Wearable/Personal Monitoring Devices Present to Future

Gaetano D. Gargiulo · Ganesh R. Naik
Editors

Wearable/Personal Monitoring Devices Present to Future

Springer

Editors
Gaetano D. Gargiulo
BioMedical Engineering
and Instrumentation
Western Sydney University
Penrith, NSW, Australia

Ganesh R. Naik
MARCS Institute for Brain, Behaviour
and Development
Western Sydney University
Penrith, NSW, Australia

ISBN 978-981-16-5323-0 ISBN 978-981-16-5324-7 (eBook)
https://doi.org/10.1007/978-981-16-5324-7

This Springer imprint is published by the registered company Springer Nature Singapore Pte Ltd.
The registered company address is: 152 Beach Road, #21-01/04 Gateway East, Singapore 189721, Singapore

Preface

This book presents recent advances in wearable health monitoring technology. The recent advances in this field yield the potentials to improve the accuracy and reliability of medical diagnoses, and their integration in the everyday personal electronic devices will re-ignite human-computer interaction enabling a more intimate relationship with electronic devices that will help to close the loop of personalized medicine empowering patients of their own health focusing on the real lifestyle and daily routine.

Our editorial goal is to provide a forum for researchers to exchange ideas and foster a better understanding of the "state of the art" in this matter; therefore, this book is intended for wearable technology experts, biomedical, computer science, and electronics engineers (researchers and graduate students) who wish to get novel research ideas and some training in wearable technologies. Additionally, the research results previously scattered in many scientific articles worldwide are collected methodically and presented in the book in a unified form.

The book is organized into four sections. Part I is devoted to Skin contact (Electrodes) based wearables. In this section, we have collected five chapters with several novel contributions. The set of chapters include ECG Lead Reconstruction Methodologies by *Gundlapalle and Acharyya. Karlsson et al.* report the state-of-the-art literature on brain-computer interface as a potential access method for communication in children with cerebral palsy. PPG-based Non-invasive Methodologies for Pervasive Monitoring of Vitals: BP and HR is reported by *Panwar et al. Arjunan and Joseph* report a wearable sensor system to monitor the status of the automobile drivers. *Das et al.* report Movement analysis of Human lower Limb using EMG Sensor for Lower Extremity. The second section focuses on the various applications of Non-contact-based wearables. We have gathered three chapters in this section, which are, respectively, Sleep Monitoring Wearables: Present to Future by *Dutta et al.*, Non-invasive Monitoring of Health Using Sensor-Rich Wearables by *Tiwari et al.*, Sleep Monitoring in Adults Using Wearables and Unobtrusive Technology by *Deviaene et al.* The third section covers Beyond wearables, namely The CueMinder Project: Patient-Driven Wearable Technology by *Kempner et al.*, and IoT and Machine Learning Algorithms for Fall Detection by *Pahuja* and *Veer*. The final section—Signal Processing for Wearable/IOT applications covers two chapters: Smart Home Automation by *Volety,* and *Geethanjali*, and a novel architecture design for complex network measures of brain connectivity aiding diagnosis by *Pal and Acharyya.*

We want to thank the authors for their excellent submissions (chapters) to this book and their significant contributions to the review process, which have helped ensure this publication's high quality. Without their contributions, it would not have been possible for the book to come successfully into existence.

Penrith, Australia
May 2021

Gaetano D. Gargiulo
Ganesh R. Naik

Contents

Part I
Skin Contact (Electrodes)-Based Wearables

Chapter 1
ECG Lead Reconstruction Methodologies for Remote Health Monitoring of Cardiovascular Diseases (CVD)

Vishnuvardhan Gundlapalle and Amit Acharyya

Abstract Cardiovascular diseases (CVD) are the prime cause of the high mortality rate across the world. Clinical CVD identification is, generally, done by exploitation of either Standard 12-Lead (S12) or Mason–Liker 12-Lead (ML12) Electrocardiogram (ECG) system. However, acquisition of these 12-lead ECG is cumbersome and time consuming process. In remote health monitoring cases, 12-lead ECG requires high bandwidth, storage and data transmission time. Thus to mitigate these, synthesis of 12-lead ECG from reduced lead set has been investigated over the past decade. This chapter identifies and review on the state of the art methodologies that exists in the literature.

1.1 Introduction

Cardiovascular Diseases (CVD) are the prime cause of the high mortality rate across the globe (CVD statistics as per WHO). Around 30% deaths all over the world is due to the CVD as per the World Health Organization (WHO) statistics. The advancement of medical industry and huge growth in IoT technology is gradually making the remote CVD monitoring a reality. Recent years have witnessed a growing interest for developing personalized and nonhospital-based care systems to improve the management of cardiac care, to reduce the time before treatment, and consequently, to reduce cardiac morbidity and mortality. Furthermore, because the population is rapidly aging, the number of cardiac patients is steadily increasing and almost 2/3 of the cardiac patients die in the prehospital phase. Therefore, greater deployment of resources for prehospital care is needed to reduce the fatality rate.

V. Gundlapalle (✉) · A. Acharyya
Department of Electrical Engineering, IIT Hyderabad, Kandi, Sangareddy 502285, Telangana, India
e-mail: ee19mtech01004@iith.ac.in

A. Acharyya
e-mail: amit_acharyya@ee.iith.ac.in

G. D. Gargiulo and G. R. Naik (eds.), *Wearable/Personal Monitoring Devices Present to Future*, https://doi.org/10.1007/978-981-16-5324-7_1

Electrocardiogram (ECG) is the only easy to use diagnosis tool useful for assessing the probability of cardiac events in home care, self-care, ambulatory, or emergency recording conditions. Clinical CVD diagnosis is, generally, carried out using Standard 12-Lead (S12) or Mason-Liker 12-Lead (ML12) system [18]. However, in such situations, recording a standard 12-lead ECG, is often difficult and impractical considering the high bandwidth, memory and data transmission time. Moreover, the reduced lead (RL) systems with 2–3 leads are generally utilized in tele monitoring application, which is not decent for diagnosis. It would thus be valuable to design a minimal, easy to use reduced lead set at the transmission end, from which the 12-lead ECG can be accurately reconstructed at the receiver end. It can also be the reconstruction of standard 12-lead (S12) system from Frank vectorcardiogram (FV). A personalized reduced 2/3 lead system is required which can offer equivalent information as contained in S12 system, so as to accurately reconstruct S12 system from RL system for diagnosis. Among the 12-lead system, selection of such RL systems suitable for personalized remote health-monitoring applications is also a research challenge, followed by its subsequent successful reconstruction of the S12 system. The accuracy and real-time performance of traditional methods needs improvement. Also if accurate Frank's vectorcardiogram (VCG) system is made available avoiding the limitations of proper electrode settings and hardware complexity, VCG will complement S12 system in diagnosis of CVDs.

There exist various methodologies that provides a solution for the aforementioned problems. Among the existing literature Dower was the first to propose a transformation matrix, known as Dower Transform (DT), to transform the Frankvectorcardiographic (FV) system to S12 system [7]. In Dawson et al. [4], Affine transformation (AT) was proposed which employed Least-square (LS) fit method on a population of patients to obtain population based transformation matrix and has been shown to outperform DT. EASI system to S12 system is another transformation proposed which uses LS fit method or DT [7]. Finlay et al. [11] posits that the transformation involving reconstruction of missing precordial leads of S12 out-performs the transformation of EASI system to S12 system. There have been several investigations involving reconstruction of missing leads, however, investigations on transformation of 3-lead subset to S12 system are very few and most of the works have focused on the transformation of 4-lead subset to S12 using either patient-specific or population specific coefficients. Horacek et al. [14] has investigated into the transformation of reduced 4-lead systems comprising of basis leads derived from ML12 system to the missing leads of ML12, S12 and 18-lead ECGs. In [18] R3L system was formed using leads I, II and one of the six precordial leads i.e. V1 to V6 resulting in a total of six such combinations and along with FV system they all were used to reconstruct S12 system. In all the works that have been carried out over the years, personalized or patient-specific transformations have outperformed population specific transformations. Recently Naresh et al. [25] proposed 2-lead to S12 system which used Lead I and V2 as the basis leads. There is a high change of research that will be carried out in this domain and even 1-lead to S12 system transformation will be possible with the advent of IoT and cloud technologies.

1.2 A Glimpse of State of the Art Methodologies

- Maheshwari et al. [19] proposed a personalized reconstruction methodology for the reconstruction of missing precordial leads of S12 and ML12 systems. A Reduced 3-Lead system (R3L) from S12 and ML12 systems at the transmission end which reduces the number of signals to 3 and then reconstruct the S12 and ML12 systems at the receiver end using the proposed personalized reconstruction methodology, thus allaying aforementioned limitations. I, II and V2 form the basis leads and the precordial leads form the target leads. Least square fit and heart vector projection theory have been used to obtain personalized transformation coefficients. Accuracy of reconstruction has been evaluated on PhysioNet PTBDB (Physikalisch-Technische Bundesanstalt database) and INCARTDB (Institute of Cardiological Technics 12-Lead Arrhythmia Database) [13], Bousseljot et al. [2], after wavelet based preprocessing, using R2 statistics, correlation (rx) and regression (bx) coefficients. Re-usability of personalized coefficients has also been investigated in the proposed work. Mean R2 values obtained from the reconstruction of target leads are 91.87% (PTBDB) 83.75% (INCARTDB). R3L system reduces the number of leads/signals from 8 to 3 and as the results indicate the possibility of reusing the transformation coefficients, the number of electrodes can be reduced from 10 to 5, thereby, increasing the comfort of patients and caregivers.
- Maheshwari et al. [18] proposed a robust and accurate method for the reconstruction of S12 system from FV system using personalized transformation (PT) matrices targeting personalized remote health-monitoring applictions. FV system is used in the 3D visualisation of heart and diagnosis and prognosis of many cardiologic disorders including myocardial infarction, Brugada syndrome etc. However, cardiologists are accustomed to S12 system pertaining to its decades-old usage and widespread acceptability and hence, it is generally used as primary ECG acquisition system and state-of-the-art inverse Dower transform (DT) and affine transform (AT) are used to obtain FV system from S12 system. Here, authors have proposed the acquisition of FV system and use PT to reconstruct S12 system from FV system. PhysioNet's PTB database after wavelet-based pre-processing to remove baseline wandering and noise has been used in this investigation. The personalised coefficients have been obtained using least-squares fit method and heart-vector projection theory and evaluation metrics used are R^2 statistics, correlation and regression coefficients. The proposed PT methodology has outperformed AT and DT by mean R^2 values of 16.36% and 26.53%, respectively. For the practical application of the proposed system, authors have investigated into the reusability of personalised coefficients which has been shown to outperform state-of-the-art AT and DT.
- Maheshwari et al. [21] proposed a personalized reduced 3-lead (R3L) system formation methodology which employs principal component analysis, thereby, reducing the redundancy and increasing SNR ratio, hence, making it suitable for wireless transmission. Accurate S12 system is made available using personalized lead reconstruction methodology, thus addressing medical constraints.

Mean R^2 statistics values obtained for reconstruction of S12 system from the proposed R3L system using PhysioNet's PTB and TWA databases were 95.63% and 96.37% respectively. To substantiate the superior diagnostic quality of reconstructed leads, root mean square error (RMSE) metrics obtained upon comparing the ECG features extracted from the original and reconstructed leads, using Time Domain Morphology and Gradient (TDMG) algorithm [24], have been analysed and discussed. The proposed system does not require any extra electrode or modification in placement positions and hence, can readily find application in computerized ECG machines

- Maheswhari et al. [22] made an attempt to provide technical methodology to the medical practitioners for selection of RL systems suitable for personalised remote health monitoring applications. Subsequently, a novel S12-lead ECG reconstruction methodology is also proposed which is shown to be more reliable than the state-of-the art lead reconstruction methodologies. In this study, along with Frank's vectorcardiographic system, reduced 3-lead systems consisting of leads I, II and one of the six precordial leads (V1–V6) leading to a total of six such reduced lead sub-systems are considered. Based on the proposed lead reconstruction methodology, these aforementioned reduced lead systems' performances are evaluated and compared comprehensively using R^2 statistics, correlation and regression coefficients. The advantages and disadvantages of using a particular RL system have been discussed in the context of remote health monitoring applications.
- Maheshwari et al. [20] proposed a methodology to construct Frank VCG from S12 system using Principal Component Analysis (PCA). They have compared the proposed work with state-of-the-art Inverse Dower Transform (IDT) [7, 9] and Kors Transform (KT) [15]. Mean R^2 statistics and correlation coefficient values obtained for CSE multi-lead database (CSEDB) and PhysioNet's PTBDB using proposed method were (73.7%, 0.869), for IDT (57.6%, 0.788) and for KT (56.2%, 0.781). From remote healthcare perspective, a reduced 2–3 lead system is desired and Frank lead system seems to be promising as shown by previous studies. However, cardiologists are accustomed to S12 system due to its widespread usage and derived Frank lead system might not be sufficient. Hence, to bridge the gap, authors have presented the results of personalized reconstruction of S12 system from derived VCG, obtained using proposed PCA-based method and compared it with results obtained when originally measured Frank leads were used.
- Naresh et al. [25] proposed a novel 2–12 lead reduction system for remote ECG monitoring applications. In the proposed method, the coefficient values of all the leads are generated as the first step using the least-square fit method and heart-vector projection. For the 12 lead ECG reconstruction, lead I and V2 are taken as the basis leads. As a part of the reconstruction, Lead II is derived initially from these two basis leads (Lead I, V2) using the HVP computation. After Lead II derivation, these three (I, II, V2) leads acts as basis leads and will derive remaining leads (V1, V3, V4, V5, V6) by repeating the HVP computation. Three categories Healthy Control (HC), Bundle Branch Block (BBB) and, Myocardial Infarction (MI) were taken from PTBDB for the proposed work. R^2 statistics, correlation and

regression coefficients were used to evaluate the performance, the mean values of the stated performance metrics obtained were 91.94%, 0.957, 0.921 for HC, 85.81%, 0.920, 0.856 for BBB and 81.42%, 0.889, 0.820 for MI respectively.

- Wang et al. [28] proposed a novel method based on convolutional neural networks (CNNs) for the synthesis of missing precordial leads. The results show that the proposed method receives better similarity and consumes less time using the PTB database. Particularly, the presented method shows outstanding performance in reconstructing the pathological ECG signal, which is crucial for cardiac diagnosis. The proposed CNN-based method is shown to be more accurate and time-saving for deployment in non-hospital situations to synthesize a standard 12-lead ECG from a reduced lead-set ECG recording. This is promising for real cardiac care.
- Atoui et al. [1] proposed a novel approach for deriving 12-lead ECGs from a pseudo orthogonal three-lead subset via generic and patient-specific nonlinear reconstruction methods based on the use of artificial neural-networks (ANNs). Authors have trained and tested the proposed ANN on a set of serial ECGs from 120 cardiac inpatients from the intensive care unit of the Cardiology Hospital of Lyon. Then assessment has been done on the similarity between the synthesized ECGs and the original ECGs at the quantitative level in comparison with generic and patient-specific multiple-regression-based methods. The ANN achieved accurate reconstruction of the 12-lead ECGs of the study population using both generic and patient-specific ANN transforms.
- Dawson et al. [5] proposed a statistical approach to transform 3-lead Frank VCG to 12- lead ECG signals and vice versa, based on Dower's pioneering work on lead transformation. This approach enables compensation of baseline shifts and other constant biases present in long ECG data streams, so that the resulting statistical transforms can be more consistent and accurate. Authors have compared the performance of the affine transform with that of Dower transform (from 3 to 12 and from 12 to 3) using the data from the PhysioNet PTB database. The results show that for both myocardial infarction (MI) and healthy control (HC) subjects, the statistical affine transform that the authors have presented maps 3-lead VCG to12-lead ECG more accurately than Dower or other lead transformation matrices of the ECG recordings.
- Nelwan et al. [26] presented the assessment on how well absent, noisy, or defective leads can be reconstructed from different lead subsets and how well lead reconstruction performs over time. A data set of 234 24-hour ECG recordings was divided into an equally sized training and test set. Precordial leads were systematically removed, and for all lead subsets including both limb leads and at least one precordial lead, the absent leads were reconstructed using general and patient-specific reconstruction templates. Reconstruction performance was measured by correlation between the original and reconstructed leads over the QRS and T waves, by average and maximum absolute ST differences, and by agreement when a clinical decision rule was applied. Reconstruction performance over time was evaluated at baseline, at 20 min, and 1, 6, 12 and 24 h after the start of each recording. Reconstruction accuracy was high (correlation $\geq$0.932, average ST difference $\leq$30 μV, agreement $\geq$94.9%) with general reconstruction

for lead sets with 1 or 2 precordial leads removed but was less satisfactory when more leads were missing. Patient-specific reconstruction performed well when up to 4 precordial leads were removed (correlation ≥0.967, average ST difference ≤26 μV, agreement ≥95.7%). Patient-specific reconstruction performance initially slightly decreased and then stabilized over time but remained much better than general reconstruction after 24 h. Accurate reconstruction of the 12-lead ECG from lead subsets is possible over time. General reconstruction allows reconstruction of 1 or 2 precordial leads, whereas up to 4 leads can be reconstructed well using patient-specific reconstruction.

1.3 Explanation of the State of the Art Methodologies

1.3.1 Accurate and Reliable 3-Lead to 12-Lead ECG Reconstruction Methodology for Remote Health Monitoring Applications

The envisaged remote health monitoring system is shown in Fig. 1.1. The proposed system [19] involves acquisition of raw ECG signal, preprocessing, coefficient generation, transmission and lead reconstruction at the receiver end. Acquisition of the raw ECG is not discussed in this approach as it is beyond the scope of the research. Hence authors assume that the raw ECG is available and the main focus is on preprocessing, coefficient generation and reconstruction of missing leads. The details of each method is summarized as follows.

In the first stage raw ECG signal is passed through a preprocessing module with comprises of removal of baseline wandering (BW) using discrete wavelet transform (DWT) and de-noising based on translation invariant wavelet transform (TIWT)

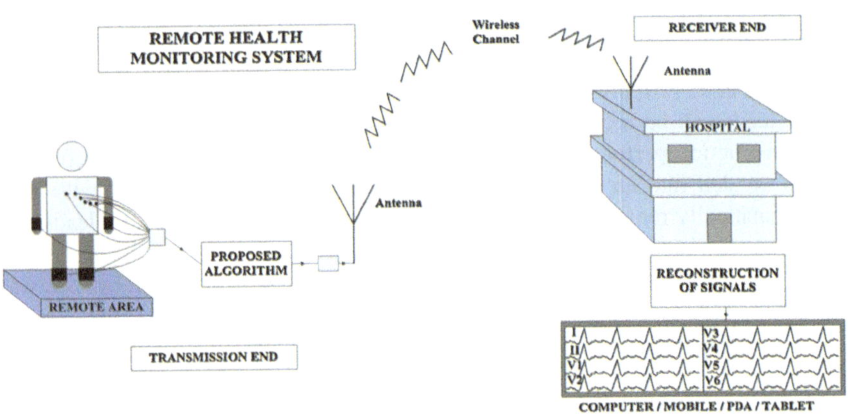

Fig. 1.1 Envisaged remote health monitoring system

[19]. The overall methodology is shown in Fig. 1.2. This pre-processed signal is then used for generation of coefficients using heart vector project (HVP) theory and least squares fit (LS) method. HVP theory [12, 16] states that heart can be approximated as single dipole vector (known as heart vector, $\vec{H}$) fixed in 3-D space whose orientation and magnitude varies during a cardiac cycle. This dipole vector is responsible for the body surface potential observed when electrodes are placed on the body. The potential at any point on the body is the projection of $\vec{H}$ on the lead vector ($\vec{L}$) which is assumed to originate from the zero-potential region in the heart and terminates on the point located on the body is given by (1.1).

$$V = \vec{H}.\vec{L} = a_1 X + b_1 Y + c_1 Z \tag{1.1}$$

where $\vec{H} = X\,\hat{i} + Y\,\hat{j} + Z\,\hat{k}$, $\vec{L} = a_1\,\hat{i} + b_1\,\hat{j} + c_1\,\hat{k}$, this linear model will be helpful to obtain the potential at any point using leads apart from the heart vector components.

In the propose method authors used leads I, II and V2 to generate the signal of some other lead. It can be seen from (1.1), (1.2) that any lead of S12 system can be generated from a set of three independent leads provided the coefficients are available. These coefficients can be generated statistically using LS fit technique upon the availability of leads appearing on the right-hand side (RHS) and the left-hand side (LHS) of (1.1), (1.2). The solution of LS fit when applied to $V_i = x_i I + y_i II + z_i V_2$ is given by (1.3).

$$V_i = x_i I + y_i II + z_i V_2 \tag{1.2}$$

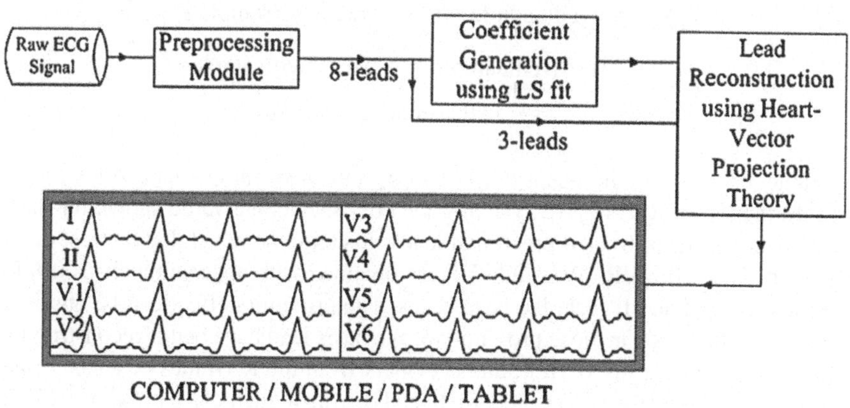

Fig. 1.2 Block level diagram of the proposed method

$$\begin{bmatrix} x_i \\ y_i \\ z_i \end{bmatrix} = \begin{bmatrix} \sum I^2 & \sum I.II & \sum I.V_2 \\ \sum I.II & \sum II^2 & \sum II.V_2 \\ \sum I.V_2 & \sum II.V_2 & \sum V_2^2 \end{bmatrix} \begin{bmatrix} \sum V_i.I \\ \sum V_i.II \\ \sum V_i.V_2 \end{bmatrix} \tag{1.3}$$

where I, II, V_2 are the basis leads and x_i, y_i, z_i are their respective coefficients. When LS fit method is applied on one patient the coefficients obtained are personalized or patient-specific and when applied on a group of patients, are known as population-specific coefficients. PTBDB and INCARTDB sampled at 1 kHz and 257 Hz have been used in this study to evaluate the proposed method. These are the widely used and accepted databases among the research community. For both PTBDB and INCARTDB, a training set of 5000 samples from the beginning of the recording of each patient was used in (1.3) for obtaining the transformation coefficients and the rest of the recording was used as the testing set. The complete work was carried out on MATLAB (Version 7.10.0.499 R2010a).

R^2 statistics has been used to evaluate the degree of association between the measured and the reconstructed signal. Perfect retracing of the measured wave by the reconstructed wave will be indicated by a value of 100%. Correlation coefficient (r_x) is a metric to estimate the similarity between two signals and the regression coefficient (b_x) fairly estimates the amplitude differences between the measured and reconstructed signals. Formulas for these mentioned metrics are given by Eqs. (1.5), (1.6) and (1.7) respectively.

$$R^2 = \left\{ \frac{\sum [\text{Derived(sample } i) - \text{Measured(sample } i)]^2}{[\sum (\text{Measured(sample } i))]^2} \right\} \times 100 \tag{1.5}$$

$$r_x = \frac{\sum [\text{Derived(sample } i) - \text{Measured(sample } i)]}{[\sum (\text{Derived(sample } i))^2 \times (\text{Measured(sample } i))^2]^{1/2}} \tag{1.6}$$

$$b_x = \frac{\sum [\text{Derived(sample } i) - \text{Measured(sample } i)]}{[\sum (\text{Measured(sample } i))]^2} \tag{1.7}$$

Table 1.1 presents the mean R^2, r_x and b_x values of various categories for the transformation of R3L to S12 system for PTBDB. Average value shown has been taken over the reconstructed precordial leads. It should be noted that the rest of the leads viz. I, II, III, aVR, aVL and aVF are obtained with approximately 100% R^2 values as leads I and II form the basis leads. The proximity effect can be observed from the table. Leads in close proximity of the basis lead have better reconstruction compared to those away from it, this is evident from the R^2 values of leads V1and V3which are higher than other precordial leads owing to their close proximity to V2, one of the chosen basis lead in this investigation. The lowest values can be found for lead V4. Table 1.2 presents the mean values of evaluation metrics for INCARTDB. We can see that reconstruction results in the case of S12 system are superior to those of ML12 system. This difference is probably because of the electrode positions in

Table 1.1 Mean R^2, r_x and b_x values of various categories for transformation of R3L system to S12 system for PTBDB

	V1			V3			V4			V5			V6			Average
	R^2	r_x	b_x	R^2	r_x	b_x	R^2	r_x	b_x	R^2	r_x	b_x	R^2	r_x	b_x	R^2
BB	90.40	0.947	0.899	95.78	0.979	0.975	87.39	0.937	0.913	88.47	0.944	0.919	92.41	0.962	0.952	90.89
HC	94.69	0.973	0.952	96.52	0.983	0.973	91.83	0.959	0.939	93.53	0.968	0.957	94.69	0.973	0.967	94.25
HY	97.44	0.987	0.979	96.23	0.981	0.968	87.45	0.934	0.897	89.90	0.945	0.899	95.74	0.978	0.957	93.35
MI	94.34	0.971	0.947	95.92	0.979	0.959	89.52	0.945	0.895	89.22	0.944	0.889	91.52	0.956	0.911	92.10
VA	93.73	0.967	0.942	94.25	0.970	0.945	89.97	0.948	0.906	91.89	0.959	0.921	93.78	0.968	0.934	92.72
ND	90.92	0.949	0.913	93.54	0.967	0.931	83.72	0.910	0.838	84.61	0.916	0.847	87.28	0.920	0.876	88.01

Table 1.2 Lead-wise mean R^2, r_x and b_x values for INCARTDB

	R^2	r_x	b_x
$V1$	86.38	0.936	0.923
$V3$	86.91	0.940	0.926
$V4$	83.61	0.922	0.901
$V5$	83.74	0.921	0.911
$V6$	78.11	0.893	0.872
Avg	83.75	0.922	0.907
Min	78.11	0.893	0.872

Table 1.3 Mean values of Personalized Transformation Coefficients obtained

PTBDB							INCARTDB		
	Healthy control			Unhealthy					
V1	−0.6767	−0.0561	0.4926	−0.6250	−0.0965	0.5006	−0.7450	−0.1431	0.5434
V3	0.7297	0.3298	0.8506	0.4449	0.3588	0.9637	−0.2267	0.5070	1.1181
V4	1.4101	0.5072	0.4149	0.9036	0.5792	0.5915	0.1140	0.7191	0.5214
V5	1.5883	0.4610	0.0314	1.0042	0.5933	0.1774	0.7173	0.6890	−0.0455
V6	1.1729	0.3618	−0.1173	0.7330	0.4710	−0.0442	0.6792	0.6080	−0.3074

ML12 which are on the torso and not on arms and legs as in S12 system. Table 1.3 provides the mean values of PT coefficients for PTBDB (healthy control [HC] and unhealthy [UH]) and INCARTDB. PTBDB database has been divided into healthy and unhealthy categories where as INCARTDB has not been categorized.

Figure 1.3 shows some of the results based on the proposed work in original vs reconstructed signal format. Figure 1.3a, b correspond to patients with mean R^2 values of 94.28% (HC) and 90.06% (UH) respectively (mean taken over 5 reconstructed missing precordial leads) which are nearly equal to the mean values given in Table 1.1. Figure 1.3c is a similar mean case subject from INCARTDB with mean R^2 value being 84.97%. Similarly, Fig. 1.3d, 3E and 3F provide the reconstruction results of subjects with minimum mean R^2 values from HC, UH and INCARTDB respectively. The range of R^2 values for re-constructed leads shown in Fig. 1.3 is from 5.512% to 99.52%. It should be noted that R^2 values of 80% and above can be considered to have high diagnostic value and R^2 values of 90% and above can be considered to be an accurate retrace of the original signal for all practical purposes. In HC category, a total of 255 (51×5) leads were reconstructed out of which 96.08% leads had more than 80% R^2 value and 88.6% leads had more than 90% R^2 value. In UH category, a total of 1130 (226×5) leads were reconstructed out of which 85.75% patients had more than 80% (51×5) value and 66% leads had more than 90% R^2 value. For the reconstruction of 340 (68×5) precordial leads of ML12 system about 73.82% had more than 80% R^2 value and 50% leads had more than 90% R^2 value. In Fig. 1.3e, f, we can see relative smoothing and stabilization of probably badly

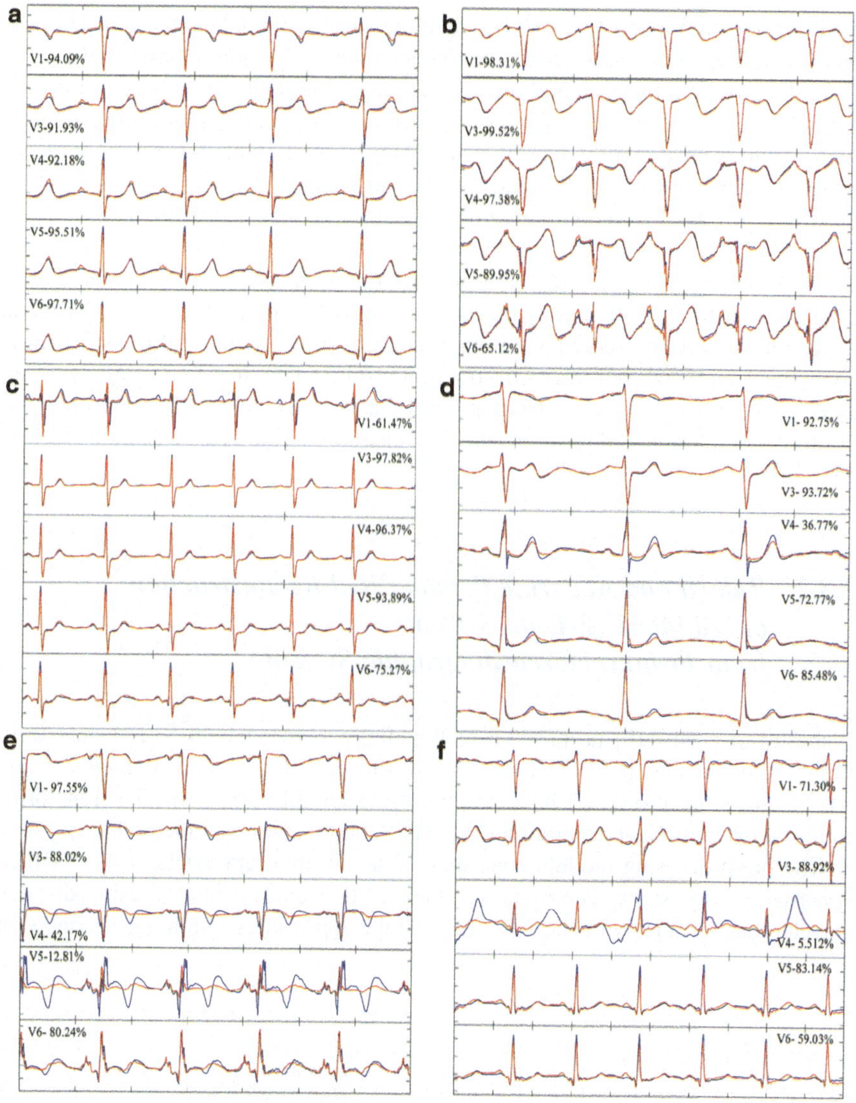

Fig. 1.3 Comparison between original (blue) and derived (red) leads of the subjects from PTBDB and INCARTDB. Sub-figures **a** and **b** show the comparison of subjects from HC and UH category respectively with the mean R^2 values near to actual means provided in Table 1.1. **c** is the corresponding mean case from INCARTDB. Sub-figures **d**, **e** and **f** correspond to minimum case of mean R^2 values for the subjects from HC, UH and INCARTDB respectively. The corresponding R^2 values of the reconstructed lead have been provided with the figure. (For interpretation of the references to color in this figure legend, the reader is referred to the web version of this article.)

positioned leads V4 and V5. This is one of the advantages of lead reconstruction as badly acquired leads can be reconstructed from the well acquired basis leads.

In the proposed method, authors have attempted to address the requirements of cardiologists and technological constraints encountered in tele monitoring applications. Authors have proposed a personalized reconstruction methodology for reconstruction of missing pre-cordial leads of S12 and ML12 systems and the results show that using proposed methodology reliable and accurate reconstruction is

possible. R3L system used in the proposed work requires no change in the standard electrode positions of S12 or ML12 system and hence its a practical option for clinical usage. Reconstruction of missing precordial leads of S12 outperforms ML12 system. Here, authors have presented an in-depth analysis of re-usability of transformation coefficients and analyzed the reconstruction results when the same coefficients are used for reconstruction. It has been inferred that if care is taken while placement of electrodes at the standard positions then reconstruction results are less likely to vary over time.

1.3.2 Robust and Accurate Personalised Reconstruction of Standard 12-Lead System from Frankvectorcardiographic System

Maheshwari et al. [18] proposed a methodology to reconstruct S12 system from Frankvectorcardiographic (FV) system. As shown in Fig. 1.4, first raw ECG and VCG signals are processed by a preprocessing module using wavelet transforms followed by coefficient generation using HVP theory and LS fit method.

The pre-processing module consists of four different approaches, all of which consisted of two steps. The first step involves the removal of BW using discrete wavelet transform (DWT) and is same in all the approaches. While the second step

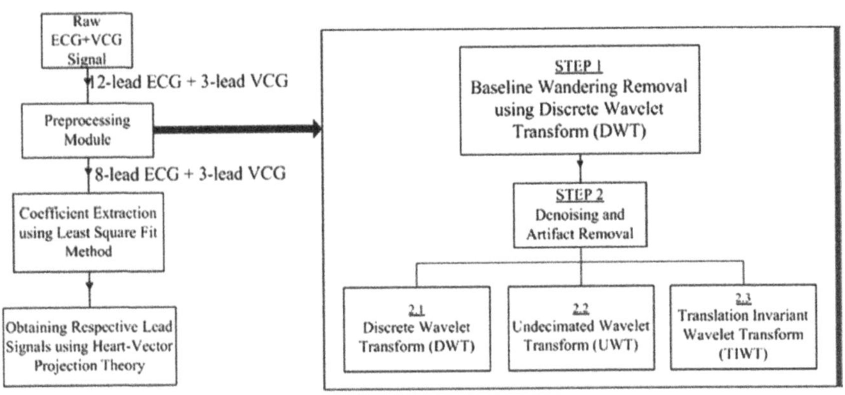

Fig. 1.4 Block level representation of the proposed system

is de-noising and is performed in the second, third and fourth approach. The details of pre-processing approaches used are as follows.

- A1: Only BW removal was performed using DWT
- A2: BW removal and de-noising both using DWT was performed
- A3: BW removal using DWT and de-noising using un-decimated wavelet transform (UWT) was performed
- A4: BW removal using DWT and de-noising using translation invariant wavelet transform (TIWT) was performed

Next step is to generate personalized transformation (PT) coefficients for each individual lead. In this work FV system's 3 leads namely, X, Y and Z are taken as the basis leads to generate the 8 independent leads in the S12 system namely, I, II, V1, V2, V3, V4, V5 and V6. For generating coefficients authors have used HVP theory and LS fit method as explained in the Sect. 3.1. R2, correlation (rx) and regression (bx) metrics are used to evaluate the performance of the proposed system. Publicly available PTBDB has been used in this study to evaluate the proposed methodology.

Table 1.4 presents the mean R2, rx and bx values for all the eight independent leads of S12 system over all the patients in two categories namely HC and UH subjects. It can be seen that very high values of evaluation metrics have resulted from the usage of PT and has incremented from A1 to A4, although not with significant improvement among themselves. From this authors have considered A4 as the preferred pre-processing module.

Figure 1.5 shows a comparison of the reconstruction performances of AT and PT for two subjects one each from HC and UH categories with their mean R2 values (obtained using AT), i.e. 83.89% (HC) and 72.82% (UH) being near to actual mean values obtained using AT, i.e. 84.48% (HC) and 72.75% (UH) as can be seen from Table 1.4. The R2 values of corresponding leads have been shown in the figure. Figure 1.5 shows the reconstructed lead (in red) overlapping the originally measured lead (in blue) of mean case AT subject belonging to HC category, with Fig. 1.5a showing the reconstructed leads obtained when AT was used and Fig. 1.5b showing the reconstructed leads obtained using PT. Similarly, Fig. 1.5c, d shows the mean AT-UH subject when reconstructed using AT and PT, respectively. The corresponding mean R2 values obtained when PT was used for the same subjects were 98.06% (HC) and 85.07% (UH). Figure 1.6 shows a similar comparison for minimum R2 value cases obtained when PT was used. The mean R2 values of the two subjects were 88.41% (HC) and 75.22% (UH), and the corresponding values when AT was used on the same subjects were 85.94% (HC) and 61.72% (UH). Figure 1.6a, b shows reconstructed lead (in red) overlapping the originally measured lead (in blue) of minimum case PT subject belonging to HC category with 3A showing the reconstructed leads obtained when PT was used and 6B showing the reconstructed leads obtained using AT. Similarly, Fig. 1.6c, d shows the minimum case UH subject when reconstructed using PT and AT, respectively.

In this work, authors have proposed a personalised lead reconstruction methodology to reconstruct S12 system from FV leads. A wavelet based pre-processing module has been introduced to remove BW and noise. Three variants of wavelet

Table 1.4 Mean R^2, rx, and bx Values Obtained on Applying the Proposed PT (Personalized Transformation), AT (Affine Transform), and DT (Dower Transform) on HC and UH Subjects

	Al			A2			A3			A4			AT			DT		
	R^2	r_x	b_x	R^2	r_x	b_x	R^2	r_x	b_x	R^2	r_x	b_x	R^2	r_x	b_x	R^2	r_x	b_x
Healthy control																		
I	92.89	0.963	0.926	93.01	0.963	0.927	93.05	0.964	0.928	93.00	0.963	0.927	76.58	0.899	0.812	66.91	0.893	0.808
II	97.36	0.987	0.975	97.43	0.987	0.976	97.46	0.987	0.976	97.38	0.987	0.975	93.84	0.976	0.927	92.28	0.973	1.028
V1	95.70	0.978	0.956	95.75	0.978	0.957	95.77	0.979	0.957	95.77	0.979	0.957	81.76	0.942	0.908	73.46	0.910	0.858
V2	93.90	0.969	0.943	93.94	0.969	0.944	93.95	0.969	0.944	93.96	0.969	0.943	75.79	0.921	0.929	63.87	0.806	0.651
V3	96.46	0.982	0.966	96.48	0.982	0.966	96.49	0.982	0.966	96.63	0.983	0.965	83.10	0.927	0.897	57.55	0.827	0.880
V4	97.42	0.987	0.974	97.44	0.987	0.974	97.46	0.987	0.974	97.65	0.988	0.976	82.34	0.937	0.925	72.56	0.921	0.972
V5	98.94	0.995	0.991	98.95	0.995	0.991	98.96	0.995	0.991	98.96	0.995	0.991	89.71	0.962	0.963	89.27	0.953	0.891
V6	98.98	0.995	0.991	99.00	0.995	0.991	99.01	0.995	0.991	99.00	0.995	0.991	92.70	0.970	0.997	90.58	0.958	0.844
Average	**96.46**	0.982	0.965	**96.50**	0.982	0.966	**96.52**	0.982	0.966	**96.54**	0.982	0.966	**84.48**	0.942	0.920	**75.81**	0.905	0.867
Unhealthy																		
I	90.04	0.948	0.909	90.34	0.950	0.913	90.39	0.950	0.913	90.40	0.950	0.913	48.32	0.866	0.884	60.23	0.867	0.681
II	92.79	0.963	0.936	92.99	0.964	0.939	93.04	0.964	0.939	93.07	0.964	0.939	85.54	0.932	0.894	78.64	0.921	0.999
V1	92.83	0.963	0.936	92.94	0.964	0.937	92.96	0.964	0.938	92.97	0.963	0.938	73.58	0.923	0.912	60.42	0.865	0.745
V2	91.06	0.953	0.916	91.14	0.953	0.917	91.13	0.953	0.917	91.16	0.953	0.918	67.18	0.893	0.867	55.62	0.807	0.612
V3	93.44	0.966	0.939	93.49	0.967	0.941	93.50	0.967	0.941	93.51	0.967	0.941	71.46	0.905	0.909	47.86	0.772	0.661
V4	94.51	0.972	0.951	94.56	0.972	0.952	94.57	0.972	0.952	94.57	0.972	0.952	72.67	0.909	0.890	46.56	0.779	0.711
V5	96.08	0.980	0.965	96.14	0.980	0.965	96.15	0.980	0.966	96.15	0.980	0.966	80.46	0.914	0.877	70.06	0.872	0.818
V6	95.45	0.976	0.957	95.55	0.977	0.958	95.57	0.977	0.958	95.57	0.977	0.959	82.81	0.943	0.917	69.36	0.903	0.832
Average	**93.28**	0.965	0.939	**93.39**	0.966	0.940	**93.41**	0.966	0.940	**93.42**	0.966	0.941	**72.75**	0.911	0.894	**61.09**	0.848	0.757

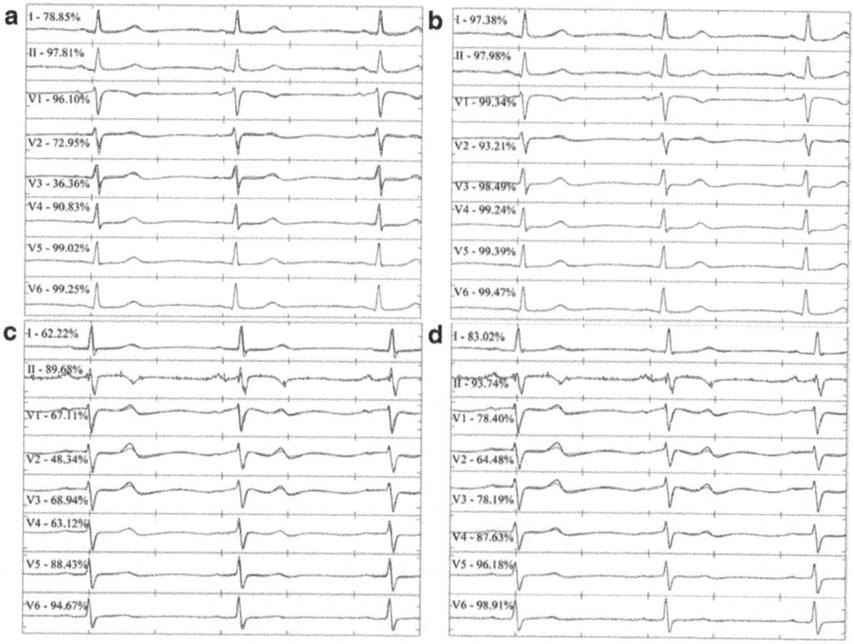

Fig. 1.5 (Colour online) Comparison between the reconstructed (red) and original (blue) signal. One patient each has been taken from HC and UH categories, and the reconstruction has been performed using AT and PT methodologies. The patients chosen have mean R2 values of 83.89% (HC) ad 72.82% (UH), obtained using AT, which is close to the overall mean using AT, i.e. 84.48% (HC) and 72.75% (UH) as can be seen from Table 1.4. **a** reconstruction of HC subject using AT; **b** reconstruction of HC subject using PT; **c** reconstruction of UH subject using AT; **d** reconstruction of UH subject using PT

transform have been evaluated, and it has been found that TIWT outperforms others, however, not with significant differences. The proposed methodology was compared with the state-of-the-art DT and AT and it has been shows that proposed PT outperformed state-of-the-art DT and AT.

1.3.3 Personalized Reduced 3-Lead System Formation Methodology for Remote Health Monitoring Applications and Reconstruction of Standard 12-Lead System

Maheshwari et al. [21] proposed a personalized reduced 3-lead (R3L) system formation methodology to accurately reconstruct the S12 system at the receiver side. Figure 1.7 shows the summary of the proposed methodology. Here authors envisaged a scenario where Standard 12-lead ECG is being acquired in a remote or hospital

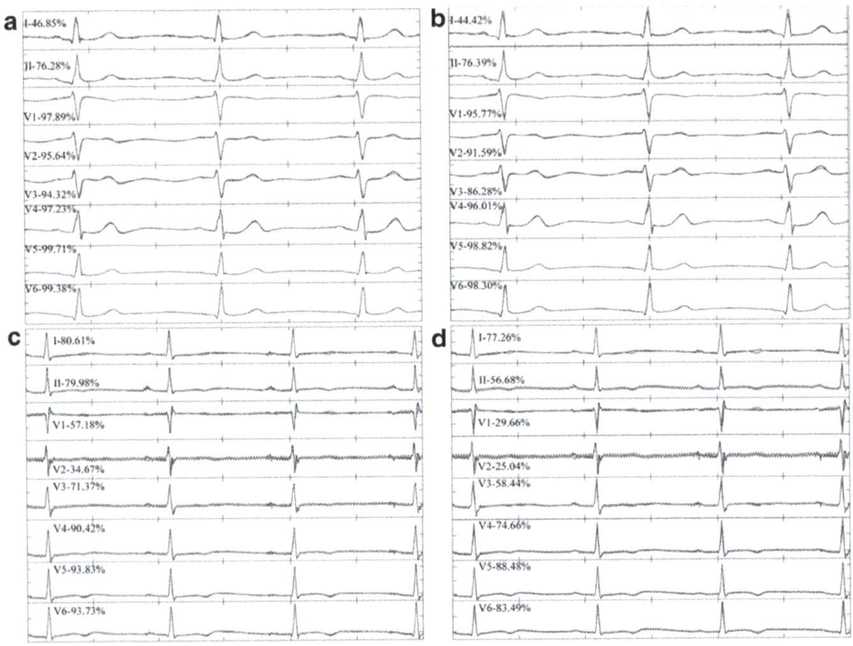

Fig. 1.6 (Colour online) Comparison between the reconstructed (red) and original (blue) signal. One patient each has been taken from HC and UH categories, and the reconstruction has been performed using AT and PT methodologies. The patients chosen have minimum mean R2 values of 88.41% (HC) and 75.22% (UH), obtained using proposed PT, and the corresponding values when AT was used on the same subjects were 85.94% (HC) and 61.72% (UH). **a** Reconstruction of HC subject using PT; **b** reconstruction of HC subject using AT; **c** reconstruction of UH subject using PT; **d** reconstruction of UH subject using AT

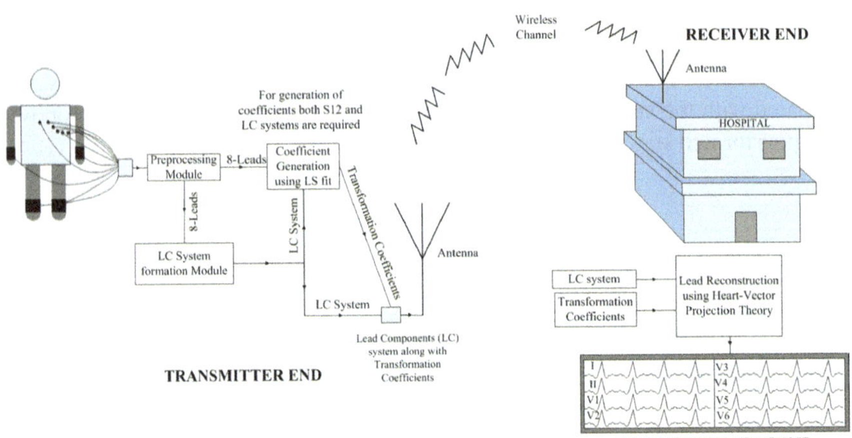

Fig. 1.7 Summary of proposed methodology

based environment. Generally, in a remote health monitoring scenario, these leads are needed to be transmitted to nearby state-of-the-art health centre for diagnosis, storage and updating of patient's health record. At the transmission end, the conventional S12 acquisitions system is used to capture the ECG. Eight independent leads of the S12 system are then passed through the Lead Component (LC) system formation module. Using LC and S12 system together the transformation coefficients are obtained by employing least-square (LS) fit method. Then, LC leads along with the transformation coefficients can be transmitted and eventually 8 independent leads of the S12 system can be reconstructed for diagnosis at the receiver's end in an optimum way.

In the proposed methodology a R3L has been obtained upon employing principle component analysis (PCA) on 8 independent leads of S12 system. PCA is a widely used methodology for noise removal and dimensionality reduction. Applying PCA on 8 independent leads of S12 results into orthogonal components and authors have selected first 3 principal components forming R3L system. From here onwards this newly formed R3L system will be taken as basis leads to reconstruct the S12 system. Next step is to generate the transformation coefficients using R3L and S12 system using HVP theory and LS fit method as explained in Sect. 3.1. Along with the proposed LC system results authors have obtained the results using a subset of S12 system which include I, II and V2 (S12S system) for comparison. To validate the proposed methodology authors have used PTBDB and TWADB. A training set of 5000 samples from the middle of the recording (training set) of each patient was used for obtaining the transformation coefficients and the whole recording was used as the testing set. It should be noted that for both LC system formation, transformation coefficient generation and S12 reconstruction only 8 independent leads of S12 system has been considered, as rest viz. III, aVR, aVL and aVF, can be readily obtained from leads I and II and are redundant.

Table 1.5 presents mean R^2, r_x and b_x values for the reconstruction of S12 system from LC and S12S systems using both PTBDB and TWADB. The average values have been calculated over 8 independent leads of S12 system. The R^2 values are higher for the precordial leads compared to bipolar leads I and II. The transformation of proposed LC system to S12 has outperformed S12S system. This confirms that the information obtained from standard leads are better represented in LC system than that of a subset. Figure 1.8 shows the comparison between the originally measured (blue) and reconstructed (red) signal. One subject each from TWADB (Fig. 1.8a, b) PTBDB (Fig. 1.8c, d). The two subjects have mean R^2 values of 91.69% (TWADB) and 88.11% (PTBDB), which is very close to the overall mean values stated in Table 1.5 for S12S to S12 transformation and the corresponding values for LC to S12 transformation were 96.71% and 92.64%. The R^2 values of respective leads have been indicated in the figure. The R^2 values in Fig. 1.8 ranges from 54.61% to 100% and provide insight into the correspondence between the quality of reconstruction and R^2 values. Table 1.6 presents the RMSE values for the features extracted using Time Domain Morphology and Gradient (TDMG) algorithm [24] from 8 independent leads of S12 system between the originally measured and reconstructed signal. Thirteen different features were extracted for both the methodologies3. Mean improvement

Table 1.5 Mean R^2, r_x and b_x values of the transformation of LC to S12 (1) and S12S to S12 (2) systems for both TWADB (72 patients) and PTBDB (first recording of 290 patients)

		TWADB									PTBDB								
		I	II	V1	V2	V3	V4	V5	V6	Average	I	II	V1	V2	V3	V4	V5	V6	Average
(1)	R^2	93.24	91.72	96.19	97.33	98.77	98.52	98.11	97.07	96.37	93.32	93.25	95.33	96.76	97.46	96.96	97.17	94.77	95.63
	r_x	0.974	0.956	0.984	0.989	0.994	0.993	0.992	0.987	0.984	0.970	0.968	0.978	0.985	0.989	0.987	0.987	0.975	0.979
	b_x	0.969	0.951	0.987	0.996	0.997	0.994	0.992	0.994	0.985	0.956	0.969	0.974	0.986	0.993	0.995	0.987	0.964	0.978
(2)	R^2	100	100	85.25	100	96.17	88.93	82.51	79.71	91.57	100	100	88.04	100	90.34	77.25	73.99	75.46	88.14
	r_x	1	1	0.954	1	0.981	0.942	0.926	0.918	0.965	1	1	0.938	1	0.955	0.902	0.886	0.887	0.946
	b_x	1	1	0.958	1	0.969	0.928	0.931	0.935	0.965	1	1	0.918	1	0.947	0.890	0.876	0.877	0.939

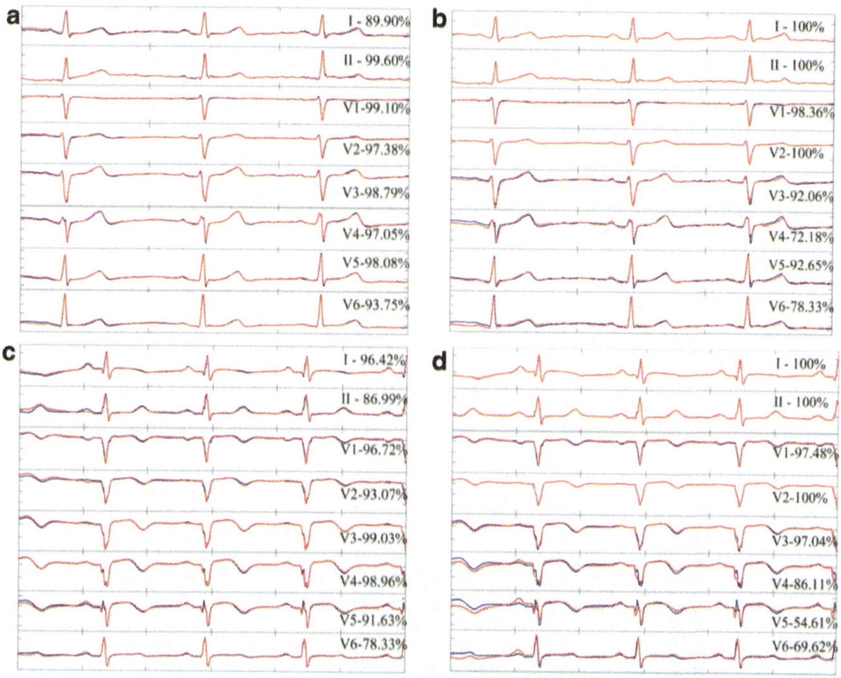

Fig. 1.8 Comparison between the two systems (LC and S12S) for the mean case S12S patient one each from TWADB and PTBDB. Sub-figures **a** and **b** shows comparison of 8 independent leads i.e. I, II, V1 to V6 for the TWADB and PTBDB and sub-figures **c** and **d** show a comparison of a mean case patient from PTBDB. The reconstructed lead (red) has been plotted over original lead (blue) corresponding R^2 values have shown on respective plots

of over 45% was observed in the proposed LC to S12 compared to S12S to S12 transformation. In this methodology authors proposed a personalized and robust reduced 3-lead system viz. Lead Component (LC) system, formation methodology targeting the emerging field of remote health monitoring to improve diagnosis and prognosis by overcoming the constraints from technical and medical science side.

It has been shown that the proposed LC system out performs the existing methodologies by various investigators for the reconstruction of standard 12-leads and has been verified on two widely accepted databases viz. PTBDB and TWADB of PhysioNet. The importance of an appropriate personalized reduced lead system has been showcased. Authors introduced PCA in this context to obtain and orthogonal 3-lead personalized system which reduces dimension of the data from 8 leads to 3 leads resulting in an increase in the SNR and Compression ratio, which favours wireless transmission of the data either to the medical practitioner's mobile or to the personalized health data repository in the cloud. It should be noted that, the personalized coefficients for reconstruction are always recalculated and retransmitted each time the contact electrodes are positioned on the patient.

Table 1.6 Mean root mean square error (RMSE) values for various features extracted using TDMG for the reconstruction methodology. Mean was taken over all the 49 patients. The leads I, II and V2 formed the basis leads of S12S, so the RMSE values for them were zero and hence, has not been reported

		LC to S12								S12S to S12				
Sl. no	**Feature (unit)**	**I**	**II**	**V1**	**V2**	**V3**	**V4**	**V5**	**V6**	**V1**	**V3**	**V4**	**V5**	**V6**
1	P duration (ms)	9.755	13.18	13.43	10.12	7.306	9.51	9.632	7.755	16.86	12.37	18	18.82	19.63
2	P height (μV)	52.89	63.59	92.53	84.82	50.52	81.96	37.58	37.03	133.3	110.1	177.7	113.4	89.27
3	PR interval (ms)	11.39	11.55	15.27	12.65	6	7.225	9.020	9.102	19.59	11.47	12.98	12.61	17.06
4	PR segment (ms)	10.78	6.449	15.14	8.163	5.551	9.388	5.265	5.020	18.33	11.35	16.78	12	16.04
5	Q peak (μV)	19.58	14.33	47.73	60.85	54.27	52.51	26.20	16.27	77.22	111.6	353.4	62.59	22.27
6	QRS length (ms)	10.65	7.184	7.633	3.755	4	6.816	5.184	6.122	9.306	9.143	11.63	10.61	10.37
7	QT interval (ms)	16.08	15.8	26.98	7.551	18.61	10	5.469	8.531	30.61	21.59	20.33	11.10	18.65
8	R height (μV)	47.26	51.03	47.12	99.95	60.18	88.54	48.41	31.72	78.21	182.4	480.4	158.9	90.75
9	S peak (μV)	9.125	10.36	19.27	0.035	0.036	2.825	9.714	10.20	28.90	2.689	4.220	19.85	17.29
10	ST interval (ms)	9.184	8.857	24.25	7.143	6.612	6.122	4.367	7.796	31.59	7.225	15.06	11.76	18.16
11	ST segment (ms)	14.16	11.92	18.61	10.37	7.184	11.27	6.490	10.90	28.86	7.633	17.76	18.08	20.37
12	T duration (ms)	43.32	64.35	107.5	114.5	139.0	105.2	73.27	43.96	161.2	199	202.5	135.7	98.76
13	T height (μV)	22.18	32.95	55.05	58.62	71.18	53.84	37.51	22.51	82.53	101.9	103.7	69.50	50.57

1.3.4 *Reduced Lead System Selection Methodology for Reliable Standard 12-Lead Reconstruction Targeting Personalised Remote Health Monitoring Applications*

Maheswhari et al. [23] proposed a technical methodology to the medical practitioners for selection of reduced lead (RL) systems suitable for personalised remote health monitoring applications. Figure 1.9 shows the two possible ways in which a patient can be registered at the health centre. During the registration process, the transformation coefficients are generated which can be eventually used to reconstruct the S12 system. The first scenario shows the online registration process and second shows the offline registration process. 'Online' means that the patient needs not to be present in the hospital/health-centre physically. In this case, the RL or Frank leads can be captured and transmitted to the hospital/health-centre for reconstruction of standard 12-leads. 'Offline' means that the patient is present in the hospital/heath-centre physically. In this case, standard 12-lead reconstruction is done at the hospital/health-centre itself. The aforementioned remote health monitoring service can benefit both ambulatory patients and the patients living in rural or remotely accessed areas. The following methods have been investigated in the proposed method.

R3L system to S12 system

If the reconstruction methodology being adopted is the transformation of R3L systems to S12 system, then for coefficient generation the acquisition of only S12 system is required. These coefficients can then be used to reconstruct S12 system on eventual readings.

FV system to S12 system

If the reconstruction methodology being adopted is the transformation of FV to S12 system, then for the coefficient generation a simultaneous acquisition of both the systems is required. On eventual monitoring, only FV system's acquisition is required which can be then transformed to S12 system.

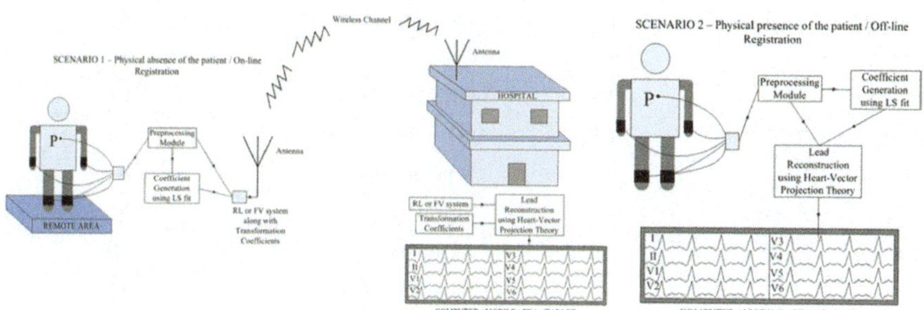

Fig. 1.9 Envisaged remote health monitoring scenarios: (1) when patient may not be physically present for registration (online) and (2) when patient is available for registration (offline)

The proposed system include pre-processing of the ECG signal using wavelet transforms. Then the generation of coefficients using HVP theory and LS fit methods. Authors have used PTBDB for evaluating the proposed methodology. Two different lead reconstruction methodologies have been investigated in this work. The first reconstruction methodology (FRM) involves transformation of six R3L systems which is comprised of leads I, II and one of the six precordial leads (V1–V6) to S12 system, and the second reconstruction methodology (SRM) involves transformation of FV (X, Y and Z) system to S12 system. The 12 leads of the S12 system have been divided into two lead sets: set 1 comprises of leads I, II, III, aVR, aVL, aVF and set 2 comprises of the precordial leads. The FRM involves a total of six R3L system and SRM involves one, i.e. FV system; therefore, a total of seven RL systems. Table 1.7 shows the denotations of these seven RL systems. Authors have used **'I'**, **'II'**, etc. in bold letters to denote R3L systems with precordial leads V1, V2, etc. as the basis leads in the R3L systems, respectively (Table 1.7). FV system has been denoted by FV in bold. Table 1.7 also shows the denotations of 13 different features that were extracted using TDMG for comparison starting from 1 to 13 in bold letters.

Table 1.8 presents the mean R^2, correlation and regression coefficient values of the FRM i.e. R3L systems to S12 system for the reconstruction of lead set 2. The last column of Table 1.8 shows the mean R^2 values over all the 12-leads of patients in the corresponding categories using FRM. As leads I and II are involved in the basis lead sets of all the six R3L systems, the resulting R^2, rx and bx values of lead set 1 when FRM is used are 100%, 1.0 and 1.0, respectively. Better results have been obtained when the precordial lead in the basis lead set are V2 or V3 followed by V1 or V4 and

Table 1.7 Denotations of R3L Systems and Various ECG Features Extracted Using TDMG Algorithm

Basis Leads of RL (Reduced Lead) System	Denotation	ECG Features (Represents Unit)	Denotation
I, II, V1	**I**	P duration (ms)	**1**
I, II, V2	**II**	P height (mV)	**2**
I, II, V3	**III**	PR interval (ms)	**3**
I, II, V4	**IV**	PR segment (ms)	**4**
I, II, V5	**V**	Q peak (mV)	**5**
I, II, V6	**VI**	QRS length (ms)	**6**
Frank leads X, Y, and Z	**FV**	QT interval (ms)	**7**
		R height (mV)	**8**
		S peak (mV)	**9**
		ST interval (ms)	**10**
		ST segment (ms)	**11**
		T duration (ms)	**12**
		T height (mV)	**13**

Table 1.8 Mean R^2, r_x and b_x values of various categories for the transformation of RL systems (V_1 –V_6: I–VI) to S12 systems

		V_1			V_2			V_3			V_4			V_5			V_6			Avg
		R^2	r_x	b_x	R^2	r_x	b_x	R^2	r_x	b_x	R^2	r_x	b_x	R^2	r_x	b_x	R^2	r_x	b_x	
BB	I	100	1	1	83.36	0.91	0.872	79.93	0.9	0.865	80.54	0.91	0.886	87.64	0.938	0.909	92.59	0.963	0.945	93.07
	II	90.4	0.947	0.899	100	1	1	95.78	0.979	0.975	87.39	0.937	0.913	88.47	0.944	0.919	92.41	0.962	0.952	95.6
	III	87.05	0.929	0.864	95.07	0.975	0.942	100	1	1	94.16	0.971	0.965	89.36	0.949	0.93	92.25	0.962	0.952	95.89
	IV	84.68	0.919	0.849	77.53	0.873	0.779	89.47	0.943	0.896	100	1	1	91.6	0.963	0.954	92.03	0.962	0.951	94
	V	78.24	0.889	0.815	64.08	0.798	0.688	67.28	0.813	0.728	74.2	0.848	0.805	100	1	1	94.24	0.971	0.963	89.23
	VI	83.18	0.911	0.832	71.52	0.847	0.757	73.5	0.867	0.804	76.13	0.885	0.839	91.15	0.954	0.913	100	1	1	90.69
HC	I	100	1	1	94.57	0.972	0.947	93.51	0.967	0.935	93.14	0.965	0.936	95.29	0.976	0.957	96.82	0.984	0.972	97.76
	II	95.79	0.978	0.959	100	1	1	97.3	0.986	0.975	94.29	0.971	0.948	95.54	0.978	0.959	97.01	0.985	0.974	98.31
	III	92.17	0.958	0.924	95.55	0.978	0.961	100	1	1	96.77	0.984	0.97	96.11	0.98	0.963	96.81	0.984	0.972	98.1
	IV	82.23	0.901	0.827	80.59	0.888	0.816	92.57	0.962	0.93	100	1	1	97.66	0.988	0.978	96.6	0.983	0.969	95.79
	V	73.18	0.846	0.74	66.38	0.794	0.678	77.04	0.868	0.777	92.18	0.959	0.924	100	1	1	97.78	0.989	0.981	92.2
	VI	78.49	0.878	0.795	74.56	0.857	0.764	79.04	0.885	0.801	87.66	0.935	0.883	96.92	0.984	0.969	100	1	1	93.04
HY	I	100	1	1	96.15	0.98	0.962	90.83	0.953	0.921	85.62	0.925	0.883	89.13	0.939	0.893	95.39	0.976	0.953	96.41
	II	97.44	0.987	0.979	100	1	1	96.23	0.981	0.968	87.45	0.934	0.897	89.9	0.945	0.899	95.74	0.978	0.957	97.22
	III	95.58	0.978	0.961	97.3	0.987	0.972	100	1	1	93.52	0.966	0.949	91.61	0.955	0.918	95.98	0.979	0.958	97.82
	IV	91.88	0.958	0.925	88.94	0.941	0.888	93.48	0.967	0.93	100	1	1	94.45	0.971	0.941	96.63	0.983	0.962	97.1
	V	85.25	0.921	0.861	78.51	0.878	0.797	77.9	0.883	0.796	86.81	0.929	0.887	96.91	0.984	0.967	100	1	1	93.77
	VI	84.73	0.916	0.855	77.05	0.861	0.778	70.86	0.83	0.721	74.65	0.842	0.76	90.32	0.947	0.906	100	1	1	91.45
MI	I	100	1	1	92.1	0.958	0.923	87.73	0.935	0.88	84.07	0.914	0.844	87.56	0.934	0.875	91.19	0.954	0.907	95.2
	II	94.34	0.971	0.947	100	1	1	95.92	0.979	0.959	89.52	0.945	0.895	89.22	0.944	0.889	91.52	0.956	0.911	96.69

(continued)

Table 1.8 (continued)

		V_1			V_2			V_3			V_4			V_5			V_6			Avg
		R^2	r_x	b_x	R^2	r_x	b_x	R^2	r_x	b_x	R^2	r_x	b_x	R^2	r_x	b_x	R^2	r_x	b_x	
	III	89.82	0.946	0.907	95.35	0.976	0.959	100	1	1	95.55	0.977	0.954	91.49	0.956	0.912	91.64	0.956	0.913	96.96
	IV	82.21	0.903	0.834	83.34	0.909	0.844	92.55	0.96	0.93	100	1	1	94.84	0.974	0.947	91.83	0.958	0.915	95.37
	V	73.83	0.852	0.75	68.54	0.818	0.701	73.96	0.851	0.753	84.38	0.907	0.85	100	1	1	94.72	0.973	0.946	91.26
	VI	73.93	0.851	0.752	64.98	0.786	0.666	65.35	0.796	0.67	70.35	0.826	0.714	89.99	0.947	0.905	100	1	1	88.69
VA	I	100	1	1	90.63	0.95	0.901	86.57	0.929	0.866	85.54	0.924	0.863	91.6	0.957	0.917	93.74	0.968	0.934	95.63
	II	93.73	0.967	0.942	100	1	1	94.25	0.97	0.945	89.97	0.948	0.906	91.89	0.959	0.921	93.78	0.968	0.934	96.93
	III	88.68	0.935	0.888	92.86	0.961	0.927	100	1	1	95.84	0.979	0.96	93.79	0.968	0.968	94.27	0.97	0.937	97.08
	IV	81.68	0.895	0.817	80.38	0.889	0.801	91.69	0.957	0.913	100	1	1	93.96	0.963	0.941	94.89	0.974	0.945	95.18
	V	74.11	0.834	0.748	68.59	0.816	0.691	77.72	0.877	0.779	90.66	0.95	0.908	100	1	1	97.14	0.985	1	92.31
	VI	78.02	0.874	0.782	69.18	0.819	0.692	72.02	0.841	0.723	80.46	0.888	0.816	96.02	0.979	0.967	100	1	1	91.27
ND	I	100	1	1	90.42	0.951	0.916	81.28	0.891	0.825	78.06	0.876	0.787	82.08	0.902	0.823	85.84	0.91	0.86	92.39
	II	90.92	0.949	0.913	100	1	1	93.54	0.967	0.931	83.72	0.91	0.838	84.61	0.916	0.847	87.28	0.92	0.876	94.44
	III	88.18	0.937	0.899	95.74	0.978	0.963	100	1	1	93.07	0.965	0.931	88.5	0.939	0.887	88.4	0.932	0.89	96.02
	IV	83.27	0.911	0.856	83.5	0.914	0.857	90.22	0.95	0.916	100	1	1	95.58	0.977	0.952	92.85	0.963	0.928	95.39
	V	76.77	0.874	0.796	68.32	0.825	0.723	67.11	0.802	0.704	90.12	0.948	0.909	100	1	1	96.89	0.984	0.969	91.52
	VI	78.51	0.883	0.816	72.83	0.853	0.766	69.69	0.831	0.728	83.32	0.911	0.843	94.97	0.974	0.951	100	1	1	91.51

V5 or V6. As for an illustration, row 1 of the 'BB' case in Table 1.8, basic lead of RL system i.e. I (comprising of lead I, II and V1 as per the definition given in Table 1.7) is used to reconstruct V1–V6 in standard 12-lead ECG system. It is apparent from the row 1 that R^2, correlation (rx) and regression (bx) for the reconstruction of V1 is 100%, 1 and 1 respectively; V2 is 83.36%, 0.91 and 0.872, respectively and so on. In this way, all rows of Table 1.8 for different diseased classes as well as Healthy Control can be read and interpreted.

Table 1.9 shows the reconstruction results from SRM, i.e. FV to S12 system with average values mentioned in the last row. From Tables 1.8 and 1.9, it can be seen that the FRM outperforms SRM for lead set 1 as leads I and II constitute the basis lead set and other four dependent leads can be derived from them using simple linear algebraic relations. However, still second methodology produces more than 90% R2 value for all these leads. For leads V5 and V6, second methodology (i.e. FV to S12) outperforms all others except where the same leads themselves form the basis leads. For rest of the four precordial leads, the performance of R3L systems are inconsistent compared to FV system.

Figure 1.10a–c present reconstructed signal (red) overlapping the original signal (blue), obtained using SRM, of three different patients with worst (71.3%), 80% and best case (99.61%) mean R^2 values for lead set 1. The R^2 values of corresponding leads are mentioned in the figure. Figure 1.10d–i presents a similar comparison for lead set 2 for both FRM and SRM, where the first six boxes in all the sub-figures correspond to FRM in the order of R3L systems with leads V1, V2, V3, V4, V5 and V6 as the precordial lead in the basis lead set and the last box corresponds to SRM. Figure 1.10d corresponds to reconstruction of V1, similarly, Fig. 1.10e corresponds to reconstruction of V2 and so on. Figure 1.11 provides the box plot of lead-wise mean RMSE for 13 different features extracted using TDMG over the complete database. Each sub-figure in Fig. 1.11, i.e. A–H, presents 13 boxes numbered 1–13 each corresponding to a particular feature. The correspondence between the labels in the horizontal axis and features has been described in Table 1.7. Figure 1.11a–f follows from the reconstruction results of RL systems for lead set 2 with basis lead following the order: A–V1, B–V2, C–V3, D–V4, E–V5 and F–V6 (all of these basis lead sets essentially contain leads I and II). Figure 1.11g, h follows from the results of SRM for lead set 2 and lead set 1, respectively. The edges of the box are the 25th and 75th percentiles, the whiskers extend to ±2.7 standard deviation and rest are plotted individually. Table 1.10 provides the mean RMSE values for all the seven methodologies. It can be seen that higher values of error has mainly occurred when height or depth of a peak was measured in comparison to the features measuring the horizontal intervals. Among the R3L systems, R3L systems with basis leads V2 and V3 have outperformed the others. SRM has outperformed FRM for lead set 2; however, the results are otherwise for lead set 1. Table 1.11 presents a comparison between FRM and SRM in the context of remote healthcare monitoring applications.

Table 1.9 Mean R^2, r_x, and b_x Values of Various Categories for the Transformation of FV System to S12 System

	BB			HC			HY			MI			VA			ND		
	R^2	r_x	b_x	R^2	r_x	b_x	R^2	r_x	b_x	R^2	r_x	b_x	R^2	r_x	b_x	R^2	r_x	b_x
I	90.89	0.952	0.898	95.99	0.98	0.96	95.34	0.976	0.959	92.41	0.961	0.93	94.15	0.971	0.955	93.61	0.968	0.927
II	94.29	0.97	0.944	99.04	0.995	0.99	98.33	0.992	0.983	96.54	0.982	0.967	97.41	0.987	0.982	93.55	0.966	0.947
III	88.29	0.916	0.88	93.18	0.965	0.929	95.41	0.976	0.952	93.55	0.967	0.94	91.99	0.958	0.918	91.84	0.952	0.92
aVR	91.06	0.953	0.914	98.49	0.993	0.985	97.71	0.989	0.979	93.86	0.969	0.945	95.59	0.978	0.966	91.6	0.955	0.918
aVL	91.05	0.953	0.904	91.75	0.957	0.918	93.63	0.967	0.937	92.18	0.96	0.926	90.28	0.945	0.914	94.24	0.971	0.963
aVF	96.01	0.98	0.961	98.09	0.99	0.98	97.34	0.987	0.974	95.99	0.979	0.96	96.41	0.982	0.966	95.6	0.977	0.959
V1	91.12	0.954	0.897	96.41	0.982	0.966	97.29	0.986	0.977	94.15	0.97	0.949	94.83	0.974	0.953	92.07	0.959	0.928
V2	89.31	0.943	0.89	94.03	0.97	0.942	96.39	0.982	0.961	92.29	0.96	0.929	91.19	0.951	0.916	91.28	0.955	0.925
V3	94.44	0.973	0.953	96.9	0.984	0.968	96.91	0.985	0.971	94.56	0.972	0.95	95.2	0.976	0.959	94.72	0.973	0.953
V4	93.75	0.97	0.961	98.06	0.99	0.982	95.82	0.979	0.974	95.66	0.977	0.959	97.36	0.987	0.984	97.04	0.985	0.967
V5	97.79	0.989	0.976	99.27	0.996	0.993	96.19	0.98	0.964	97.53	0.988	0.975	98.8	0.994	0.991	96.64	0.983	0.964
V6	98.7	0.994	0.991	99.51	0.998	0.996	99.24	0.996	0.992	97.54	0.988	0.974	98.74	0.994	0.988	98.12	0.991	0.983
Average	93.06	0.9623	0.931	96.73	0.983	0.967	96.63	0.983	0.968	94.69	0.973	0.950	95.16	0.975	0.958	94.11	0.969	0.9436

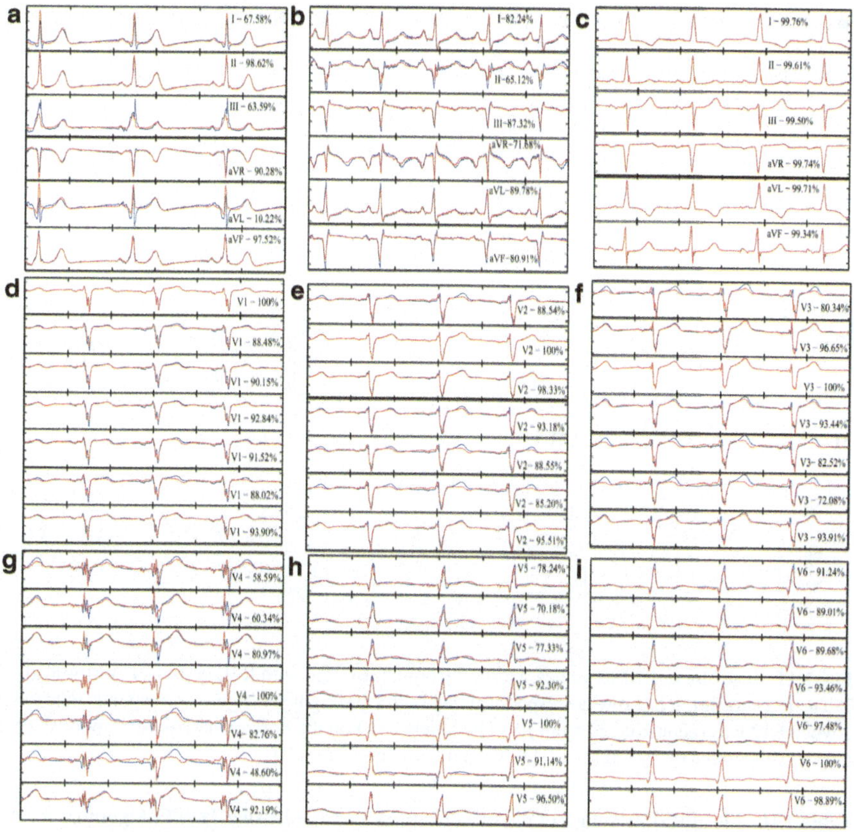

Fig. 1.10 (Colour online) Comparison between the reconstructed (red) and the measured (blue) ECG signal. **a–c** Shows the reconstruction of lead set 1 using SRM, i.e. FV to S12 for the worst, 80% and the best case mean R^2 values. **d–i** show the reconstruction of lead set 2 (V1 –V6) for all the RL systems with the basis leads starting from V1 (top) to V6 and FV system (bottom)

1.3.5 Frank Vectorcardiographic System from Standard 12 Lead ECG: An Effort to Enhance Cardiovascular Diagnosis

Figure 1.12 shows the summary of the methodology followed in this investigation [20]. First the databases are de-noised using the pre-processing module followed by PCA module for the construction of FV from S12 system. IDT and KT were also used to reconstruct Frank's leads. All the three sets of derived Frank leads were then compared with actually recorded Frank leads. Finally, the S12 leads was reconstructed from the derived FV leads, obtained using the proposed PCA-based methodology and were compared with the originally measured S12 leads. The personalized reconstruction of S12 system from derived FV system further includes a module for

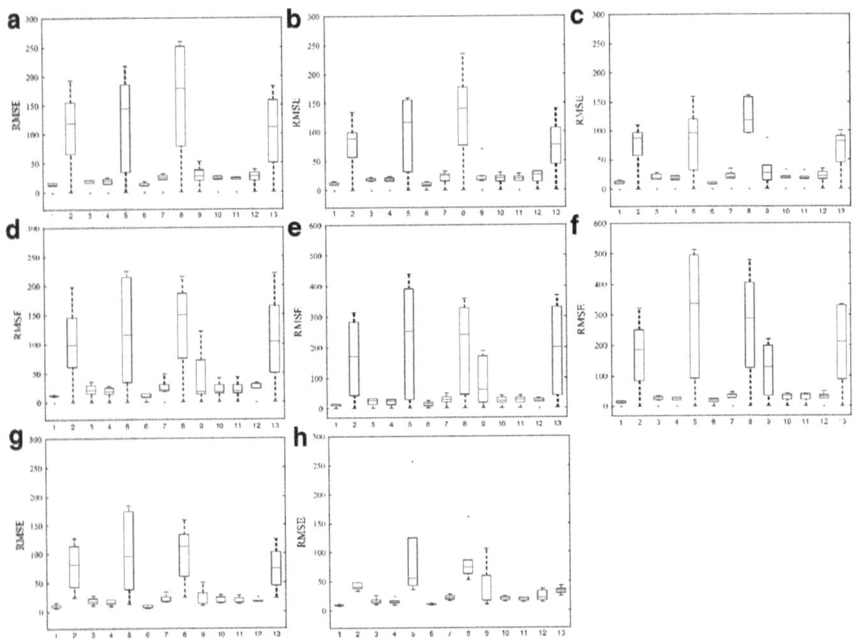

Fig. 1.11 Box plot of RMSE values. Starting from left to right with R3L systems I–VI for lead set 2. The left sub-figure last row corresponds to lead set 2 and right sub-figure corresponds to lead set 1 when S12 was reconstructed from FV. The labels 1–13 on the horizontal axis corresponds to the respective features extracted from TDMG as mentioned in the text

Table 1.10 Mean RMSE Values of the ECG Features Extracted Using TDMG Algorithm for All the RL Systems

Sr.	Feature (in Unit)	I	II	III	IV	V	VI	FV
1	P duration (ms)	6.700	5.536	5.423	5.644	5.888	6.357	9.70
2	P height (mV)	55.47	40.04	37.30	51.00	82.70	86.57	60.24
3	PR interval (ms)	9.058	8.639	9.281	10.96	11.80	12.27	17.20
4	PR segment (ms)	8.982	8.567	8.595	9.36	10.68	11.45	15.81
5	Q peak (mV)	63.60	51.33	44.34	61.49	116.4	150.6	97.75
6	QRS length (ms)	5.997	4.710	4.355	4.966	7.407	8.527	9.40
7	QT interval (ms)	11.97	11.26	10.64	12.41	14.38	15.98	21.41
8	R height (mV)	79.96	65.48	54.85	65.47	102.3	133.1	92.78
9	S peak (mV)	14.25	12.87	16.67	20.79	42.10	59.59	29.73
10	ST interval (ms)	10.94	9.889	9.629	10.94	12.69	14.72	19.32
11	ST segment (ms)	10.45	9.542	9.316	10.96	13.36	15.00	18.49
12	T duration (ms)	13.33	11.85	11.03	12.04	12.21	14.88	21.16
13	T height (mV)	52.42	38.26	33.79	54.17	95.76	98.51	53.61

Table 1.11 Comparison of FRM and SRM in Context of Remote Health-care Applications

	R3L system	FV system
1	Five electrodes system	Eight electrodes system
2	Three leads	Three leads
3	Inconsistent reconstruction of precordial leads	Consistent reconstruction of precordial leads
4	Comparatively bad reconstruction of V5 and V6	Comparatively better reconstruction of V5 and V6
5	Leads I, II, III, aVR, aVL, and aVF are obtained with approximately no information loss	Comparatively less accurate and information is lost in the reconstructed signal
6	Not much change in already existing acquisition system is required	A different system is required which can acquire both ECG and VCG
7	Online and offline registration possible	Online registration difficult

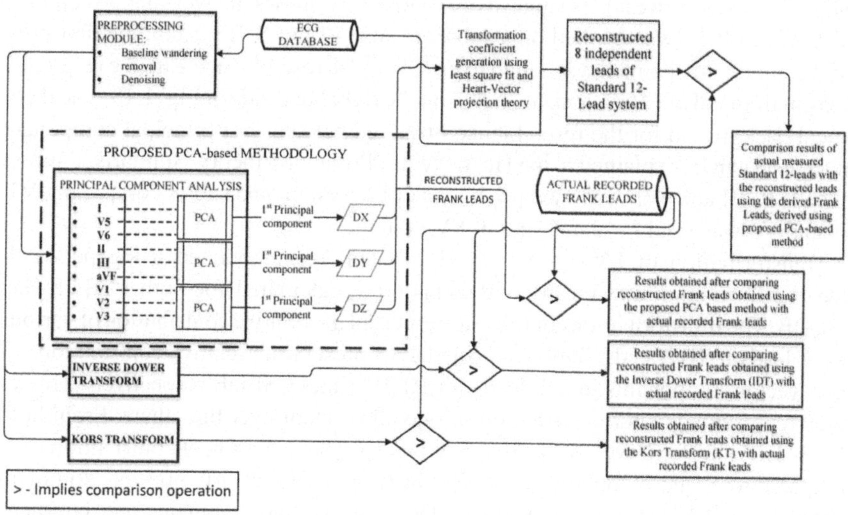

Fig. 1.12 Summary of the proposed methodology for reconstruction of Frank system from Standard 12- lead system and then using the derived Frank leads (DX, DY and DZ) to reconstruct standard 12-leads using personalized transformation employing least square fit method and Heart-Vector projection theory

generation of transformation coefficient. PTBDB sampled at 1000 Hz and CSEDB sampled at 500 Hz were used in this investigation.

Pre-processing involves two steps namely, baseline wandering removal and de-noising of the signals. These signals are then used to derive FV system leads. PCA is mainly used for dimensionality reduction. It obtains the best set of the points which is the best representation of the dataset, known as principal component. The 1st principal component has the maximum variance, hence maximum information,

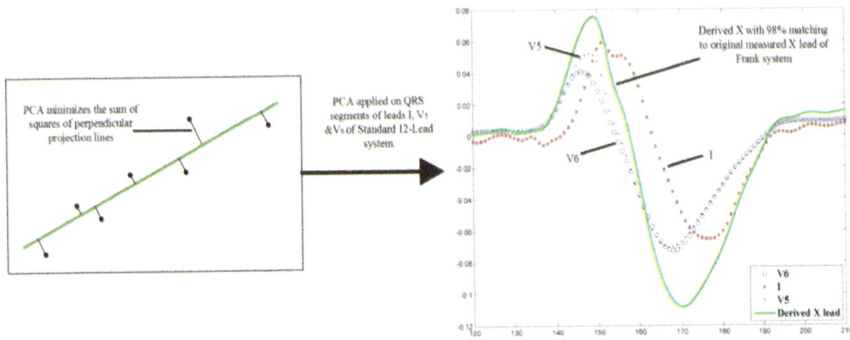

Fig. 1.13 Effect of PCA on a data set and when applied on 3-lead subset of standard 12-lead system. The subset in the figure includes I, V5 and V6. The resulting first principal component is shown as a continuous plot in green whose resemblance to originally measured Frank's X lead is 98.69%

and it decreases with 2nd, 3rd and so on. Figure 1.13 shows the working principal of PCA. When PCA was applied on a subset (I, V5, V6) of S12 system, the first principal component obtained was found to have 98.69% resemblance with the originally measured X lead of FV system, as shown on the right hand side of Fig. 1.13. The theoretical background for the resemblance observed between 1st principal component and X lead can be explained using Heart-Vector Projection theory. Similarly, a subset of leads II, III and aVF was used for Y lead of FV system and subset comprising V1, V2 and V3 was used for the Z lead of FV system.

Transformation of FV system to S12 system were obtained using personalized transformation and mean coefficients were calculated for both PTBDB and CSEDB. The transformation coefficients represent the relative contribution of various Frank leads to S12 leads. Tables 1.12 and 1.13 present the relative contributions of heart vector components in S12 leads for PTBDB and CSEDB respectively. Higher value represents greater contribution of one component over the others. From both Tables 1.12 and 1.13, we can see that I, V5 and V6 have dominant contribution of X compared to Y and Z components. Leads II, III and aVF, clearly presents dominant contribution of Y compared to X and Z. The balanced and dominant contribution of Z from Tables 1.12 and 1.13 is not readily visible however, leads V1, V2 and V3 can be found to lead among the rest.

The derived Frank leads, obtained using three different methodologies viz. the proposed PCA-based method, IDT and KT have been compared independently with the originally recorded Frank lead using the evaluation metrics in Table 1.14. It shows the number of patients (in %) in various ranges of mean R^2 values. Mean was taken over derived X, Y and Z leads of FV system and all the patients of PTBDB and CSEDB were included. All the transformations have performed relatively well for CSEDB compared to PTBDB. For PTBDB, PCA-based has significantly outperformed IDT and KT transformations. However, for CSEDB, KT outperforms the PCA-based method followed by IDT. Superior results obtained for KT on CSEDB can be explained upon considering that it was derived using population based linear

Table 1.12 Fractional Content of Heart Dipole Components in S12 Leads for PTB Database (PTBDB)

Lead	x/y	x/z	y/x	y/z	z/x	z/y
I	7.338	10.06	0.136	1.370	0.099	0.73
II	0.164	8.155	6.112	49.85	0.123	0.02
III	0.589	1.296	1.699	2.202	0.771	0.45
aVR	0.129	0.245	7.767	1.904	4.079	0.52
aVL	0.797	3.560	1.255	4.468	0.281	0.22
aVF	0.355	1.583	2.821	4.465	0.632	0.22
V1	0.677	2.079	1.477	3.071	0.481	0.32
V2	2.773	3.352	0.361	1.209	0.298	0.82
V3	1.239	8.521	0.807	6.876	0.117	0.14
V4	0.834	17.23	1.199	20.67	0.058	0.04
V5	0.445	5.094	2.254	11.48	0.196	0.08
V6	0.996	4.884	1.004	4.905	0.205	0.20

Table 1.13 Fractional Content of Heart Dipole Components in S12 Leads for CSE Database (CSEDB)

Lead	x/y	x/z	y/x	y/z	z/x	z/y
I	7.207	13.79	0.139	1.914	0.072	0.523
II	0.315	2.829	3.178	8.990	0.354	0.111
III	0.540	2.927	1.850	5.417	0.342	0.185
aVR	1.141	126.5	0.877	110.9	0.008	0.009
aVL	1.082	7.522	0.924	6.953	0.133	0.14
aVF	0.149	0.833	6.722	5.594	1.202	0.19
V1	2.637	0.592	0.379	0.224	1.691	4.45
V2	0.295	0.086	3.385	0.290	11.65	3.43
V3	3.301	0.574	0.303	0.174	1.744	5.75
V4	81.77	1.401	0.012	0.017	0.714	58.36
V5	6.204	3.472	0.161	0.560	0.288	1.78
V6	3.155	14.89	0.317	4.718	0.067	0.22

regression method from CSE database itself, which has since been used as the universal transformation matrix. However, KT fails to replicate its performance in PTBDB. It should be noted that KT produces a positive R^2 values only in 56.47% of patients for PTBDB, for all other patients it critically fails. The proposed PCA-based method is personalized compared to database based Kors Transform. IDT's performance fluctuates with respect to KT, for positive R^2 values IDT outperforms KT for PTBDB, however, for R^2 values greater than 50% KT has outperformed IDT. IDT is based on the assumption of homogeneity and fixed body characteristics, however, KT

Table 1.14 Number of subjects (in %) with various values of reconstruction accuracy of frank system from Standard 12-Lead system for both PTBDB and CSE database using the Proposed Methodology (PCA-Based), Inverse Dower Transform (IDT), and Kors Transform (KT)

Mean R^2 values	No. of Patients in CSEDB (in %)			No. of Patients in PTBDB (in %)		
	PCA-Based	IDT	KT	PCA-Based	IDT	KT
>0%	100%	98.8%	99.6%	97.63%	73.04%	56.47%
>50%	90.4%	94%	97.2%	73.95%	47.91%	47.18%
>80%	68%	66.8%	75.2%	40.98%	30.05%	34.97%
>90%	48.8%	40%	55.6%	19.85%	11.66%	19.31%
Overall mean R^2 value	81.6%	80.93%	85.52%	65.77%	34.15%	26.89%
Overall mean correlation coefficient	0.8289	0.6708	0.6344	0.9080	0.9046	0.9276

Table 1.15 Fraction (in %) of Subjects PTBDB with various reconstruction accuracy values for reconstruction of Frank System from Standard 12-Lead System for Healthy Control (HC), Unhealthy (UH) for 290 Records, and Remaining Records

Mean R^2 values	Healthy control: 52	Unhealthy: 238 (%)	Remaining records
>0%	100%	99.2%	96.14%
>50%	92.31%	92.44%	58.69%
>80%	82.69%	67.65%	17.76%
>90%	63.46%	47.06%	2.703%
Overall mean R^2 value	86.63%	80.90%	52.84%
Overall mean correlation	0.933	0.904	0.764

has been obtained from real patients, hence, includes the effects due to heterogeneity and other body characteristics i.e. body fat distribution, size and shape of the body.

Table 1.15 presents a comparison in performance of proposed PCA-based method on the first recordings of patients in PTBDB, which excluded patients with pace makers and other therapies, and the remaining recordings. First recordings were further subdivided into healthy control and unhealthy subjects. A remarkable difference of 5.73% (R^2) in performance for healthy subjects can be seen compared to unhealthy subjects and 29.09% (R^2) for first recording compared to rest. Even if electrode placement error is considered, it does not explain the considerable difference in reconstruction result, since, 290 first recordings and 259 further records were considered. Condition of heart can severely affect the reconstruction process. This comparison supports the theoretical basis behind the PCA-based reconstruction methodology. Table 1.16 presents the reconstruction results of S12 system from derived FV system. The Frank leads used for reconstruction of S12 system have been synthesized using the proposed PCA-based method. Previously, it has been shown that personalized transformations outperformed state-of-the-art Dower transform and Affine transform, here, in Table 1.16 a comparison between the personalized reconstruction of S12 system from derived FV leads and original FV leads has been

Table 1.16 Mean R^2 and Correlation coefficient (CC) values for the reconstruction of Standard 12-Lead System from Derived Frank Leads, derived using PCA-based methodology and its comparison with the reconstruction result using originally measured Frank Leads

Derived frank leads			PTBDB				CSEDB	
			Original Frank Leads		Derived Frank Leads		Original Frank Leads	
Leads	R^2 (%)	CC (r_x)	R^2 (%)	CC (r_x)	R^2(%)	CC (r_x)	R^2 (%)	CC (r_x)
I	48.85	0.527	46.39	0.494	92.06	0.9646	89.91	0.950
II	97.57	0.987	92.21	0.959	98.36	0.9928	95.29	0.977
V1	87.54	0.930	92.91	0.961	92.45	0.9609	91.61	0.961
V2	97.32	0.986	83.87	0.912	97.67	0.9884	87.41	0.954
V3	96.56	0.983	85.73	0.920	96.28	0.9840	92.86	0.967
V4	94.82	0.972	88.34	0.935	93.49	0.9676	95.41	0.977
V5	97.25	0.986	90.42	0.947	97.15	0.9871	97.46	0.988
V6	97.31	0.987	95.39	0.975	97.54	0.9912	97.46	0.988
Mean	89.65	0.9198	84.41	0.8878	95.62	0.9795	93.43	0.970

presented. Derived FV leads outperform original FV leads by 5.24% and 2.19% for PTBDB and CSEDB respectively.

Figure 1.14 shows the comparison between derived (red) and originally measured (blue) FV leads for three different patients in PTBDB. The mean R^2 values of the three patients lie close to mean, median and maximum R^2 values respectively, obtained using PCA-based method. Corresponding reconstruction performances for IDT and KT have also been shown. Figure 1.14a–c presents mean case patient's FV leads (X, Y and Z) reconstruction using PCA-based method, IDT and KT respectively. Similarly, Fig. 1.14d–f shows the median case and Fig. 1.14g–i shows the maximum case for PCA-based method, IDT and KT respectively.

Figure 1.15 presents a similar comparison as mentioned in previous paragraph for mean, median and maximum case patients in CSEDB using PCA-based method and its comparison with IDT and KT. Figure 1.15a–c presents mean case patient, Fig. 1.15d–f presents median case and Fig. 1.15g–i presents maximum case. Figures 1.14 and 1.15, contains a range of R^2 values (20.94% to 99.51%) and presents a relation between R^2 values and its corresponding reconstruction. From the wide variety of evaluation results and consultation with two co-authoring experienced cardiologists, authors conclude that the R^2 value of 80% and above can be assumed to be diagnostically accurate for all practical purposes.

Vectorcardiography in general and Frank system, in particular, were postulated and realized six decades ago in pre-informatics period devoid of modern computing capabilities and low-power low-cost electronic devices. The authors, hence, strongly believe that possible advantages of orthogonal leads were not fully explored. Unavailability of hassle-free Frank VCG was among the primary reasons behind dropping it from routine tests. In this chapter, authors have attempted to address this issue.

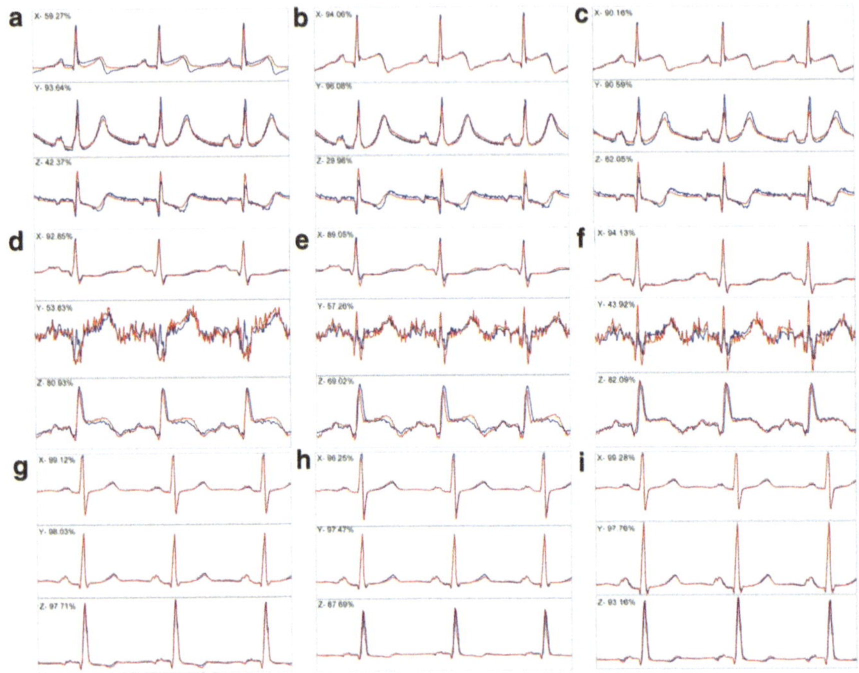

Fig. 1.14 Comparison between original (in blue) and derived (in red) signals of Frank system when constructed from Standard 12-lead system. **a**, **d** and **g** shows the construction using the proposed PCA-based method for the subjects which had mean, median and maximum R^2 value in PTBDB. **b**, **e** and **h** shows reconstruction for Inverse Dower Transform and **c**, **f** and **i** shows reconstruction for Kors Transform for the same subjects

1.3.6 A Novel 2 Lead to 12 Lead ECG Reconstruction Methodology for Remote Health Monitoring Applications

Naresh et al. [25] presented a novel solution for 2-lead to 12lead ECG reconstruction methodology that improves the remote health monitoring applications. In this study authors have used PTBDB database sampled at 1000 Hz. Three categories Healthy Control (HC), Bundle Branch Block (BBB), and Myocardial Infarction (MI) are chosen to validate the proposed methodology. HC consists of 80 patient's data, BBB consists of 16 patient's data and MI consists of 333 patient's data, each patient data holds 1 min recording. As a part of pre-processing, artifacts are removed from the raw ECG signal by passing through 4th order butter worth band pass filter having cut-off frequencies as 0.05 Hz–40 Hz, as most of the ECG signal frequencies lies in that range. Median filter is used to estimate baseline which is then subtracted from the filtered signal. The pre-processed data is then used for further processing throughout the work.

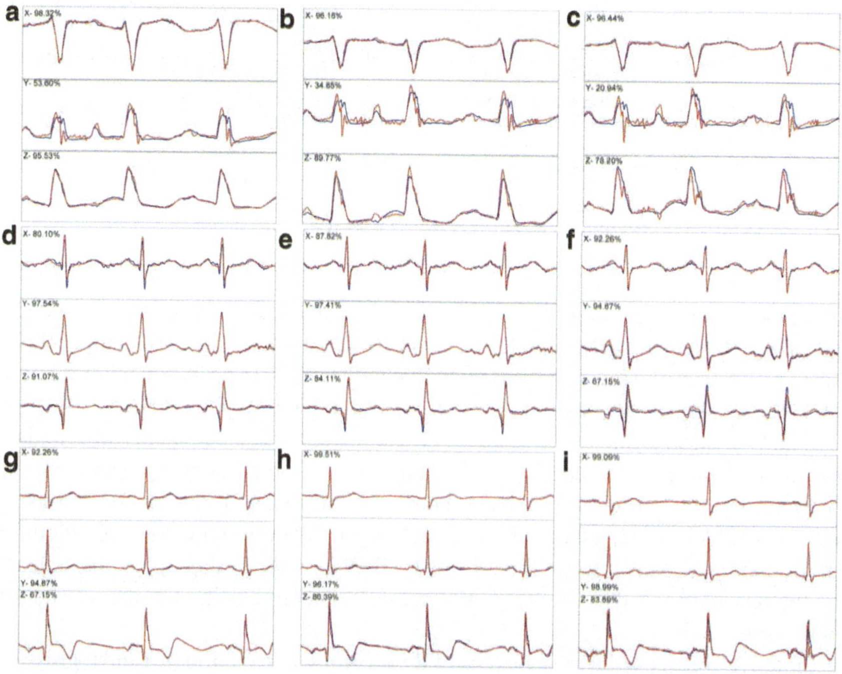

Fig. 1.15 Comparison between original (in blue) and derived (in red) signals of Frank system when constructed from Standard 12-lead system. **a**, **d** and **g** shows the construction using our proposed PCA-based method for the subjects which had mean, median and maximum R^2 value in CSEDB. **b**, **e** and **h** shows reconstruction for Inverse Dower Transform and C, F and I shows reconstruction for Kors Transform for the same subjects

The noise free 12 lead ECG signals obtained from the pre-processing block are then used for generating the personalized coefficients using Least Square (LS) fit method and Heart-Vector Projection (HVP) theory as shown in Fig. 1.16. These

Fig. 1.16 Generation of the coefficients for all the leads with the existing data

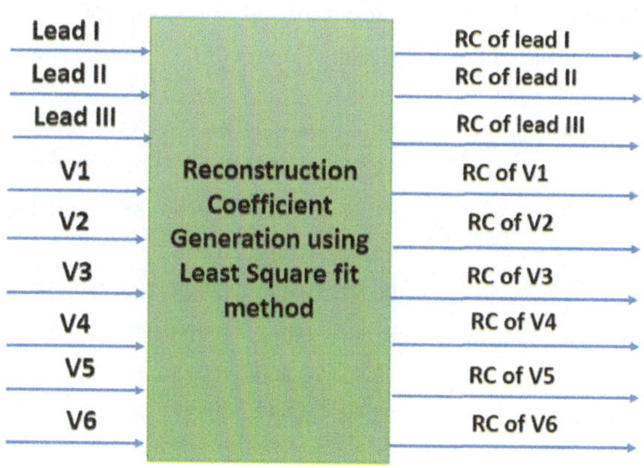

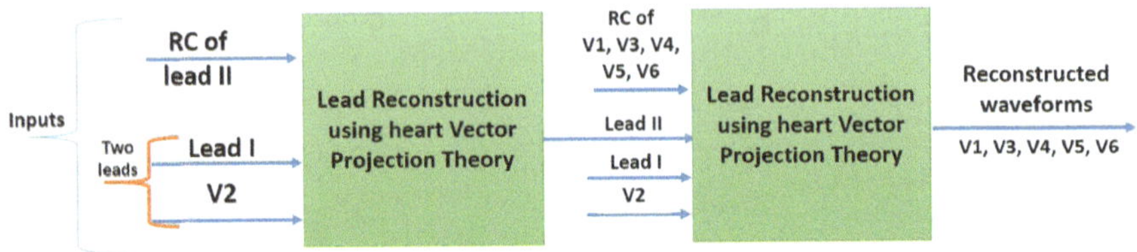

Fig. 1.17 Reduced lead system to reconstruct the 12 lead system using 2 lead ECG

coefficients along with Lead I and V2 can be transmitted to server. The reduced 2 lead system along with the generated personalized coefficients are used at the receiver side to reconstruct the missing leads as shown in Fig. 1.17. In proposed method, Lead I and V2 are taken as the basis leads to reconstruct the missing leads (II, V1, V3, V4, V5 and V6). Lead III, aVR, aVL and aVF can be reconstructed using the standard procedure (Lead III = Lead II-Lead I) using Lead I and II. Hence, these leads are not reconstructed using the proposed method. Firstly lead II will be derived using these two basis leads (I, V2) and reconstruction coefficients in the proposed method. After obtaining Lead II, these three leads (I, II, V2) will be taken as the basis leads and the remaining leads (V1, V3 to V6) will be generated using the reconstruction coefficients.

Table 1.17 presents the mean values of R^2 statistics, correlation (r_x) and regression (b_x) coefficients for the three categories HC, BBB and MI. The mean values of the stated performance metrics obtained were 91.94%, 0.95, 0.921 for HC, 85.81, 0.920, 0.856 for BBB and 81.42, 0.889, 0.820 for MI respectively. From the Table 1.17, it is evident that the maximum reconstruction performance obtained for lead V3 for all the categories while the minimum reconstruction performance obtained for lead II. And also the results show that HC reconstruction performance is greater than that of BBB and MI. This indicates that the accuracy of reconstruction depends on many factors which affect the values of personalized coefficients and it can be interpreted that the performance in incremented and decremented due to the variation in the

Table 1.17 Performance results of 2 to 12 lead reconstruction

Reconstructed Leads	Healthy Control (80)			Bundle Branch Block (16)			Myocardial Infarction (333)		
	R^2	r_x	b_x	R^2	r_x	b_x	R^2	r_x	b_x
II	88.21	0.935	0.882	78.18	0.874	0.782	63.17	0.768	0.632
V1	91.77	0.955	0.918	88.84	0.937	0.888	92.59	0.960	0.923
V3	95.77	0.978	0.958	93.79	0.968	0.938	93.97	0.968	0.940
V4	90.81	0.950	0.908	84.71	0.918	0.847	82.98	0.904	0.830
V5	92.97	0.962	0.930	82.51	0.903	0.825	78.29	0.871	0.783
V6	92.11	0.960	0.927	86.82	0.921	0.856	77.54	0.865	0.814
Average	**91.94**	**0.957**	**0.921**	**85.81**	**0.920**	**0.856**	**81.42**	**0.889**	**0.820**

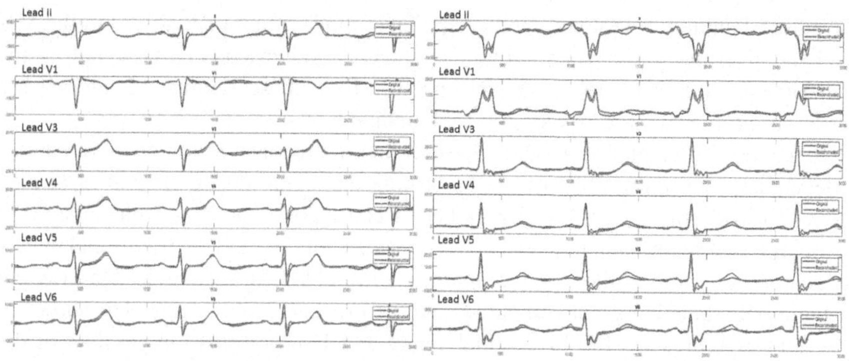

Fig. 1.18 Comparison analysis of original (blue) and reconstructed (red) ECG signals

placement of electrodes while recording the ECG data, as the electrodes placement changes for every recording on the body. Other factors like age, sex, size, shape, fat distribution and any arrhythmias are less likely to change the coefficient variation. Figure 1.18 shows the originals vs reconstructed signal plots. It illustrates that most of the clinical features are retained in the reconstructed signal for all the missing leads.

1.3.7 A Novel Method Based on Convolutional Neural Networks for Deriving Standard 12-Lead ECG from Serial 3-Lead ECG

Wang et al. [28] presented an approach to synthesize missing precordial leads from S12 system using convolutional neural networks (CNN). Authors have used PTBDB for the validation of the proposed approach. Authors have divided each patients data into 2 pairs each consisting of the following.

- A 3 lead ECG consisting of I, II and V2 as the input database
- Other 9 lead of S12 system as the target database.

To evaluate the results, authors have divided the study population into two subsets, DS1 and DS2. Dataset DS1 consists of the 161 patients (116 male, 45 female; mean age $\pm$SD $= 58.55 \pm 14.25$). Dataset DS2 consists of the remaining 129 patients (93 male, 36 female; mean age $\pm$ SD $= 55.51 \pm 15.19$).

In the proposed method first all the signals are de-noised using DWT and soft thresholding. The Daubechies6 wavelet is used to decompose the ECG stream to eight levels, and only 3rd/4th/5th detail coefficients are used to reconstruct the new signal for removing motion artifacts. Then the authors have segmented the signals beat wise using Pan and Tompkins algorithm (Pan et al. 1985). Since the V4-lead signal of this lead has the best geometric characteristics of the R wave, authors have used Pan and Tompkins' method to detect the location of R waves in the V4-lead signal in the study. Then authors segment the other leads ECG signal by the $[R_i-(R_i-R_{-1})/2, R_i + (R_{i+1}-R_i)/2]$ interval. Then these segmented interval are further used in next steps in the proposed method.

Since the ECG signal is a 1D data vector and the reconstruction process is time-independent, the first problem faced is turning the signal into a 2D matrix of data points. Considering 1D-ECG to 2D-ECG conversion, reference has been made to the gradient-based method on image/shape reconstruction [10] and the traditional matrix reconstruction method based on slipping insertion [17]. In this study, the gradient of the input signal has been computed, and then repeated and arranged as shown in Fig. 1.19 (1 represents the #1 original signal and 1' represents the gradient of the #1 signal). Overall, a total of 150000 images are obtained. Authors have used 120 000 images as a training set to train the CNNs, and for validation, 30000 images as a testing set to evaluate the proposed method. The aim of this transformation is not only to satisfy the input requirements, but also to obtain more details of the input signal.

Figure 1.19 is the structure schematic of the proposed method and the details of the CNNs are given in Fig. 1.20. The convolutional and pooling layers compose each CNN stage. The aim of the convolutional layer is to represent more information from the previous layer, and the pooling layer is used to merge similar features in each stage. In this study, back-propagation method has been used to train the multi-layer CNN, which applies the chain rule for derivatives to compute the gradient of a predefined objective function relative to all the neuron parameters. Learning rate has been set to 0.1, the batch size as 100, and the number of training epochs as 1.

As shown in Fig. 1.21a, first convolution layer C1 consisting of six feature maps, which are computed using overlapping 4 × 4 kernels on the input 9 × 9 signal data. The convolution operation can enhance the original signal feature and reduce the noise. At layer S2 (first subsampling layer) (Fig. 1.21b), there are six feature maps, each of size 3 × 3 obtained by subsampling based on max-pooling using a 2 × 2 kernel on the output of the Cl layer. According to the principle of image local correlation, subsampling of the image can reduce the amount of data processing while retaining useful information. Each unit in the feature map in layer S2 is connected to the 2 × 2 neighbourhood of the corresponding feature map in layer C1. A sigmoid function

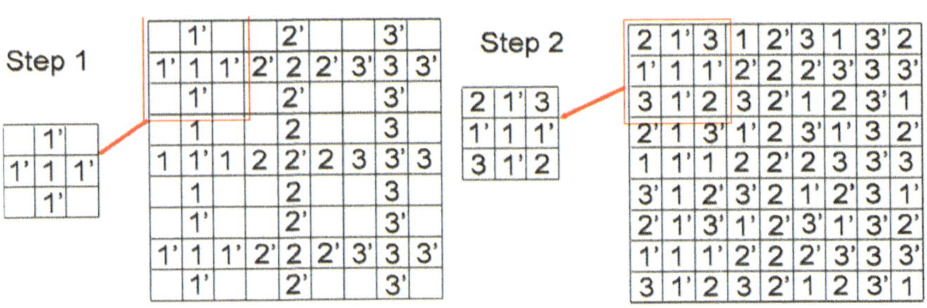

Fig. 1.19 The structure of the input signal

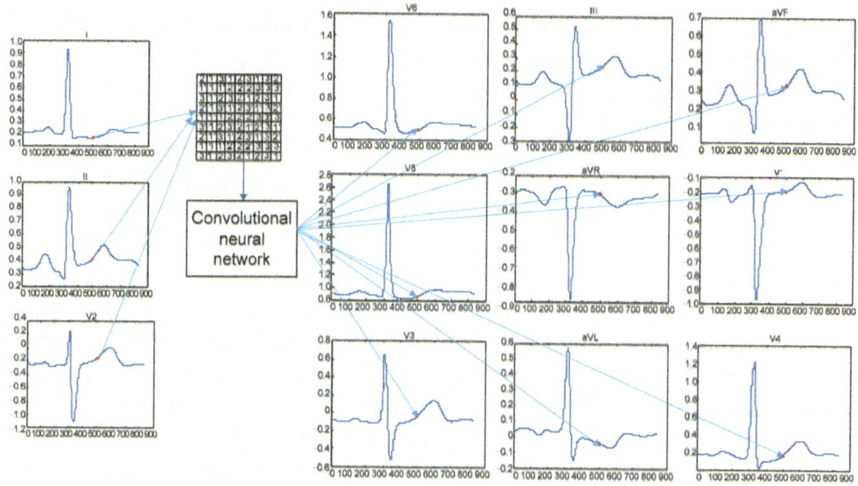

Fig. 1.20 The schematic of the proposed method

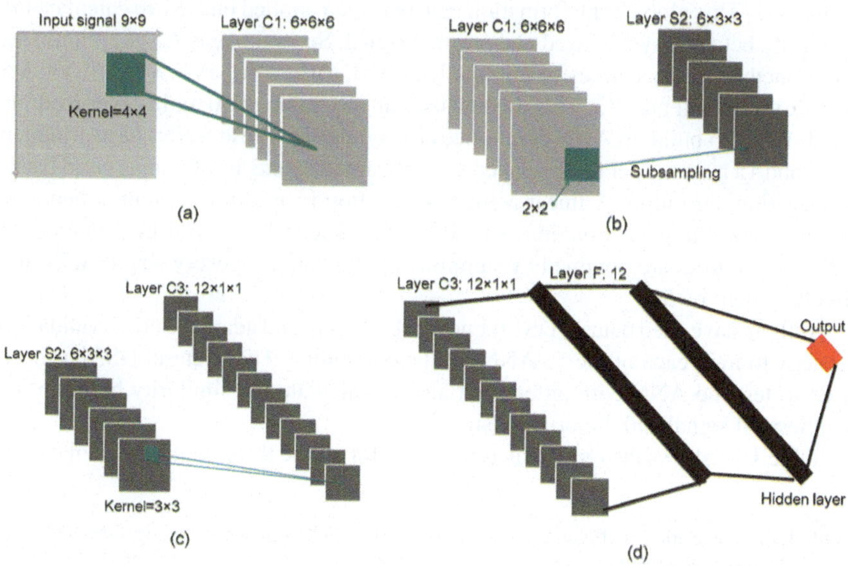

Fig. 1.21 Actions at different layers of the CNN: **a** computation from the input signal to layer C1; **b** computation from layer C1 to layer S2; **c** computation from layer S2 to layer C2; **d** computation from layer C2 to output

is used to compute the final output. So, the mapping from a layer to the next layer can be seen as a convolution operation. The

S-layer can be seen as a fuzzy filter, playing a secondary feature extraction role. The spatial resolution between the hidden layers decreases, and the number of planes contained in each layer increases. This can be used to detect more feature information. Similarly, at layer C3 (2nd convolution layer) (Fig. 1.21c), there are 12 feature maps which are computed using overlapping 3×3 kernels on the output of layer S2. Then, 12 features at layer F have been obtained that are computed from these 12 kernels of size 1×1 (Fig. 1.21d). Finally, a sigmoid function has been used as the networks output instead of the traditional one. When training, the feature maps of C3 are concatenated into a feature vector, feeding into the fifth layer and the final layer, which consists of N output neurons corresponding to the point number of the ECG sample. The proposed CNN has been used with stochastic gradient descent (SGD) on the full subset DS1. The batch size is set as 100 and the fixed learning rate is set as 0.1. Weights have been initialized using the stochastic weight method.

Authors have used two other methods for synthesis comparison. First one is multiple regression based method where a global transformation matrix is obtained using DS1. This global transformation matrix is then applied on DS2 to calculate the similarity between synthesized and original signal. Second one is ANN-committees based method [1] to synthesize missing leads (III, aVR, aVL, aVF, V1, V3, V4, V5 and V6) from I, II and V2. This method uses an ensemble of 50 multi-layer feedforward ANNs to build up ANN committees for synthesis. These ANNs have 1 hidden layer and 15 neurons per hidden layer and are trained by means of a supervised back-propagation algorithm. A linear activation function is used for the output neurons, and a sigmoid transfer function is used for the hidden layer. The final outputs of ANN committees are obtained by summing up the output of every single ANN and dividing them by 50.

Authors have used dataset DS1 for network training and adapt the cross-validation strategy to train each of the 50 ANNs of the committee. Then dataset DS2 has been used to test the ANN committees and also to calculate the similarity between the synthesized signal and the original signal.

Table 1.18 shows the reconstruction results for all the three methods and it can be

Table 1.18 Correlation coefficient r between the original ECG signal and the reconstructed ECG signal obtained using the three methods

Lead	r (%)		
	CNN	ANN-committees	Multiple regression
III, aVR, aVL, aVF	100	100	100
V1	94.40	91.40	91.42
V3	94.48	95.66	96.51
V4	93.80	89.74	89.20
V5	94.65	94.54	93.01
V6	97.44	97.15	96.58

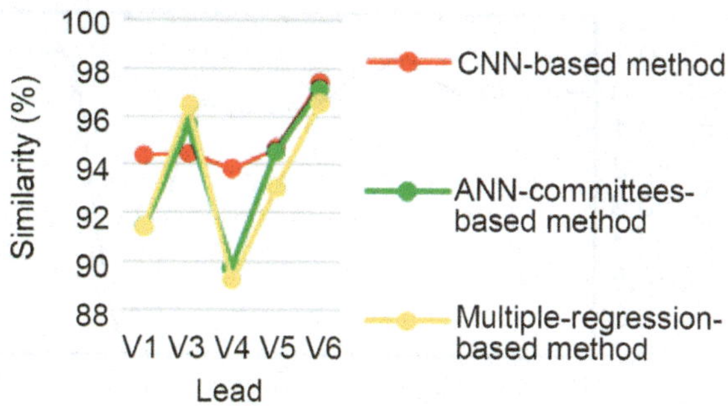

Fig. 1.22 Line chart of similarity obtained using three methods

observed that the proposed CNN-based method has better reconstruction results than the ANN-committees-based method and multiple-regression-based method. Figure 1.22 shows the line chart of the results for all the three mthods.

The results show that the proposed method performs better in the reconstruction of leads V1, V4, V5, and V6. Note that the proposed method increases the similarity between the reconstructed signal and the original signal of the V4 lead by over 4%. As can be seen, the ANN-committees-method has worse performance when the ECG is for patients with pathological changes. Figure 1.23b presents a comparison of the reconstructed signal of the V4-lead signal using the

ANN-committees based method and based on the original pathological changes. It is clear that the reconstructed signal does not displace such meaningful pathological changes (The direction of R-wave in the reconstructed signal is contrary to that of the real signal, which may lead to misdiagnosis). Figure 1.23a shows that the signal reconstructed using the proposed CNN-based method displaces an invert R wave as the original signal. The correlation coefficient is 35.63% for the ANN-committees-based method and 91.04% for the CNN-based method. Therefore, the proposed method has better performance in the reconstruction.

The authors have also compared the time consumption of the proposed method with that of ANN committees-based method. As shown in Fig. 1.24, the elapsed time of the proposed CNN-based method is only 7% of that of the ANN committees based method.

In this study, authors have investigated the convolutional neural networks method in detail, and applied it to the reconstruction of an ECG signal.

Compared to linear regression and ANN, proposed method has not only better similarity between the reconstructed and original ECGs, but also less time consumption. Authors also believe that the generic synthesis method can be further modified using a higher volume training database.

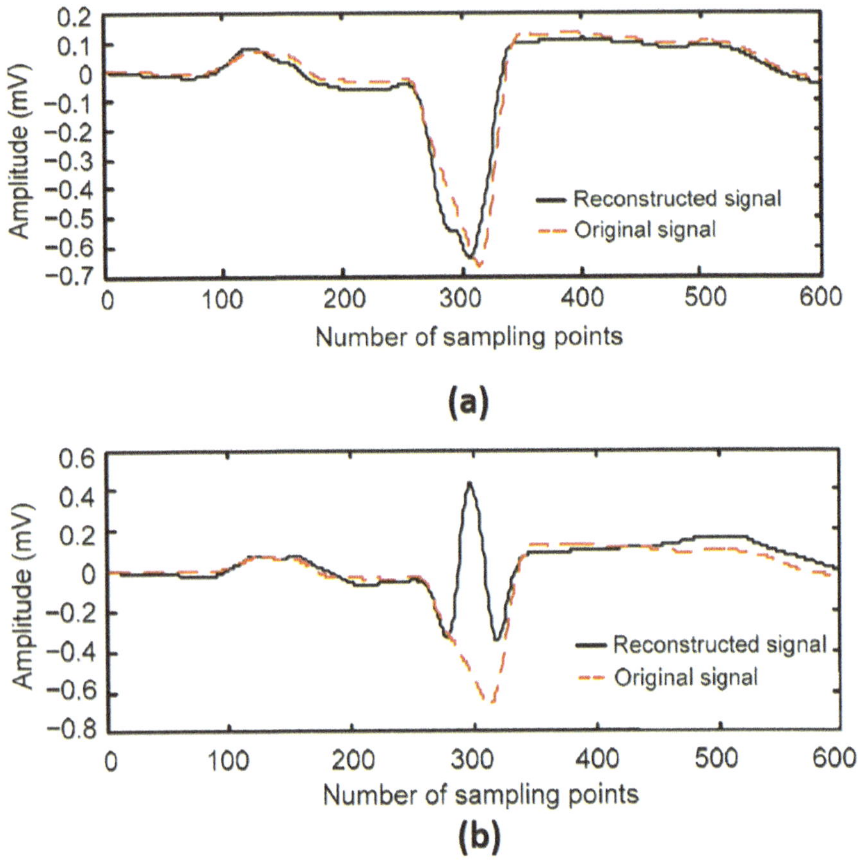

Fig. 1.23 Comparison of the original V4-lead signal and the reconstruction signal using **a** proposed CNN based method **b** ANN-committees based method

1.3.8 A Novel Neural-Network Model for Deriving Standard 12-Lead ECGs from Serial Three-Lead ECGs: Application to Self-Care

Atoui et al. [1] proposed a solution based on Artificial Neural Networks (ANN) to derive standard 12-lead ECG from serial 3-lead ECG. TO achieve this transformation, authors have determined easy-to-use pseudo orthogonal set of ECG leads, based on four active electrodes located in the Mason Likar positions and in V2, respectively, which have proven to be adequate for home care or ambulatory ECG recording. In this study authors have used database of 10 s ECGs that have been recorded on cardiac inpatients from the intensive care unit of the Cardiology.

Hospital of Lyon. Each patient had one or two pairs of ECGs, each consisting of the following:

Fig. 1.24 Running time of the proposed CNN-based method and the ANN-committees-based method

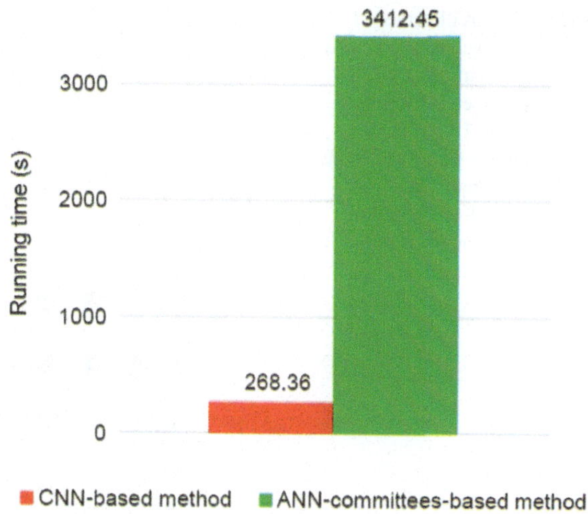

- A standard 10 s 12-lead ECG, hereafter called reference ECG, recorded by using a standard digital electrocardiograph (Cardiette);
- A 10 s three-lead PEM (Personal ECG Monitor) ECG based on I, II, and V2.

The PEM ECGs were recorded in average at less than half an hour interval of the recording of the reference ECGs.

The study population thus consists of a series of 157 pairs of digital ECGs (sampling rate: 500 samples/s; resolution: 5 μV) collected on 120 patients. Acute coronary syndrome (ACS) was diagnosed in 46 (29.3%) of the 157 standard 12-lead ECG tracings. The ECGs were processed to extract a 1-s-duration representative cycle from every original 10 s ECG record and determined the corresponding P, QRS, and T onsets and offsets. Finally, the ECG signal from the representative cycle was used in the interval for computing the reconstruction transforms. The study population was divided into two subsets DS1 and DS2 of 83 and 74 pairs of ECGs, respectively. Dataset DS1 consists of the 83 patients of the study population who had only one pair of ECGs (21 ACS ECGs and 62 control ECGs). Dataset DS2 consists of the remaining 37 patients who had two pairs recorded in average at three (range: 1–21) days interval (25 ACS ECGs and 49 control ECGs).

To synthesize the missing (V1, V3, V4, V5, and V6) ECG signals from the recorded (I, II, and V2) 12-lead ECG subset, an ensemble of N multilayer feed-forward ANNs have been used trained by means of a supervised back-propagation algorithm. Each individual ANN consists of one input layer with three input neurons (one for each recorded signal), one output layer with five output neurons (one for each derived signal), h = 1 hidden layer and n = 15 neurons per hidden layer, as shown in Fig. 1.25. The relations between the input and output layers are expressed through the weight and biases of the neurons of the hidden layer. These weight and biases are chosen randomly at the beginning of the training phase and are then iteratively optimized during the back-propagation process. The hidden layers number, the

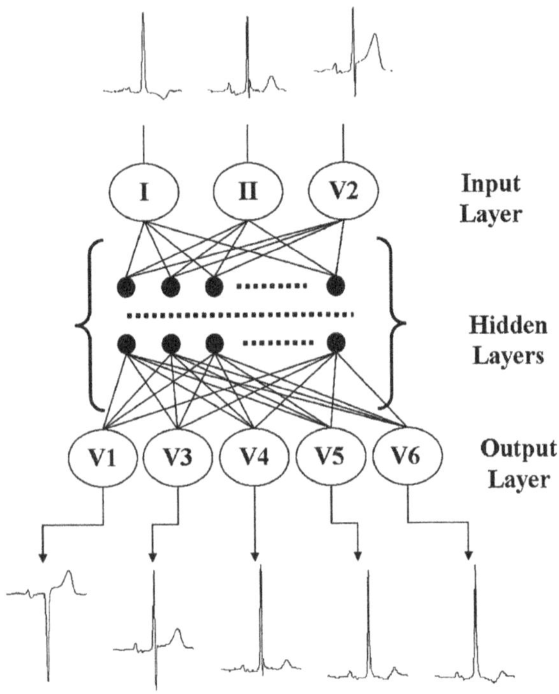

Fig. 1.25 Architecture of the ANNs used to synthesize the five missing V1, V3, V4, V5, and V6 leads (output layer) of the 12-lead ECG using a three-lead ECG (I, II, and V2) as input layer, *h* (typically 1 or 2) hidden layers and *n* neurons per hidden layer

activation function type, and the numbers of neurons per layer were chosen on the basis of previous experiments. As usual in approximation tasks, a linear activation function is used for the output neurons. A sigmoid transfer function was chosen for the hidden layer.

To overcome the performance limitations inherent to the training process, such as the random selection of the initial settings of the ANN weights and biases of the individual networks, a solution consisting of building up ANN committees of N = 50 individual ANN of the Fig. 1.25 type has been adopted. The five outputs of the ANN committees are obtained by summing up and dividing by N, the outputs of the N individual ANNs. It is thus expected that, as an application of the central limit theorem, the performance of the ANN committee surpasses each individual ANN performance.

Generic Neural-Network Training

To train each of the N = 50 individual, generic ANN of the committee, a cross-validation-like strategy has been adopted as follows. The 83 standard 12-lead ECGs of DS1 were randomly divided into $m = 5$ subsets. Each individual network of the committee was trained m times, each time leaving out one of the training subsets. The ECGs of the $m-1$ subsets were juxtaposed to form one continuous ECG recording. Stopping the learning process was performed by monitoring the reconstruction error (rms) on the remaining subset, where the ECGs were also juxtaposed to form one continuous ECG recording. This procedure was repeated for each of the N different networks biases and weights initializations and for each of the *m* subsets. A fixed number of iterations *K1* was then determined by recording, for each training case, the number of iterations k and by averaging these values over the *N X m* trainings. Finally, each individual network of the committee was trained over the 83 12-lead ECGs of DS1 using the fixed number of iterations *K1* as an early stopping point.

Patient Specific Neural-Network Training

Ideally, three serial 12-lead ECGs per patient would be needed for deriving a patient-specific ANN ensemble: one for training the network (ECG-L), one for stopping the learning process in order to avoid network overdesign (ECG-S), and one for testing the generalization capacity of the trained network (ECG-T). Since the database comprised at maximum only two serial standard 12-lead ECGs per patient (ECGs L and T), to overcome this difficulty, instead of stopping the learning process after a variable number of iterations that should ideally be determined by ECG-S, a fixed number of iterations *K2* that was determined in a similar fashion, but on a different, independent serial database. The learning process was then performed by training each of the N individual ANN on the first reference ECG (ECG-L) of each of the 37 patients of DS2.

Once trained, the generic and patient-specific ANN transforms are tested on a different dataset than the one used to train the generic and patient-specific committees. This test is intended to assess the ANN models performance by simulating their use in real-life situations and measuring the reconstruction accuracy of both transformation models. The simulation process however substantially differs from one reconstruction method to the other. In the generic approach, the second standard 12-lead and PEM ECG pairs of dataset DS2, i.e., ECG-T and the corresponding PEM ECG from DS2 will be used as a test set. The five "missing" leads of every standard 12-lead and PEM ECG are synthesized, as shown in Fig. 1.26, by the generic ANN committee designed, and the reconstructed V1, V3 to V6 leads are then compared with the original leads of the corresponding standard 12-lead ECG of DS2. In the patient-specific approach, a patient-specific committee is trained on the first standard 12-lead ECG (ECG-L) of the first ECG pair of each of the 37 patients of DS2. The patient-specific committee is then tested on each second ECG pair of the corresponding patient by deriving the five "missing" leads for both the Cardiette (ECG-T) and the PEM ECGs. Finally, the reconstructed V1, V3 to V6 leads are compared with the original leads of the corresponding Cardiette ECG (ECG-T). This procedure is applied in turn for each case of the validation set DS2.

Table 1.19 displays the rms values and the correlation coefficients between the first and second Cardiette ECGs of DS2, after removal of the 5% extreme values (N = 35). The average rms signal difference between the standard I, II, V1, V3, V4, V5 and V6 leads is 107 μV. The average correlation coefficient is 0.92. Table 1.20 displays the rms values and the correlation coefficients for the generic ANN committee and multiple-regression based models after removal of the two highest rms and the two lowest correlation coefficients. It shows a slight superiority of the ANN committee, with an average rms value of 122 μV between the original and the reconstructed

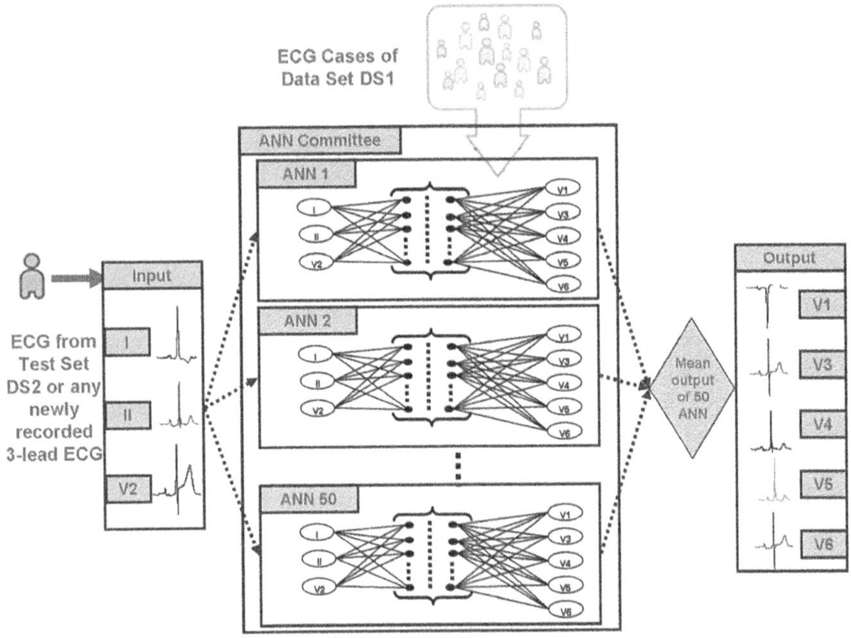

Fig. 1.26 Reconstruction of a standard 12-lead ECG out of the three-lead PEM ECG. The synthesis of the five missing V1, V3, V4, V5, and V6 leads is obtained by averaging the outputs of a committee of 50 neural networks

Table 1.19 RMS (in micro Volts) and Correlation Coefficient (r_x) between the original I, II and V2 leads of the first and the second standard 12-lead lead ECGs recordings for every patient in DS2 (N = 35)

Lead	RMS (±SD)	r_x (±SD)
I	60 (±30)	0.93 (±0.09)
II	73 (±32)	0.85 (±0.18)
V2	125 (±58)	0.95 (±0.07)
V1	82 (±44)	0.95 (±0.08)
V3	135 (±57)	0.95 (±0.04)
V4	136 (±56)	0.92 (±0.08)
V5	139 (±59)	0.90 (±0.13)
V6	100 (±41)	0.91 (±0.10)

leads versus an average rms of 126 μV for the regression-based synthesis method. However, the correlation coefficients between the original and

the reconstructed leads score almost the same values for the ANN committee and the regression-based approaches, with an average value of 0.93. Table 1.21 displays the rms values and the correlation coefficients for the patient-specific ANN committee and multiple regression based models. In contrary to the generic approach, it shows a more pronounced superiority of the ANN committee, with an average rms of 94 μV between the original and the reconstructed leads versus an average

Table 1.20 RMS (in micro Volts) and Correlation Coefficient (r_x) between the original and reconstructed V1, V3 to V6 leads for the Generic ANN committee (ANN-S) and regression based models (REG-S) for DS2 (N = 35)

Method	Lead	RMS (±SD)	r_x (±SD)
ANN-S	V1	87 (±42)	0.97 (±0.002)
	V3	108 (±50)	0.96 (±0.003)
	V4	171 (±77)	0.88 (±0.014)
	V5	152 (±65)	0.91 (±0.009)
	V6	96 (±44)	0.93 (±0.007)
REG-S	V1	88 (±42)	0.97 (±0.007)
	V3	111(±49)	0.96 (±0.006)
	V4	176 (±73)	0.88 (±0.014)
	V5	157 (±63)	0.90 (±0.010)
	V6	98 (±45)	0.92 (±0.009)

Table 1.21 RMS (in micro Volts) and Correlation Coefficient (r_x) between the original and reconstructed V1, V3 to V6 leads for the Patient Specific ANN committee (ANN-S) and regression based models (REG-S) for DS2 (N = 35)

Method	Lead	RMS (±SD)	r_x (±SD)
ANN-S	V1	68 (±30)	0.98 (±0.002)
	V3	93 (±45)	0.98 (±0.003)
	V4	114 (±45)	0.95 (±0.014)
	V5	112 (±42)	0.95 (±0.009)
	V6	84 (±34)	0.95 (±0.007)
REG-S	V1	75 (±33)	0.98 (±0.002)
	V3	99 (±46)	0.97 (±0.002)
	V4	140 (±55)	0.92 (±0.007)
	V5	132 (±46)	0.93 (±0.005)
	V6	93 (±40)	0.94 (±0.006)

rms of 108 μV for the regression-based synthesis method. The same yields for the correlation coefficients between the original and the reconstructed leads, which score 0.961 in average for the patient-specific ANN committee versus 0.948 in average for the patient-specific regression model. The median correlations are 0.975 and 0.967, respectively.

In this study Patient-specific transforms have proven to be more accurate. Nevertheless, generic reconstructions remain particularly attractive in situations, where no standard 12-lead ECG could be recorded and/or no patient-specific transform could be computed and/or retrieved, prior to the recording of a reduced ECG lead set. Also, generic reconstructions are less sensitive to artifacts and electrode displacements, since their coefficients were computed from a very large set of ECGs of different origins.

1.3.9 Linear Affine Transformations Between 3-Lead (Frank XYZ Leads) Vectorcardiogram and 12-Lead Electrocardiogram Signals

Dawson et al. [4] proposed an approach to to transform 3-lead Frank VCG to 12-lead ECG signals and vice versa. In this study PTBDB database has been used. The database used consists of 80 HC subjects and 361 MI patients. The sampling rate was 1 kHz, and the samples were typically gathered for one minute durations. The 15-lead ECG signals were used in the following investigations:

- Compare the overall accuracies of Dower and affine transforms for correctly deriving the 12-lead from the known 3-lead Frank XYZ, separately for HC and MI recordings as well as for combined HC and MI recordings.
- Compare the overall accuracies of the Dower and the affine transforms presented here for correctly deriving the 3-lead Frank XYZ from the known 12-lead ECG for MI and HC subjects separately as well as combined.
- Evaluate using statistical hypothesis tests, the diagnostic capability of the transform matrices.

Figure 1.27 gives the flow chart of the overall approach used in this investigation. For deriving the 12-lead ECG from the known Frank XYZ leads, 25 randomly chosen recordings of the available 80 HC recordings have been used for training and the remaining for testing. From each recording, 2000 points were extracted from the mid portion of each signal, and all the selected recordings for training are concatenated to generate an input matrix of known lead values for a least squares fit. A linear

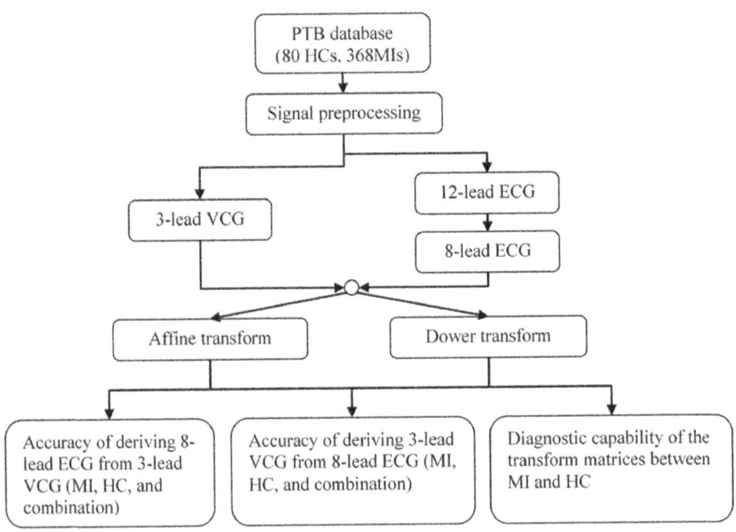

Fig. 1.27 Summary of the overall approach

regression fitting process was used to yield the affine transform matrix to derive the 8 leads from the known 3 Frank (XYZ) VCG leads. As stated in the foregoing statement, an affine transform is commonly used for deriving multiple regression models. These models can compensate for constant bias and inconsistent baselines in the ECG signals.

Linear regression model assumes that every derived lead (here, the individual 8-lead values denoted as Y = [y1, y2, ……y8]T, where T denotes a transpose) can be obtained from a linear combination of the 3 Frank VCG values (here, the 3 leads are denoted as X = [× 1, × 2, × 3]), i.e.,

$$Y = AX + e = a0 + a1 \times 1 + a2 \times 2 + a3 \times 3 + e \quad (1.8)$$

where a0, ai…an are the columns of the transform coefficients, and A is the transform matrix and e is the error.

Table 1.22 presents the R^2 values for derived (affine) and Dower transforms for HC subjects and MI patients for the 12-lead ECG from 3-lead VCG. For both MI patients and HC subjects, the affine transformation matrix presented here yields improved accuracy over Dower transform. For example, for the HC subjects, R^2 average is 84.79 for affine and 77.31 for Dower. For MI, it is 80.77 for affine and 72.09 for

Table 1.22 Distribution of R^2 values for derived (affine) and Dower transforms for HC subjects and MI patients for the 12-lead ECG from 3-lead VCG

Leads	HC		MI		Combined HC and MI	
	R^2		R^2		R^2	
	Derived	Dower	Derived	Dower	Derived	Dower
I	78.30	73.27	71.84	67.64	70.95	70.46
II	92.27	90.65	89.64	86.51	91.91	88.58
III	78.76	67.17	77.34	69.02	72.69	68.09
aVR	89.68	88.80	81.38	79.37	85.96	84.09
aVL	63.99	48.37	70.24	63.08	61.50	55.73
aVF	89.50	84.52	87.14	80.64	87.05	82.58
V_1	82.38	77.81	79.06	71.86	77.83	74.83
V_2	81.91	66.98	73.10	59.71	79.45	63.34
V_3	84.45	63.98	78.09	59.31	82.85	61.64
V_4	89.06	83.09	81.77	58.38	86.57	70.74
V_5	93.28	92.04	87.97	83.76	92.24	87.90
V_6	93.90	91.08	91.71	85.76	94.10	88.42
Total	1017.48	927.78	969.29	865.04	983.11	896.41
Average	84.79	77.31	80.77	72.09	81.93	74.70
Maximum	93.90	92.04	91.71	86.51	94.10	88.58
Minimum	63.99	48.37	70.24	58.38	61.50	55.73

Dower. While the maximum values for affine and Dower were very close (93.9 vs. 92.04) for HC and (91.71 vs. 86.51) for MI, there is a significant difference in the minimum values (63.99 vs. 48.37) for HC and (70.24 vs. 58.38) for MI. Figure 1.28 shows the ECGs measured (blue), derived (red), and Dower (green) of lead II, V4, and V6 obtained for (A) HC subjects and (B) MI patients. Table 1.23 presents the R^2 values derived (affine) and Dower transforms for HC subjects and MI patients for the 3-lead VCG from 12-lead ECG. Figure 1.29 shows the VCG and XYZ leads measured (blue), derived (red), and Dower (green) from 12-lead ECG to 3-lead VCG for HC subjects and MI patients, respectively.

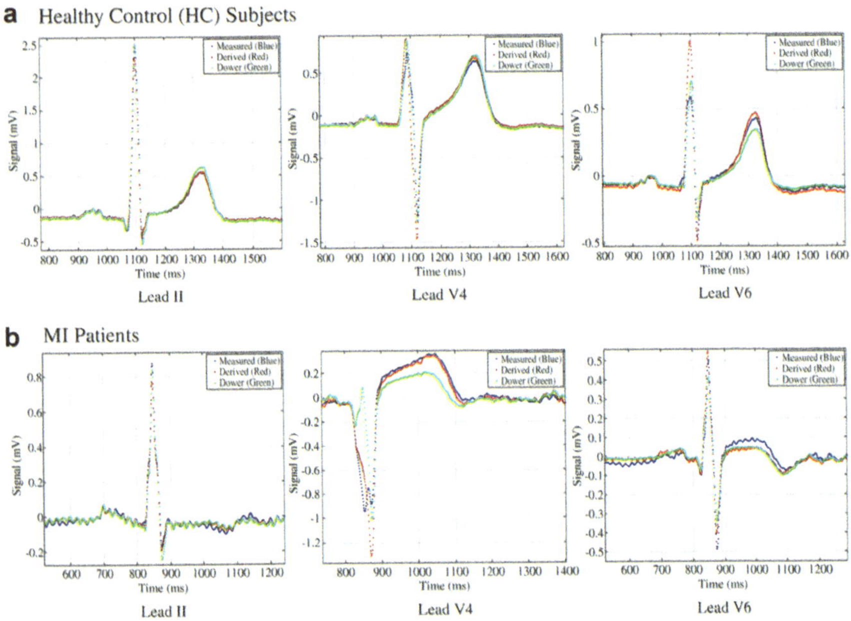

Fig. 1.28 Electrocardiograms (measured (blue), derived (red), and Dower (green)) of lead II, V4, and V6 obtained for (**a**) HC subjects and (**b**) MI patients Thus, from knowing the input lead values, the corresponding coefficient vectors can be used to derive each of the 8 leads. It has been found that at least 25 recordings are needed to provide statistically consistent distribution of transform matrix values. Authors have repeated the random drawing of training and testing patterns to derive 25 different realizations of the transform matrix. The same study is repeated next for MI recordings taking 25 random recordings for training and the remaining for testing. For deriving the three Frank XYZ leads from eight independent ECG leads (out of 12, because the augmented leads are calculated from leads I, II, and III), the fitting system becomes overdetermined, i.e., the same XYZ lead values can be obtained from a given 12-lead values using completely different transform coefficients. This is because the eight leads carry redundant information. For example, leads II and III can have highly redundant information, and so do V5 and V6. This redundancy needs to be reduced to achieve consistent 8-lead to 3-lead transformation. Principal component analysis (PCA) is widely used in engineering to reduce such data redundancy. This technique was used in this investigation to determine a consistent transformation matrix to derive three Frank XYZ lead values for HC, MI, and combined HC-MI recordings as described in Results

Table 1.23 Distribution of R^2 values for derived (affine) and Dower transforms for HC and MI for the XYZ leads from 12-lead ECG

Leads	HC		MI		Combined	
	R^2		R^2		R^2	
	Derived	Dower	Derived	Dower	Derived	Dower
b	95.78	91.46	95.42	83.02	96.69	90.42
Y	91.51	85.93	90.33	83.07	91.07	86.48
Z	91.07	49.42	86.82	19.32	85.60	37.84
Total	277.99	226.81	272.58	185.41	273.37	214.74
Average	92.66	75.60	90.86	61.80	91.12	71.58
Maximum	96.76	91.46	95.43	83.07	96.68	90.42
Minimum	89.72	49.42	86.82	19.32	85.60	37.84

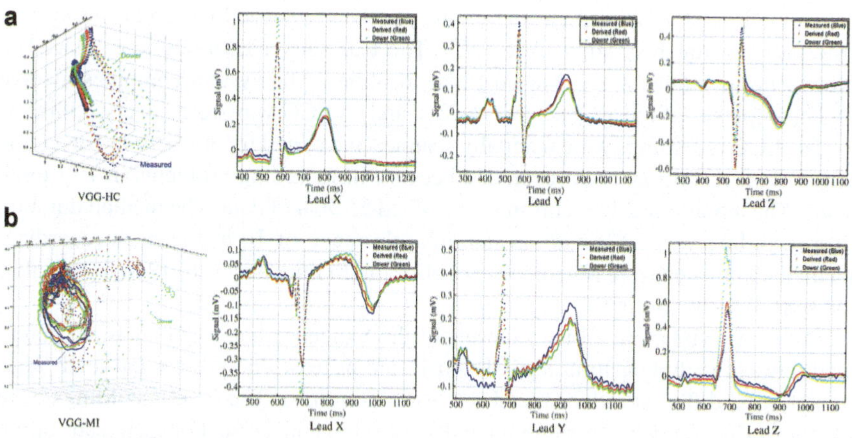

Fig. 1.29 A and B, VCG and XYZ leads (measured (blue), derived (red), and Dower (green)) from 12-lead ECG to 3-lead VCG for HC subjects and MI patients, respectively

In this investigation, an affine transformation is presented to derive statistically consistent transformations between 3-lead VCG and 12-lead ECG and vice versa. Compared to a geometry-based Dower transform, the statistical affine transform is found to yield improved accuracies (in terms of R^2 values) for both MI patients and HC subjects. The accuracy improvement is found to be more pronounced in the case of MI subjects. This is perhaps because under pathological conditions, single dipole assumptions that underlie geometric transform will not hold. The coefficients of the affine transforms derived for HC subjects were found to be statistically different from those from MI subjects. As such, the coefficients can serve as discriminating features for diagnostic purposes.

1.3.10 Reconstruction of the 12-Lead Electrocardiogram from Reduced Lead Sets

Nelwan et al. [26] investigated on how well absent, noisy, or defective leads can be reconstructed from different lead subsets and how well lead reconstruction performs over time. In this study 236 continuous 12-lead ECG recordings from patients enrolled in the PURSUIT ECG ischemia monitoring have been taken. In summary, patients were eligible if they presented within 24 h of an episode of ischemic chest pain (>10 min) and had transient ST-segment elevation (>50 μV), transient or persistent ST-segment depression (>50 μV), T-wave inversion (>100 μV). The study protocol excluded patients with persistent (>30 min) ST-segment elevation and those with ECG abnormalities interfering with ST-segment interpretation, such as left bundle branch block, third-degree atrioventricular (AV) block, persistent arrhythmias, or pacemakers. There were no restrictions regarding age. Patients were monitored for 24 h with a continuously updated 12-lead ECG recording system (Mortara Instruments, Milwaukee, WI). The extremity electrodes were placed in monitoring positions at the Mason-Likar locations. The system computes median ECG complexes of the 12 leads every 20 s, and an ECG was stored every 5 min.

The ECG recordings were manually scanned and edited for artifacts and detection errors. Two recordings were excluded because of recording problems in 1 or more leads. The remaining 234 recordings contained 5,796 h of data. The median duration of a recording was 24 h with an interquartile range of 10 h. For each recording, median complexes were obtained at baseline, at 20 min, and 1, 6, 12 and 24 h after the start of the recording.

In this investigation, all lead subsets of the 8 independent ECG leads are generated, always including limb leads I and II and at least one pre-cordial lead. Thus a total of 62 different reduced lead sets were examined. The ECG recordings were randomly divided into a learning set and a test set of equal size. For each lead subset, general reconstruction coefficients to synthesize the absent leads were derived from the baseline ECGs in the learning set using linear regression. These coefficients were subsequently applied on the ECGs of the test set. Patient-specific coefficients were derived from each baseline ECG in the test set. These reconstruction coefficients were applied to the ECGs at baseline, 20 min, and 1, 6, 12, and 24 h after the start of the recording.

For each lead subset, reconstruction accuracy of the missing leads was evaluated on 3 different levels. First, overall waveform similarity for each ECG was assessed by averaging, over the reconstructed leads, the correlations between the originals and their reconstructions. The correlations were based on all samples in the QRS-T complex. Second, reconstruction accuracy was determined for a clinically important measurement, the amplitude of the ST segment at 60 ms after the J-point (ST60). The average of the absolute differences between the original and reconstructed ST60 amplitudes have been determined. In addition, the maximum absolute difference at ST60 was determined. Third, reconstruction accuracy was assessed at the diagnostic level. A clinically accepted decision rule for detection of myocardial ischemia and evolving myocardial infarction to the original ECG and to the reconstructed ECGs was applied and determined the agreement defined as the percentage of cases in which decisions concurred. The decision rule holds true when 2 or more contiguous leads show ST elevations of ≥0.2 mV in leads V1, V2, or V3, or ≥0.1 mV in other leads.

Table 1.24 shows the median and range of the general and patient-specific results of lead subsets of a given size on the baseline recordings of the test set. Patient-specific reconstruction performed better than general reconstruction. For general reconstruction, median correlations of 0.978 were obtained for lead subsets with 1 or 2 precordial leads removed; considering the best lead subsets, removal of even 3 precordial leads yielded similar high correlations. With patient specific reconstruction, lead subsets with 4 precordial leads removed had median correlations of 0.989 or higher. The best lead subset with 5 precordial leads removed still had a correlation of at least 0.964. The average and the maximum absolute ST amplitude differences for general reconstruction were about twice as high as the differences for patient-specific reconstruction. By and large, general reconstruction results for a lead subset of a given size were comparable with patient-specific results for a subset with 2 more precordial leads removed. For example, using general reconstruction, the average absolute ST difference for lead subsets with 2 precordial leads removed had a median (range) of 16 (13–30) μV, whereas patient specific reconstruction for lead subsets with 4 leads removed gave differences of 13 (9–26) μV. For the clinical decision rule, high to very high median agreements were obtained for any lead subset using both general and patient-specific reconstruction. In lead subsets with less than 5 precordial leads removed, the best lead subsets of a given size had a minimal agreement of at least 95.1% for general reconstruction and 95.7% for patient-specific reconstruction, whereas maximal agreement was 100% regardless of the reconstruction method.

Table 1.25 shows the best lead subsets of a given size based on the highest correlation. The same subsets were almost always the ones with the lowest average and maximum absolute ST difference and with the highest agreement. When only 1 or 2 precordial leads were removed, several subsets showed a performance almost identical to the best. The second best lead set with 3 precordial leads removed was the other set with nonadjacent precordial leads (V2, V4, V6), performing nearly as well as the best lead set (V1, V3, V5). In lead sets with 4 precordial leads removed, the combination of V2 and V6 was second best, whereas in lead sets with 5 precordial leads removed, V2 performed clearly better than other leads.

In this study, authors have investigated how well the 12-lead ECG can be reconstructed from different lead subsets. Both general and patient-specific reconstructions were evaluated on a well-defined set. General reconstruction allows accurate reconstruction of 1 or 2 missing precordial leads, whereas up to 3 to 4 precordial leads can be reconstructed well by patient-specific reconstruction parameters. Reconstruction of the 12-lead ECG over time remained accurate using general and patient-specific reconstruction. In all cases, patient-specific reconstruction performance was better than general reconstruction performance.

Table 1.24 Performance Results of General and Patient-Specific Reconstruction Methods on the Baseline ECGs of the test set (n = 117), for Lead Sets with Increasing Number of Precordial Leads Removed

Number of precordial leads removed	General				Patient Specific			
	Correlation	Average ST60 difference (mV)	Maximum ST60 difference (mV)	Agreement (%)	Correlation	Average ST60 difference (mV)	Maximum ST60 difference (mV)	Agreement (%)
1	0.988 (0.966–0.995)	15 (10–19)	15 (10–19)	100 (95.7–100)	0.999 (0.994–0.999)	7 (6–9)	7 (6–9)	100 (99.1–100)
2	0.978 (0.932–0.988)	16 (13–30)	32 (18–40)	100 (94.9–100)	0.998 (0.991–0.999)	8 (7–11)	12 (10–16)	100 (97.4–100)
3	0.965 (0.844–0.977)	20 (16–48)	31 (28–78)	96.6 (94.9–100)	0.995 (0.980–0.997)	10 (8–24)	17 (13–37)	99.1 (95.7–100)
4	0.938 (0.807–0.964)	24 (18–51)	43 (34–89)	95.7 (91.5–100)	0.989 (0.967–0.994)	13 (9–26)	24 (18–45)	98.3 (95.7–100)
5	0.854 (0.642–0.912)	37 (26–48)	66 (46–93)	94.5 (93.2–96.9)	0.964 (0.921–0.978)	24 (16–33)	47 (29–64)	96.1 (94.1–99.1)

Note Correlation values denote median (range)

Table 1.25 Best lead subsets of a given size based on the highest average correlations found in the baseline ecg recordings of the test set

Number of precordial leads removed	Best lead set	Correlation	
		General	Patient specific
1	I, II, V1, V2, V3, V4, V6	0.995 (0.986–0.998)	0.999 (0.999–1.0)
2	I, II, V1, V2, V4, V6	0.988 (0.966–0.994)	0.999 (0.997–0.999)
3	I, II, V1, V3, V5	0.977 (0.952–0.988)	0.997 (0.994–0.999)
4	I, II, V2, V5	0.964 (0.934–0.979)	0.994 (0.988–0.996)
5	I, II, V2	0.912 (0.858–0.950)	0.978 (0.964–0.988)

Note Correlation values denote median (interquartile range)

1.4 Conclusion

A 12-lead ECG is one of the most important cardiac diagnostic tools. The need for synthesis of 12-lead ECG from reduced leads increases dramatically as current portable devices provide often limited electrodes. In addition to this, higher memory, transmission power and bandwidth are required for standard 12 lead ECG. In this chapter, we have identified and have reported various studies, which provided solutions to the aforementioned problems. Frank Vectorcardiographic (FV) system to Standard 12-lead system (S12) and vice versa have been investigated in some of the studied. In some of the studies, a subset of S12 system is used to reconstruct the remaining independent leads of the S12 system. A new orthogonal reduced lead set was derived from S12 system using PCA and this derived reduced lead set was used to reconstruct the S12 system in one of the study. Most of these investigations have used Heart vector projection theory and linear regression models represent the target leads from the basis lead sets. Then reconstruction coefficients were generated using Least squares fit method. These reconstruction coefficients were then used for synthesizing the target leads. It was found that patient specific reconstruction outperformed the population based reconstruction in most of the investigations. There were also some studies in recent times which have used machine learning techniques for the reconstruction of S12 system from reduced lead system. These studies have used Artificial Neural Networks (ANN) and Convolutional Neural Networks (CNN) for reconstruction, where a large dataset has been used to train and test the performance. All these investigations have been summarized in this chapter along with their promising results.

References

1. Atoui H, Fayn J, Rubel P (2010) A novel neural-network model for deriving standard 12-lead ECGs from serial three-lead ECGs: Application to self-care. IEEE Trans Inf Technol Biomed 14(3):883–890
2. Bousseljot R, Kreiseler D, Schnabel A (1995) Nutzung der EKG-Signaldatenbank CARDIODAT der PTB überdas Internet. Biomed Tech 40(1):S317
3. CVD statistics as per WHO." [Online]. Available: http://www.who.int/cardiovascular_diseases/en/
4. Dawson D, Yang H, Malshe M, Bukkapatnam STS, Benjamin B, Koman-duri R (2009) Linear affine transformations between 3-lead (Frank XYZ leads) vectorcardiogram and 12-lead electrocardiogram signals. J Electrocardiol 42(6):622–630
5. Dawson D, Yang H, Malshe M, Bukkapatnam STS, Benjamin B, Komanduri R (2009) Linear affine transformations between 3-lead (Frank XYZ leads) vectorcardiogram and 12-lead electrocardiogram signals. J Electrocardiol 42(6):622–30
6. Dower GE, Machado HB, Osborne JA(1980) On deriving electrocardiogram from vectorcardiographic leads. Clin Cardiol 3(2):87–95
7. Dower GE (1968) A lead synthesizer for the Frank system to simulate the standard 12-lead electrocardiogram. J Electrocardiol 1(1):101–16
8. Dower GE, Yakush A, Nazzal SB, Jutzy RV, Ruiz CE (1988) Deriving the 12-lead electrocardiogram from four (EASI) electrodes. J Electrocardiol (Suppl.):182–7
9. Edenbrandt L, Pahlm O (1988) Vectorcardiograms synthesized from 12-lead ECGs: Superiority of inverse Dower matrix. J Electrocardiol 21(4):361–367. https://doi.org/10.1016/0022-0736(88)90113-6
10. Ettl S, Kaminski J, Knauer MC, Häusler G (2008) Shape reconstruction from gradient data. Appl Opt 47(12):2091. https://doi.org/10.1364/ao.47.002091
11. Finlay DD, Nugent CD, Kellet JG, Donnelly MP, McCullagh PJ, Black ND (2007) Synthesizing the 12-lead electrocardiogram: trends and challenges. Eur J Intern Med 18(8):566–70
12. Frank E (1954) General theory of heart-vector projection. Circulation 2:258–70
13. Goldberger AL, Amaral LA, Glass L, Hausdorff JM, Ivanov PC, Mark RG, Mietus JE, Moody GB, Peng CK, Stanley HE (2000) PhysioBank, PhysioToolkit, and PhysioNet: Components of a new research resource for complex physiologic signals. Circulation 101(23):e215–e220
14. Horacek BM, Warren JW, Wang JJ (2008) On designing and testing transformations for derivation of standard 12-lead/18-lead electrocardiograms and vector cardiograms from reduced sets of predictor leads. J Electrocardiol 41(3):220–9
15. Kors JA, Van Harpen G, Sittig AC, Van Bemmel JH (1990) Reconstruction of the Frank vectorcardiogram from standard electrocardiographic leads: diagnostic comparison of different methods. Eur Heart J 11(12):1083–1092
16. Levkov CL (1987) Orthogonal electrocardiogram derived from the limb and chest electrodes of the conventional 12-lead system. Med Biol Eng Comput 25(2):155–64
17. Lu C, Wang Z, Zhou B (2017) Intelligent fault diagnosis of rolling bearing using hierarchical convolutional network based health state classification. Adv Eng Inf 32:139–151. https://doi.org/10.1016/j.aei.2017.02.005
18. Maheshwari S, Acharyya A, Puddu PE, Mazomenos EB, Schiarti M, Maharatna K (2014) Robust and accurate personalised reconstruction of standard 12-lead system from Frank vectorcardiographic system. Comput Methods Biomech Biomed Eng Imaging Vis. https://doi.org/10.1080/21681163.2014.931029
19. Maheshwari S, Acharyya A, Rajalakshmi P, Puddu PE, Schiariti M (2013) Accurate and reliable 3-lead to 12-lead ECG reconstruction methodology for remote health monitoring applications. IEEE Healthcom 15:208–212
20. Maheshwari S, Acharyya A, Schiariti M, Puddu PE (2016) Frank vectorcardiographic system from standard 12 lead ecg: an effort to enhance cardiovascular diagnosis. J Electrocardiol 49(2):231–242

21. Maheshwari S, Acharyya A, Schiariti M, Puddu PE (2015) Personalized reduced 3-lead system formation methodology for remote health monitoring applications and reconstruction of standard 12-lead system. Int Arch Med, 8. https://doi.org/10.3823/166
22. Maheshwari S, Acharyya A, Schiariti M, Puddu PE (2014) Reduced lead system selection methodology for reliable standard 12-lead reconstruction targeting personalized remote health monitoring applications. Comput Methods Biomech Biomed Eng Imaging Vis 2(2):107–120
23. Maheshwari S, Acharyya A, Schiariti M, Puddu PE (2014) Reduced lead system selection methodology for reliable standard 12-lead reconstruction targeting personalized remote health monitoring applications. Comput Methods Biomech Biomed Eng Imaging Vis 2(2):107–20
24. Mazomenos EB, Chen T, Acharyya A, Bhattacharya A, Rosengarten J, Maharatna K (2012) A time domain morphology and gradient based algorithm for ECG feature extraction. ICIT 2012: IEEE international conference on industrial technology, Athens, Greece, March 2012, pp. 117–122. https://doi.org/10.1109/icit.2012.6209924
25. Naresh V, Vishnuvardhan G, Acharyya A, Chakravarti B (2019) A Novel 2-lead to 12-lead ECG Reconstruction Methodology for Remote Health Monitoring Applications", Computing in Cardiology (CINC), Singapore, 8–11 September 2019
26. Nelwan SP, Kors JA, Meij SH, Bemmel JH, Simoons ML (2004) Reconstruction of the 12-lead electrocardiogram from reduced lead sets. J Electrocardiol 37:11–19
27. Pan JP, Tompkins WJ (1985) A real-time QRS detection algorithm. IEEE Trans Biomed Eng, BME 32(3):230–236. https://doi.org/10.1109/TBME.1985.325532
28. Wang Lu-di, Zhou W, Xing Y, Liu N, Movahedipour M, Zhou XG (2019) A novel method based on convolutional neural networks for deriving standard 12-lead ECG from serial 3-lead ECG. www.springerlink.com, https://doi.org/10.1631/fitee.17e0413

Chapter 2
Brain–Computer Interface as a Potential Access Method for Communication in Non-verbal Children with Cerebral Palsy: A State-of-the-Art Review

Petra Karlsson, Silvia Orlandi, Haifeng Zhao, and Alistair McEwan

Abstract Cerebral palsy is an umbrella term that encompasses a group of disorders affecting movement and posture. The primary motor disorder is often accompanied by associated impairments of sensation, cognition, communication, perception, behavior, and/or seizure disorder. One in two children with cerebral palsy have a speech disorder, and one in three are non-verbal. Speech and motor difficulties may mask children's cognitive capabilities, and recognition of their strengths and difficulties, which may limit the provision of opportunities for learning and development. A Brain–Computer Interface (BCI) may hold the key to unlock the potential in children with disabilities who today have limited means to learn, play, and communicate with currently available technology. Although BCI has been successfully applied on able-bodied adults, little information is currently available on BCI use in children with severe motor impairments who may need technology for supporting their communication. The aims of this state-of-the-art review are to describe the growing research field of BCIs and to share typical clinical applications as well as to provide evidence from the literature supporting the BCI application in children with cerebral palsy. The main engineering, assistive technology and neurorehabilitation databases were searched. Information related to inclusion criteria and study protocols was critically evaluated to determine BCI's successful implementation. This chapter highlights the five factors that can be attributed to the effective use of BCI: characteristics of the

P. Karlsson (✉)
Cerebral Palsy Alliance, The University of Sydney, Sydney, Australia
e-mail: pkarlsson@cerebralpalsy.org.au

S. Orlandi
Bloorview Research Institute, Holland Bloorview Kids Rehabilitation Hospital, Toronto, ON, Canada
e-mail: sorlandi@hollandbloorview.ca

H. Zhao · A. McEwan
The University of Sydney, Sydney, Australia
e-mail: haifeng.zhao@sydney.edu.au

A. McEwan
e-mail: alistair.mcewan@sydney.edu.au

G. D. Gargiulo and G. R. Naik (eds.), *Wearable/Personal Monitoring Devices Present to Future*, https://doi.org/10.1007/978-981-16-5324-7_2

BCI, individual characteristics of the BCI user, type of feedback and instruction, and evaluation and ethical considerations and reports on the current use of BCI in children.

Keywords Assistive technology · Disability · Communication · Brain–machine interface · Complex communication needs

2.1 Introduction

Worldwide, 785 million people report a disability, almost 40 million of whom are children [1–3]. A common neurological disability includes cerebral palsy, an umbrella term that encompasses a group of disorders affecting movement and posture, with a prevalence of 1.4–2 per 1000 live births [4]. Cerebral palsy is caused by injuries or developmental anomalies that may involve the cerebral cortex of the developing, immature brain. Cerebral palsy is characterized by activity limitation due to the difficulty controlling motor functions and posture. The motor disorders are often associated with challenges with perception, sensation, cognition, communication, behavior, and epilepsy, affecting a child's ability to explore, speak, learn, and gain independence.

One in two children with cerebral palsy has some speech problems [4, 5]. Among those who are non-verbal, language skills can be relatively preserved [6]. Research shows that the development of non-verbal intellectual ability in children (5–9 years) with less severe cerebral palsy has a similar trajectory to typically developing children. However, more limited development is evident in children with more severe cerebral palsy and those with dyskinetic cerebral palsy [6]. Intellectual disability is estimated to occur in approximately 50% of people with cerebral palsy [7], with one in five people having a moderate to severe intellectual disability [8]. Gross motor function in children with cerebral palsy determines the developmental trajectories of mobility but not daily activities. However, children and youth without an intellectual disability but with a severe motor disorder may reach outcomes close to their typically developing peers [9].

A review of access solutions for communication in children with cerebral palsy concluded that the most common access pathways were pressure-sensing micro-switches, autonomic nervous system sensors, accelerometers, video cameras used for image processing, thermal cameras, and eye-track control technology [10]. However, these access pathways may not meet the needs of many children who do not have voluntary control of their movements and cannot successfully and reliably operate these access solutions. By bypassing physical movement, in a collaboration between the brain and a device, where an interface translates signals from the brain to direct external activity, brain–computer interfaces (BCIs) hold the potential as an exciting new access solution [11–13]. BCI can be used to control mainstream technology, such as computers for communication [14] and environmental control [15], and innovative technology such as robotics [16] and exoskeletons [17] for mobility. Furthermore, the pioneering work conducted by neurotechnology research groups such as those behind *BrainGate*, *Stenode*™, *Neuralink*, and *Facebook* holds the potential to further the access to communication and mobility. BCI research field is indeed a fast-growing area in neural engineering, an intersection of neuroscience, computer science, electrical engineering [18], robotics, and all other possible stakeholders (families, support network, educators, speech pathologist, and occupational therapists) [19, 20].

2.1.1 The Need for Brain–Computer Interface Research for Communication in Children with Cerebral Palsy

Equal access to education and health care/services is not only a basic human right but is also essential for an individual's strengths to be recognized in the classroom and workplace and to enable them to reach their full potential [21, 22]. Although we know that about one in two people with cerebral palsy has an intellectual impairment [7], less than one-third cannot undergo formal cognitive assessment. Standard assessment tools require manual manipulation, verbal responses, and/or are often timed, presenting a significant challenge to obtaining an accurate picture of the true potential of children with severe cerebral palsy [23]. As such, people with motor and/or speech impairment are unable to complete the tasks as precisely or efficiently as their typically developing peers, leading to an underestimation of ability. Furthermore, emerging research points to the need for young children at risk for complex communication difficulties, to receive appropriate augmented language experience early, thus having a chance to follow a more normalized developmental communication skills pathway [24].

With verbal comprehension often exceeding expressive capacities in children with complex communication needs [25] and without a means to tap into these abilities, there is an urgent need for innovative technology solutions to bridge this gap, which may pose the following questions:

- If BCI was effectively used by children with cerebral palsy, would we have a new access method to identify their intellectual ability?
- With a means to assess the child's ability of expressive and receptive language, and with the ability to engage children as communication partners more easily with them, would their language and academic abilities develop and mature like their peers?

2.2 Outline, Scope, and Limitations of the Review

Although the knowledge is currently available about BCIs to facilitate communication, mobility, and participation for adults, in particular for able-bodied adults, there is little information on how to provide BCI access to communication and assistive technologies for children [11, 26].

The purpose of this state-of-the-art review is to share information about how BCI has been used in laboratory and research environments and to extrapolate these findings to provide a justification for the development and application of BCI for communication in young children with cerebral palsy.

The review discusses challenges that must be overcome for BCI applications for communication in children to be more widely adopted in research and translated into clinical practice. The use of BCI applications in related areas such as motor rehabilitation, gaming, and a more in-depth account for the characteristics of BCI, such as mathematical formulas for signal analysis and classifications, is outside the scope of this paper. For a more comprehensive methodology review of non-invasive BCI components for Augmentative and Alternative Communication (AAC), the authors recommend the excellent review of Akcakaya et al. [11].

2.3 Research on Brain–Computer Interfaces in Adults

This section highlights the four aspects that can be attributed to the successful use of BCI:

- Characteristics of the BCI;
- Individual characteristics of the BCI user;
- Type of feedback and instruction;
- BCI performance evaluation.

2.3.1 Characteristics of the BCI

BCI systems have been developed with different modalities where brain activity is monitored by electrophysiological activity, e.g., electroencephalography (EEG) or hemodynamic activity, e.g., functional magnetic resonance imaging (fMRI), near-infrared spectroscopy (NIRS), or when the purpose is functional imaging, and functional near-infrared spectroscopy (fNIRS) [27]. The introduction of portable EEG and fNIRS devices made these techniques suitable for BCI applications. Unlike EEG and fNIRS, fMRI has excellent spatial resolution and whole-brain coverage, so it can accurately locate activation areas in the brain. With this advantage, fMRI-BCI can extract the activation levels of specific anatomical locations as feedback, allowing a better understanding of the brain anatomy and its functionality during a specific task [28]. Hybrid BCIs is a third modality in which a combination of two or more systems such as NIRS/fNIRS-EEG and EEG-electrocardiogram (ECG) are involved [29, 30] either simultaneously or sequentially.

Depending on the latency of the signals, brain activity is divided into endogenous and exogenous groups. Endogenous signals can further be classified as invasive or non-invasive. An example of an invasive endogenous signal is slow cortical potentials (SCP) and non-invasive sensorimotor rhythms which are spontaneous. Exogenous

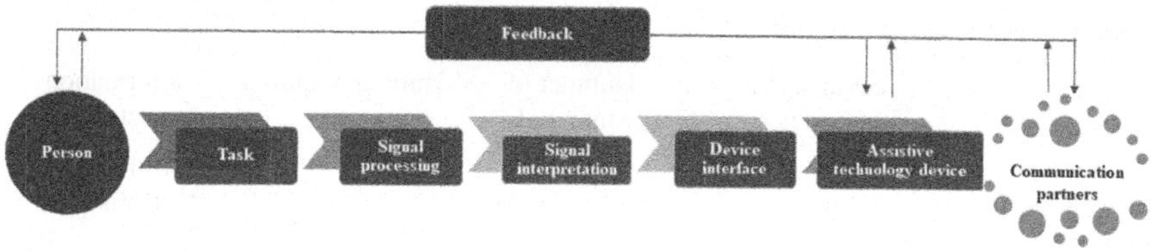

Fig. 2.1 The BCI application process

signals are non-invasive and are evoked (P300 and visual evoked (VEP)). Invasive BCIs use activity recorded by brain implanted micro- or macro-electrodes, whereas non-invasive BCIs use brain signals recorded with sensors placed on the person's scalp [31, 32].

The focus of this review will be on non-invasive systems and signals most commonly used in BCI research with the aim to facilitate communication. Figure 2.1 summarizes the BCI application process [33] and the following section will address (1) signals acquired from the human brain ("person and task"), which are then (2) filtered (signal processing), (3) classified (signal interpretation), and (4) transferred to an output device interface controlling the communication device. The feedback of the consequences of the brain activity carried out to control the assistive technology device, as well as feedback and responses from communication partners will be discussed later in the chapter.

1. **Signal acquisition**

- Electrical activity

Signals most commonly used in non-invasive BCI approaches are those generated by the electrical activity of the brain and recorded by EEG, where the electrodes are placed directly on the scalp, as reported in Table 2.1.

The different BCI systems offer different levels of complexities in terms of choices or commands, which have implications to what tasks the BCI system can be set up with and how the user, in turn, interacts with the BCI system.

The most used event-related-potential component in BCI systems is called P300 [34, 35]. P300 can be elicited with auditory and visual stimuli between 250 and 750 ms, easy to control for most users, and therefore does not require long training periods [33].

Volitional cortical potentials (VCP) such as slow cortical potentials (SCP) with gradual changes in the EEG, and sensorimotor rhythms (mu and beta rhythms), found in the motor cortex [32], are also a frequently used control mechanism based on a spontaneously generated signal [12]. VCP-based BCI normally requires long training periods for the user. Research is conducted to improve the machine learning algorithm to provide a more user-friendly experience [11]. Other common signals are steady-state evoked potential (SSEP) where evoked responses are made by flickering lights or fluctuating auditory stimuli.

Table 2.1 EEG control signals

Signal	Brain signal occurrence	Number of commands	Training required	Information transfer rate
VEP steady-state evoked potentials (SSEP) such as steady-state visual evoked potentials (SSVEP)	Visual cortex	High	No	60–100 bits/min
SCP	Slow voltage shifts	Low	Yes	5–12 bits/min
P300	Positive peaks	High	No	20–25 bits/min
Sensorimotor rhythms (mu and beta rhythms)	Synchronization to motor activities	Low	Yes	3–35 bits/min

VEP = Visual Evoked Potentials; SCP = Slow Cortical Potential; P300 = P300 Evoked Potentials

EEG-based BCIs are portable, cost-effective, and attractive to use considering the high-temporal resolution shown in the latest developments [11]. However, the type of headsets and the number of EEG channels vary between studies. Traditional EEG sensors (electrodes) require a conductive gel or paste to act as a conductive path, which limits the more widespread use of EEG as users have to wash their hair after each EEG setup and setup time can take up to 1 h. Therefore, sensors based on water [36] and dry-electrode systems are under development for clinical practice.

– Hemodynamic activity

NIRS is an imaging technique that measures light absorbance to estimate oxyhemoglobin and deoxyhemoglobin providing an indirect measure of brain activity. When NIRS is applied in brain function research, where changes in hemodynamic activity on the cortical surface are monitored, the system is called fNIRS. The infrared light is sent through the skull where the intensity of the light allows alterations in oxyhemoglobin and deoxyhemoglobin concentrations to be measured. It is a safe, relatively inexpensive non-invasive technology. Adults can learn to regulate hemodynamic activity within a few training sessions [37–39].

A recent study conducted by Chaudhary et al. [40] demonstrated an important development for individuals in a locked-in state. The participants reliably generated physiological signals to control an fNIRS-BCI enabling them to provide 'yes' and 'no' responses. Despite the positive findings, the slower transfer rate in which the hemodynamic response occurs and the susceptibly to interference from head movements or hair [37, 41] are the most challenging factors for researchers and developers to overcome when using NIRS/fNIRS. However, the slow transfer rate may be enhanced by using multiple channels and mental tasks [42]. Furthermore, NIRS systems involve minimal setup time, use an unobtrusive headband, and do not require gel, and therefore possess advantages over EEG-BCIs [42]. As such, NIRS/fNIRS hold promise for future BCI applications and may play an important role in the future combined hybrid BCI with EEG system aimed at improving accuracy [43].

- Hybrid BCI

The main purpose of a hybrid BCI is to limit the disadvantages of individual BCI systems, improving accuracy, and reducing errors [34]. Input signals used in simultaneous hybrid BCIs can be two different brain signals or one brain signal and another input such as an electromyogram (EMG). In sequential hybrid BCIs, the output of one system is used as the input of the other system. This may lead to the user finding the system more complicated. Despite this, a review of hybrid BCIs showed that in studies in which P300, steady-state visually evoked potentials (SSVEP), event-related desynchronization (ERD)-NIRS, ERD-electrooculography (EOG), ERD-SSVEP, and EEG-EMG had been combined, all studies demonstrated improved performance in a hybrid modality compared to each modality on its own [34].

Electrodermal activity (EDA) or skin conductance activity measures the resistive properties of the skin, which a person can learn to control by the anticipation and recognition of events, music, and meditation [44] to mention a few. EDA is a novel modality that has yet to be extensively studied as an alternative BCI interface. However, emerging work with able-bodied adults demonstrates that a bi-directional control of an EDA signal feature can be achieved through a mental or breathing task [44]. Fleury et al. [45] concluded in their scoping review of e-textiles in clinical rehabilitation that numerous physiological and biomechanical occurrences can already be measured. However, further research and development is required to explore e-textiles' full potential. This research will be interesting to build on to explore the possibilities to reduce the EEG setup times.

2. **Signal processing**

After the correct number and location of EEG channel selection are identified, the first step of signal processing is the filtering, to reduce artifacts and noise due to several reasons (e.g., horizontal and vertical eye movements, and eye blinking). A band-pass filter is usually applied in BCI systems where artifacts are filtered out from the recordings without compromising the relevant data. Akcakaya et al. [11] describe the most common processing approaches: down-sampling, moving average filtering; independent component analysis (ICA); common spatial patterns (CSP); xDAWN algorithm; and principal component analysis (PCA) [11]. A more extensive list of artifact removals has been outlined by Lakshmi et al. [46].

Examples of filters for NIRS and fNIRS are Chebyshev infinite impulse response (IIR) [42], Butterworth low-pass filters, wavelet-based filters [47], and Chebyshev Type II filter [48] and low-pass filter [49]. Furthermore, Akcakaya et al. [11] concluded that research suggests that BCI performance is enhanced when, in addition to selecting the Fz, Cz, and Pz channels, a channel selection according to the International 10–20 system, posterior and occipital areas is carried out.

After filtering, several algorithms may be applied to extract the main signal characteristics. This procedure is called feature extraction, and it plays a key role in a BCI system, as the performance of the next classification stage depends on the selected features. Several linear methods (e.g., Fast Fourier Transform (FFT), Autoregressive Model (AR), and Wavelet Transform (WT)) and non-linear approaches (e.g., Fractal Dimension (FD), Higher-Order Spectra (HOS), and Correlation Dimension (CD)) have been applied in BCI studies in an attempt to improve classification performance. The choice of using one of the methods mentioned depends on the paradigm applied (e.g., P300, SSVEP, motor imagery, and neurofeedback) [50].

3. **Classification**

After feature extractions, the signals are classified to detect and discriminate the response following each mental task or stimulus. Developing an effective algorithm for the EEG signal classification is critical, and [51] proposed to use an error-evoked potential for detecting and avoiding errors. Akcakaya et al. [11] list the most common classification approaches in their methodology review and also point to the challenge of identifying one 'best' classifier as it is dependent on the filtering and signal processing procedures.

fNIRS-based BCI classification has been shown to provide stronger ERD brain signals and better classification accuracy [52]. The frequently used classification methods are linear discriminant analysis (LDA)-based classifiers, support vector machine (SVM), genetic algorithms, logistic linear regression, neural networks, matched filters, Pearson's correlation method, and regularized discriminant analysis (RDA) [11]. Lakshmi et al. [46] provide a more detailed description of classifiers and also outline advantages and disadvantages with each method.

4. **Output device interface—BCI-controlled application**

The three current possibilities of output from a BCI-controlled application for communication are text, text-to-speech, and speech [11]. Direct speech output, wherein the person is using speech motor imagery to produce vowel sounds, with the end goal of developing BCI-capable device for real-time synthetic speech has been explored by [53, 54]. The commonly researched P300 Speller supported by the BCI2000 system has been validated for text output tasks (spelling, email, or Internet browsing) [55]. Text-to-speech can be achieved by enabling the synthesizer on a personal computer. Furthermore, Thompson et al. [56] developed a BCI keyboard that could be used to operate a commercially available Augmentative and Alternative Communication (AAC) speech-generating device. They showed that, like AAC in general, BCI-elicited speech was slower than natural speech (4–6 letters a minute) [33], can be difficult to learn, and in addition requires substantial training and motivation. Despite this, BCI can play a unique and important role in the field of assistive technology by means as an access method for those who do not have another reliable access method [11, 33, 40, 56].

An example of a virtual letter grid controlled by auditory evoked potentials was reported in the single-case study by Käthner et al. [57] in which eye-tracking was compared with auditory BCI in an individual in a locked-in state. Different tones were performed in random order to select one of five letters in alphabetic order. Results showed that with effort and depending on the user's concentration level that day, the user was able to successfully control the BCI to communicate [57].

BCIs for communication are challenged with many of the same difficulties as other AAC. However, in a study seeking perspectives of BCI use by people with Amyotrophic Lateral Sclerosis (ALS), participants considered that despite some challenges, access to BCI with the EEG-cap and low speed for communication was a priority [58, 59].

To facilitate the development of BCI, in particular around the translation of research findings from able-bodied participants to people with disabilities such as cerebral palsy, Wolpaw et al. [32] recommended a multi-disciplinary approach with experts in signal processing, computer science, clinical teams of neuroscientists and communication specialists, rehabilitation experts, and users. This recommendation calls for shared protocol guidelines and funding opportunities to provide an opportunity for a range of stakeholders to work together across the boundaries of agencies and disciplines.

In summary, BCIs represent one of the most innovative areas in assistive technology. However, despite technology advances current BCI research still experiences a number of challenges that limit their usefulness. These challenges are linked to the time required to reach an effective output (signal-to-noise ratio, signal processing, and classifications techniques) [60], optimizing the number of stimuli [61], the overall design of the BCI system, [62], headsets, setups [63], effectiveness [64], efficiency, and satisfaction [65, 58]. With any further advancement, it is important to adopt a multi-disciplinary approach and not only engage the end-user in its design but also evaluate its function.

2.3.2 Individual Characteristics of the BCI User

The research group of Fried-Oken et al. [66] identified the following skills as essential for successful use of BCI that is visual ERP-based: (1) hearing and auditory ability to respond correctly to stimuli, (2) ability to follow instructions; (3) vision, visual perception, and sustained visual attention ability to identify letters on the screen and attend to task; (3) literacy and spelling skills to correctly recognize letters and words and compose written messages; (4) working memory and ability to sustain attention. Like [67, 66], further recommendation is that in addition to pain and use of medications, motor functions should be assessed to determine optimal positioning to minimize the risk of signal interference.

For other factors to consider, Weyand and Chau [42] recommended that in addition to anatomical brain data, physiological and observational measures of the person's enthusiasm and arousal throughout the study should be noted. The authors found that this could contribute to a better understanding of the particular user's mood and nature [42].

Daly et al.'s [67] study on 14 individuals with cerebral palsy identified that due to uncontrolled neck muscles and/or their head supports, many users were observed to exhibit poor signal quality on occipital channels mainly due to their need for head support which put pressure on those electrodes. This points to the unique needs of individuals with severe motor impairments. Consequently, when designing new access solutions, a user-centered approach with the involvement of individuals with motor impairments prior to task development [42] gathering their perspectives of the technology as end-users and as partners in the design process [68] is imperative in future study designs.

Furthermore, in a review of the literature for EEG-based BCIs as access pathways, 3–19% of able-bodied individuals were not able to generate EEG patterns to control a specific BCI system, issues not yet established in individuals with a disability [12]. Depending on the stimulus' flashing rate/pattern, size, and color, it is found that in 0.3–3% of people it is possible that an evoked potential will trigger an adverse effect in photosensitive people and as such, for safety reasons, are not deemed eligible for BCI studies [69]. This is the main safety issue raised in the BCI literature. Moreover, with some of the challenges with EEG-based BCI, hybrid BCI systems and fMRI-BCI may warrant further research into the possibility of providing personalized and bespoke solutions for people with disabilities.

Like with all assistive technology devices, a thorough assessment of the match between the technology and the person is recommended. However, the literature around suitable assessments matching the person (user training, language development) and their environment (communication partners, settings in which they prefer to use the BCI, and portability) with the most suitable BCI system is scarce [70]. With the underlying framework built on the International Classification of Functioning, Disability and Health (which involves both individual factors, e.g., client's physical and cognitive capacities and environmental factors, e.g., technology, the physical environment, and the communication partners), the Human Activity Assistive Technology (HAAT) and the Matching Person and Technology (MPT) models promote and facilitate a user-centered approach [71]. HAAT focuses on human performance in a given task, within a specific situation [71]. MPT addresses the user's needs and preferences, the functions and features of the appropriate technology, and the environmental factors which are important to consider to ensure successful use [71].

In summary, individual characteristics to consider with an individual who may be a candidate for BCI use include developmental, psychological, physiological, and neurobiological factors [70]. Frameworks built on the International Classification of Functioning, Disability and Health which involves both individual factors and environmental factors, such as the Human Activity Assistive Technology (HAAT) and the Matching Person and Technology (MPT), may assist when matching the technology to the individual and facilitate a user-centered approach.

2.3.3 Type of Feedback and Instruction

Another essential part of a successful BCI is the feedback of the consequences of the brain activity carried out to control the computer or device [49]. For the user to modulate their brain activity, feedback of sensory stimuli, such as changes in color [72] or auditory [73], and verbal response (such as 'yes' or 'no') have been found to enhance the performance of P300 brain–computer interfaces [74]. Also, participants want to know when they are doing something wrong and what to do to change it [20]. An example of the effect on mental tasks was demonstrated in a study by Schudlo and Chau [75] where eight out of 10 participants adjusted their mental strategies when receiving feedback. A further example using NIRS was immediate feedback of real-time changes in blood oxygenation levels, conceptualized for the participant by providing a ball that rose and fell with the average change during the session [42].

As previously mentioned, user characteristics that may influence BCI performance [11, 42] are fatigue [76], mood [77], and attention [78] which may be affected by training factors including frequency of sessions, task durations, and session length. Hence, to achieve a more accurate and faster BCI control, a user-centered training protocol, in which individual adjustments and optimizations made possible, has been suggested. This concept was explored by Friedrich et al. [79] where the participants tried out seven mental tasks (mental rotation, word association, auditory imagery, mental subtraction, spatial navigation, motor imagery of hand, and motor imagery of both feet), and the task providing the best 4-class combination, in addition to a reactive frequency band, was chosen for the online control. Their protocol showed promising results and the researchers' final recommendations included not only providing a selection of mental tasks based on classification results but also considering the user's personal preferences for specific mental tasks [79].

Moreover, motivation has been identified as imperative to engage users for ongoing BCI training [12, 20]. Training protocols using more engaging training paradigms by adopting virtual environments [76] and BCI games may increase user motivation, improve training effectiveness, and reduce training times. P300 BCI games can be useful for introducing individuals to the BCI systems, as no training is required to start playing. It has been suggested that the P300 BCI games may be included in a software package for the consumer of a BCI gaming device, to allow people who purchase the device to experience the BCI control without extended training [80]. Most tasks using the P300 signal involve spelling, however, Kaplan et al. [80] took a routine BCI protocol activity and turned it into six different research games, which showed that interest generated by the P300 game ensured participation throughout the sessions. Such training regimes would, in particular, be important to explore for children.

Mindfulness meditation training with able-bodied participants with the aim to improve BCI performance has been explored. Significant improvement in accuracy was achieved compared to a control group [81, 82]. In another study, mindfulness meditation training showed significantly larger P300 amplitudes compared to control [83]. Positive results of meditation can be demonstrated after only a few days of training [84] and may be an important component of any future BCI training and setup protocols for both adults and children.

In summary, BCI use is a skill that requires engaging user training [76]. A personalized, engaging training protocol with a mindfulness meditation activity incorporated has separately shown to increase BCI performance. An adaptive training protocol will be useful for feedback and instruction to BCI users, in particular, to assess each individual's optimal (number of channels, frequency band mental task selection, and individual classifier updates.

2.3.4 BCI Performance Evaluation

Aspects that a group of individuals with severe motor impairment find most important in respect of BCI are functionality, independent use, and ease of use [59]. This corresponds with the guidelines for the ISO 9241-201, a multi-part standard from the International Organization for Standardization (ISO) [85] that covers ergonomics of human–computer interaction. The standards define usability as the "extent to which a … product … can be used by specified users to achieve specified goals with (1) effectiveness, (2) efficiency, and (3) satisfaction in a specified context of use" [86].

1. **Effectiveness**

Effectiveness, also referred to as accuracy, is often measured according to correct selections compared to the total number of selections. Accuracy is the most common metric reported for the performance evaluation of communication-based BCIs according to a review of papers between 2005 and 2012 [87].

2. **Efficiency**
- Information transfer rate

The accuracy, number of different targets, and the time needed per selection together govern the Information Transfer Rate (ITR). ITR was first defined by Wolpaw et al. [32] and is the most commonly used performance measure in BCI research [88].

The threshold of stimuli necessary to constitute a consistent and accurate response is also identified, and the level cited as being necessary for effective BCI communication is a 70% threshold [89, 90], however, users have voiced a preference for trailing devices with selection accuracy around a 90% threshold [11].

- Workload

The workload is defined as physical, mental, and temporal demands, and performance, effort, and frustration. A measure developed for human–machine systems to assess workload is the NASA task load index (TLX), a six-scale self-report questionnaire, which quantifies the workload for each task and identifies its sources [91]. The NASA TLX is a standardized assessment, and the score for each factor is weighted by the perceived contribution to total workload and summed to arrive at a final task load index [91]. Originally developed for the field of aviation, easy and quick to administer on paper, verbally or electronically, it has been used in a number of BCI studies [92–95].

3. **Satisfaction**

The Canadian Occupational Performance Measure 3rd Edition (COPM) [96], Goal Attainment Scaling (GAS) [97], and the Individually Prioritized Problem Assessment (IPPA) [98] are instruments often adopted to identify what a user wants to achieve with the use of the technology and then to measure the user's perceptions of the outcome. COPM and GAS are examples of the few instruments that are evidence-based and sensitive to measure change [99, 100]. COPM, GAS, and IPPA have the advantage that they can be used with participants of all ages, any disability, trialing any form of BCI-controlled applications or assistive technology.

A visual analogue scale (VAS) [101] ranging from 0 to 10 (not at all—absolutely satisfied) was included in Zickler et al.'s study [59] to evaluate the overall satisfaction with the device used. The VAS scale has also been used to assess motivation and mood [77]. A more specific measure of motivation the Questionnaire for Current Motivation (QCM) [102, 103] has been adapted for BCI research. The QCM-BCI comprises 18 items divided into four subscales: incompetence fear, mastery confidence, interest, and challenge, which are rated on a seven-point Likert scale [77].

In summary, in addition to effectiveness and efficiency, the user's satisfaction is one of the main aspects to measure successful BCI performance. The literature provides examples of a combination of instruments that can offer researchers additional evidence-based solutions for assessment and outcome measures in BCI-controlled applications or assistive technology.

2.4 Research on Brain–Computer Interfaces in Children for Communication and Receptive Language

A search of Medline, Embase, Google Scholar, and PubMed databases was conducted to capture the state of the art in research for BCI in children with cerebral palsy. The search revealed six studies where BCI had been used for communication or assessment purposes. Two studies included participants with cerebral palsy, as reported in Table 2.2. Alcaide-Aguirre et al. [104] conducted a verbal comprehension test using a P300 system with children with cerebral palsy, where the participant was asked to match a spoken word to one of four images on the screen using a setup similar to an AAC system. The age range in the named study was 8–29 years of age and the authors reported that the youngest group 10.6 ± 2.9 years old struggled with attention and to complete the task. The findings showed that controlling BCI in children and adults with cerebral palsy is possible, however, it may take 4 times longer compared to typically developing [104]. Perego et al. [105] assessed non-verbal, abstract, and

Table 2.2 BCI studies with children and youth addressing communication and assessment purposes

Articles	Population (sample size; age)	Diagnosis	Signal acquisition	Task (action)/Training/Feedback	Accuracies
Alcaide-Aguirre et al. [104]	N = 30; 8–29 years • $n = 19$ CP, 10 incl. in analysis • $n = 11$ able bodied	CP	P300	Peabody picture vocabulary test: selection of matching illustration to spoken words was made by focusing attention on the corresponding selection box of each illustration	82.34 ± 0.97%
Breshears et al. [26]	N = 6; 9–15 years	Intractable epilepsy	ECoG; 48–64 electrodes	Move cursor to target on screen (tongue movement = articulation; hand movement)/9 min training/visual feedback	70–99%
Ehlers et al. [106]	N = 51; 6–33 years • group 1: $n = 11$ ($\bar{x} = 6.73$) • group 2: $n = 12$ ($\bar{x} = 8.08$) • group 3: $n = 14$ ($\bar{x} = 9.86$) • group 4: $n = 14$ ($\bar{x} = 22.36$)	Able-bodied right-handed	SSVEP;8 electrodes: P_z, PO_3, PO_4, O_1, O_z, O_2, O_9, and O_{10}	Move cursor; Speller (focusing on lit LEDs)/selected letters visible on the screen	Medium frequency stimulation group 1: < 60% 2: < 60% 3: < 80% 4: < 80%

(continued)

Table 2.2 (continued)

Articles	Population (sample size; age)	Diagnosis	Signal acquisition	Task (action)/Training/Feedback	Accuracies
Lim et al. [107]	N = 20 • $n = 10$; 7–12 years medicated • $n = 10$; non-medicated matched	ADHD	EEG; FP1, FP2; dry electrodes	Attention training game; worksheet/20 training sessions/feedback unknown	Unknown
Lim et al. [107]	N = 19; 7–12 years	ADHD non-medicated	EEG; FP1, FP2; dry electrodes	Attention training game; worksheet/24 training sessions/feedback unknown	Unknown
Perego et al. [105]	N = 26; 10–41 years ($\bar{x} = 21.8 \pm 10.48$) • $n = 18$ CP • $n = 8$ SMA III, SCI, MS, hereditary spastic paraplegia	CP, SMA III, SCI, MS, hereditary spastic paraplegia	SSVP	Matching game; Raven Colored Progressive Matrices (RCPM)	Unknown

CP = Cerebral Palsy; ECoG = Electrocorticography; SSVEP = Steady-State Visual Evoked Potential; ADHD = Attention Deficit Hyperactivity Disorder; SMA III = Spinal Muscular Atrophy type 3; SCI = Spinal Cord Injury; MS = Multiple Sclerosis; N = #sample size; n = #subgroup sample size cognitive functioning, using SSVP, and they found that of the 18 participants with cerebral palsy, 10 were able to complete the test and of those 6 were children with the mean age of 12.5 years (range 11–15 years). They did not find a relationship between gross motor severity and the participants' cognitive abilities that had been assessed prior. However, due to the small sample size, caution is needed in any interpretation. It is worth noting that in these two studies, for about 50% of the children with cerebral palsy involved in the research projects, BCI was deemed as a possible access method.

Research also shows that children older than six can operate a BCI using auditory SSVEPs in spelling tasks adapted to their age-related development [108]. In their paper, Ehlers et al. [106] conducted a study using SSVEPs with able-bodied children and adults. Results demonstrated that mean accuracy rates depend on age and the frequency of stimulation (10–11 Hz).

Although not comparable with standard EEG, Roland et al. [109] found in their study that higher frequency bands showed a significant correlation with age and suggested that for Electrocorticography (ECoG)-BCI use, it would provide a more stable signal platform.

The search conducted in this paper found the forging study by Breshears et al. [26], the only study of an invasive ECoG-BCI of six children, aged 9–15 years, who required invasive monitoring for pediatric epilepsy. The participants reached 70–99% accuracy within 9 min of training. Their results demonstrate that children's level of performance is comparable to those of the adults. The results, indicating that children's signals can be decoded like those from an adult brain, show important proof of concept that decoding of brain signals from the pediatric cortex is possible [26].

Lastly, Lim et al. [107] showed improved attention in children with attention-deficit hyperactivity disorder (ADHD) through non-invasive EEG BCI-based games. In a small pilot with 10 children and 10 controls, a non-significant improvement between groups in attention after 20 sessions was noted [107]. However, when the authors followed up with 19 children who were non-medicated, they found that after an intensive training program of 3 sessions per week for 8 weeks, parents reported behavioral improvements that were sustained for 3 months post treatment [107]. Lim et al. [107] reported that two of the 19 participants had experienced headaches from their sessions, however, these were well tolerated and had not hindered them from continuing the treatment program.

Further to the questions raised in the previous section of this paper, questions around how we deliver instruction and provide feedback, how BCI can be optimally used for communication, and how we evaluate performance and user satisfaction are important to address. This state-of-the-art review raises the question:

- If we were to start applying some of the findings that are available on adults, what modifications to setup and training protocols will our research finding tell us about the match between user characteristics and BCI characteristics in children?

Research incorporating the EEG characteristics of children into machine learning and signal processing methods has been suggested to be an avenue to explore in increasing the translation of BCI applications to children [110].

2.5 Ethical Considerations

Popular media commonly and incorrectly report that science is now capable of reading people's thoughts. However, most BCI studies described in these articles are working on obtaining a binary signal (i.e., activating or deactivating a scanning system of flashing letters on a virtual keyboard) that a person can use to control an assistive technology device. The media statements pose a challenge for researchers as they may raise false expectations for participants seeking participation in future research studies. This and other ethical issues were identified by Nijboer et al. [111] who summarized the literature on emerging neuroethical topics associated with BCI. Nijboer et al. [111] concluded with a list of 17 issues: (1) obtaining informed consent from people who have difficulty communicating; (2) risk/benefit analysis needs to be done on an individual basis until more empirical data has been collected; (3) shared responsibility of BCI teams, providing one clear message to the user; (4) the consequences of BCI technology for the quality of life of the person and their families; (5) potential side-effects (such as possible long-term influences of BCI on a young developing nervous system); (6) personal responsibility and its possible constraints; (7) issues concerning personality and personhood (BCI may induce changes to cognitive capacities); (8) therapeutic applications, including risks of excessive use; (9) questions of research ethics regarding the transition from animal experimentation to application in human subjects; (10) mind-reading and privacy; (11) mind-control; (12) selective enhancement and social stratification; (13) human dignity; (14) mental integrity; (15) bodily integrity; (16) regulating safety; and (17) communication to the media, the need for BCI researchers to check statements and when needed actively speak out about incorrectly reported details.

Nijboer et al. [111] surveyed 166 attendees of the 4th International BCI conference regarding some of these neuroethical topics. The respondents were mostly male (73%), had 3–13 years of experience in BCI research (64%), and lived in Europe (42%) or North America (51%). Survey questions related to informed consent, risks and benefits, team responsibility, consequences of BCI for the person and their family, and communication with the media. The informed consent procedure is time-consuming and may be complex when the person cannot communicate. Despite the challenges, most respondents (76%) agreed that when supported by their legal representative, people who cannot talk should be given the opportunity to participate in non-invasive BCI. Nijboer et al. [111] argued that the way information on potential outcomes, benefits, and risks is presented to the legal representative may influence their decision and expectations. The weighing of the risks and benefits is an essential component of participation in any research study. Despite the challenges of achieving sufficient speed and accuracy of communication using BCI, the respondents reported they believed the potential benefits of providing communication to people with severe disabilities outweigh any risks with non-invasive BCI. Respondents recommended that the research team have a common view and work as a unified group that delivers clear messages to the person and their family. Some respondents considered that families could experience additional burdens as successful BCI use resulted in a user gaining more control over decisions about their daily living. BCI researchers expressed the importance to be able to review media releases to control for errors in reporting. Whilst having the duty to tell facts, still being enthusiastic about the field, and able to share a vision [111].

Specker Sullivan and Illes [112] conducted a review of the literature to establish components of modern ethical rationale statement to consider for BCI research. They proposed that the following four components should be included when reporting studies on human participants:

1. A statement explaining the choice of subject.
2. A statement of risk–benefit.
3. Citations supporting claims about target population desires, needs, and/or values.
4. Recognition of the multiple possible uses of the BCI device being tested, including related opportunities and caveats.

Questions in relation to BCI research with children may be raised around the following:

- Whether brain signal acquisition is safe for a developing brain and;
- Whether there are long-term effects.

Emerging research shows that age seems to affect lower frequency rhythms, while gamma rhythms appear to be set in regards to anatomical distribution and amplitude modulation [109]. To date, it has only been reported that photosensitivity may be an adverse effect [113].

In summary, ethical issues have been identified in the literature and recommendations entail more awareness of ethical aspects in the BCI field, which involves shared responsibility of collaborative multi-disciplinary BCI teams [114]. From an ethics perspective, it is therefore imperative that the concepts of user-centered design, participatory action research, and patient-centered outcomes should be applied to ensure that BCIs will meet the needs of users. Exploration of BCIs with a wide variety of physical, cognitive, and linguistic limitations and abilities should be encouraged.

2.6 Conclusions and Recommendations

A state-of-the-art review uncovered six studies where BCI had been used for communication or assessment purposes in children, where two studies included participants with cerebral palsy [104, 105]. Although challenges with BCI in adults still exists, it is imperative to expand research beyond adults and able-bodied and involve end-users, including children, to ensure that any new developments of systems are optimized from the perspective of the individual, considering differences in brain structure, topography, cognitive processing pathways, and psycho-behavioral predisposition [42]. The studies identified in this review (Table 2.2) all achieved accuracy levels beyond the gold standard 70% [89, 90] and as such show encouraging performance for future research.

BCI technology brings many promising applications in a variety of clinical applications of assistive technology, but unfortunately, the speed of writing and the EEG-cap itself were raised by users as a feature that would benefit from further development [58]. To faciliatate BCI use, engaging user training is required [12, 76]. Moreover, effectiveness, efficiency, and user satisfaction are important aspects to measure to establish successful BCI performance. Ethical issues have been identified in the literature, which among other things may call for a multi-disciplinary team collaboration. Excitingly, a small number of studies do provide encouraging evidence for continued research in children with studies demonstrating children as young as 7 years [106] able to learn to control their brain activity to perform activities on the computer.

BCI for children may hold the key to unlock the potentials in children with cerebral palsy who today have no means to learn, play, and communicate. If we are to help children and adolescents with cerebral palsy develop to their fullest potential, we must understand what works and why. To accomplish this, future research must move to the next level. We close with the following recommendations:

- Assemble multi-disciplinary and user-involved collaborative teams to translate best-available BCI evidence from existing research with adults and children to individuals with cerebral palsy across all ages. Evidence is available regarding five aspects of BCI: (1) individual characteristics of the BCI user, (2) characteristics of the BCI, (3) type of feedback and instruction, (4) the BCI-controlled application, and (5) evaluation. These collaborations should aim to identify clear clinical and research directions to accelerate the development of effective and efficient BCI solutions for communication difficulties which aim to enhance the participation of people with complex communication needs.
- Multi-disciplinary and user-involved collaborative publications, workshops, and conference presentations.
- Implementation of active surveillance programs to monitor the long-term implications of BCI on the person's medical, social, psychological, and physical situation.
- Introduction of a multi-disciplinary and user-involved collaborative research summit to advance the global priority research agenda for BCI development for individuals with cerebral palsy (and other disabilities) across all ages: shared research, training, and reporting protocols to enable future cross-study comparisons to advance the field.

References

1. Krahn GL (2011) WHO World Report on Disability: a review. Disabil Health J 4(3):141–142
2. Odding E, Roebroeck ME, Stam HJ (2006) The epidemiology of cerebral palsy: incidence, impairments and risk factors. Disabil Rehabil 28(4):183–191
3. Rosenbaum P, Paneth N, Leviton A, Goldstein M, Bax M, Damiano D, Dan B, Jacobsson B (2007) A report: the definition and classification of cerebral palsy April 2006. Dev Med Child Neurol Suppl 109:8–14
4. Australian Cerebral Palsy Register (2018) Report of the Australian Cerebral Palsy Register Birth years 1995–2012. Cerebral Palsy Alliance, Sydney
5. Nordberg A, Miniscalco C, Lohmander A, Himmelmann K (2013) Speech problems affect more than one in two children with cerebral palsy: Swedish population-based study. Acta Paediatr 102(2):161–166
6. Smits D, Ketelaar M, Gorter J, Van Schie P, Becher J, Lindeman E, Jongmans M (2011) Development of non-verbal intellectual capacity in school-age children with cerebral palsy. J Intell Disabil Res 55(6):550–562
7. Fennell EB, Dikel TN (2001) Cognitive and neuropsychological functioning in children with cerebral palsy. J Child Neurol 16(1):58–63
8. Novak I, Hines M, Goldsmith S, Barclay R (2012) Clinical prognostic messages from a systematic review on cerebral palsy. Pediatr:2012-0924
9. Vos RC, Becher JG, Ketelaar M, Smits D-W, Voorman JM, Tan SS, Reinders-Messelink HA, Dallmeijer AJ (2013) Developmental trajectories of daily activities in children and adolescents with cerebral palsy. Pediatrics 132(4):e915–e923
10. Myrden A, Schudlo L, Weyand S, Zeyl T, Chau T (2014) Trends in communicative access solutions for children with cerebral palsy. J Child Neurol 29(8):1108–1118
11. Akcakaya M, Peters B, Moghadamfalahi M, Mooney AR, Orhan U, Oken B, Erdogmus D, Fried-Oken M (2014) Noninvasive brain–computer interfaces for augmentative and alternative communication. IEEE Rev Biomed Eng 7:31–49
12. Moghimi S, Kushki A, Marie Guerguerian A, Chau T (2013) A review of EEG-based brain-computer interfaces as access pathways for individuals with severe disabilities. Assist Technol 25(2):99–110
13. Zhang J, Jadavji Z, Zewdie E, Kirton A (2019) Evaluating if children can use simple brain computer interfaces. Front Human Neurosci 13:24
14. Chaudhary U, Birbaumer N, Ramos-Murguialday A (2016) Brain–computer interfaces for communication and rehabilitation. Nat Rev Neurol 12(9):513
15. Zickler C, Di Donna V, Kaiser V, Al-Khodairy A, Kleih S, Kübler A, Malavasi M, Mattia D, Mongardi S, Neuper C (2009) BCI applications for people with disabilities: defining user needs and user requirements. In: Assistive technology from adapted equipment to inclusive environments, AAATE, vol 25, pp 185–189
16. Ajiboye AB, Willett FR, Young DR, Memberg WD, Murphy BA, Miller JP, Walter BL, Sweet JA, Hoyen HA, Keith MW (2017) Restoration of reaching and grasping movements through brain-controlled muscle stimulation in a person with tetraplegia: a proof-of-concept demonstration. Lancet 389(10081):1821–1830
17. Bogue R (2015) Robotic exoskeletons: a review of recent progress. Ind Rob 42(1):5–10
18. Mak JN, Wolpaw JR (2009) Clinical applications of brain-computer interfaces: current state and future prospects. IEEE Rev Biomed Eng 2:187–199
19. Chavarriaga R, Fried-Oken M, Kleih S, Lotte F, Scherer R (2017) Heading for new shores! Overcoming pitfalls in BCI design. Brain-Comput Interf 4(1–2):60–73
20. Taherian S, Davies TC (2018) Caregiver and special education staff perspectives of a commercial brain-computer interface as access technology: a qualitative study. Brain-Comput Interf 5(2–3):73–87
21. UNCPRD (2006) United Nations convention on the rights of persons with disabilities

22. United Nations. Educational, Scientific and Cultural Organization (2000) World education report 2000: the right to education: towards education for all throughout life. UNESCO, Paris, France
23. Geytenbeek JJ, Heim MM, Vermeulen RJ, Oostrom KJ (2010) Assessing comprehension of spoken language in nonspeaking children with cerebral palsy: application of a newly developed computer-based instrument. Augment Alternat Commun 26(2):97–107
24. Branson D, Demchak M (2009) The use of augmentative and alternative communication methods with infants and toddlers with disabilities: A research review. Augment Altern Commun 25(4):274–286
25. Sevcik RA (2006) Comprehension: an overlooked component in augmented language development. Disabil Rehabil 28(3):159–167
26. Breshears JD, Gaona CM, Roland JL, Sharma M, Anderson NR, Bundy DT, Freudenburg ZV, Smyth MD, Zempel J, Limbrick DD (2011) Decoding motor signals from the pediatric cortex: implications for brain-computer interfaces in children. Pediatr 128(1):e160–e168
27. Herold F, Wiegel P, Scholkmann F, Mueller NG (2018) Applications of functional near-infrared spectroscopy (fNIRS) neuroimaging in Exercise-Cognition science: a systematic Methodology-Focused review. J Clin Med 7(12):466
28. Wang Y, Wu D (2017) Real-time fMRI-based brain computer interface: a review. International conference neural information processing. Springer, pp 833–842
29. Pfurtscheller G, Allison BZ, Bauernfeind G, Brunner C, Solis Escalante T, Scherer R, Zander TO, Mueller-Putz G, Neuper C, Birbaumer N (2010) The hybrid BCI. Front Neurosci 4:3
30. Zephaniah PV, Kim JG (2014) Recent functional near infrared spectroscopy based brain computer interface systems: developments, applications and challenges. Biomed Eng Lett 4(3):223–230
31. Donoghue JP, Nurmikko A, Black M, Hochberg LR (2007) Assistive technology and robotic control using motor cortex ensemble-based neural interface systems in humans with tetraplegia. J Physiol 579(3):603–611
32. Wolpaw JR, Birbaumer N, McFarland DJ, Pfurtscheller G, Vaughan TM (2002) Brain–computer interfaces for communication and control. Clin Neurophysiol 113(6):767–791
33. Bamdad M, Zarshenas H, Auais MA (2015) Application of BCI systems in neurorehabilitation: a scoping review. Disabil Rehabil Assist Technol 10(5):355–364
34. Amiri S, Fazel-Rezai R, Asadpour V (2013) A review of hybrid brain-computer interface systems. Adv Hum-Comput Interact 2013:1
35. He B, Gao S, Yuan H, Wolpaw JR (2013) Brain–computer interfaces. In: Neural engineering. Springer, pp 87–151
36. Kristo G, Höhne J, Ortner R, Reuderink B, Ramsey N (2015) BNCI Horizon 2020.
37. Coyle SM, Ward TE, Markham CM (2007) Brain–computer interface using a simplified functional near-infrared spectroscopy system. J Neural Eng 4(3):219
38. Power SD, Falk TH, Chau T (2010) Classification of prefrontal activity due to mental arithmetic and music imagery using hidden Markov models and frequency domain near-infrared spectroscopy. J Neural Eng 7(2):026002
39. Sitaram R, Zhang H, Guan C, Thulasidas M, Hoshi Y, Ishikawa A, Shimizu K, Birbaumer N (2007) Temporal classification of multichannel near-infrared spectroscopy signals of motor imagery for developing a brain–computer interface. Neuro Image 34(4):1416–1427
40. Chaudhary U, Xia B, Silvoni S, Cohen LG, Birbaumer N (2017) Brain–computer interface–based communication in the completely locked-in state. PLoS Biol 15(1):e1002593
41. Power SD, Kushki A, Chau T (2012) Automatic single-trial discrimination of mental arithmetic, mental singing and the no-control state from prefrontal activity: toward a three-state NIRS-BCI. BMC Res Notes 5(1):141
42. Weyand S, Chau T (2017) Challenges of implementing a personalized mental task near-infrared spectroscopy brain–computer interface for a non-verbal young adult with motor impairments. Dev Neurorehabil 20(2):99–107
43. Mihara M, Miyai I (2016) Review of functional near-infrared spectroscopy in neurorehabilitation. Neurophotonics 3(3):031414–031414

44. Blain S, Mihailidis A, Chau T (2008) Assessing the potential of electrodermal activity as an alternative access pathway. Med Eng Phys 30(4):498–505
45. Fleury A, Sugar M, Chau T (2015) E-textiles in clinical rehabilitation: a scoping review. Electronics 4(1):173–203
46. Lakshmi MR, Prasad DT, Prakash DVC (2014) Survey on EEG signal processing methods. Int J Adv Res Comput Sci Softw Eng 4(1)
47. Falk TH, Guirgis M, Power S, Chau TT (2011) Taking NIRS-BCIs outside the lab: towards achieving robustness against environment noise. IEEE Trans Neural Syst Rehabil Eng 19(2):136–146
48. Faress A, Chau T (2013) Towards a multimodal brain–computer interface: combining fNIRS and fTCD measurements to enable higher classification accuracy. Neuroimage 77:186–194
49. Power SD, Kushki A, Chau T (2011) Towards a system-paced near-infrared spectroscopy brain–computer interface: differentiating prefrontal activity due to mental arithmetic and mental singing from the no-control state. J Neural Eng 8(6):066004
50. Vaid S, Singh P, Kaur C (2015) EEG signal analysis for BCI interface: a review. In: 2015 fifth international conference on advanced computing & communication technologies. IEEE, pp 143–147
51. Schalk G, Wolpaw JR, McFarland DJ, Pfurtscheller G (2000) EEG-based communication: presence of an error potential. Clin Neurophysiol 111(12):2138–2144
52. Kübler A, Holz E, Kaufmann T, Zickler C (2013) A user centred approach for bringing BCI controlled applications to end-users. Brain-Comput Interf Syst-Recent Progr Fut Prospects 1:19
53. Brumberg JS, Nieto-Castanon A, Kennedy PR, Guenther FH (2010) Brain–computer interfaces for speech communication. Speech Commun 52(4):367–379
54. Guenther FH, Brumberg JS, Wright EJ, Nieto-Castanon A, Tourville JA, Panko M, Law R, Siebert SA, Bartels JL, Andreasen DS (2009) A wireless brain-machine interface for real-time speech synthesis. PloS One 4(12):e8218
55. Farwell LA, Donchin E (1988) Talking off the top of your head: toward a mental prosthesis utilizing event-related brain potentials. Electroencephalogr Clin Neurophysiol 70(6):510–523
56. Thompson DE, Gruis KL, Huggins JE (2014a) A plug-and-play brain-computer interface to operate commercial assistive technology. Disabil Rehabil Assist Technol 9(2):144–150
57. Käthner I, Kübler A, Halder S (2015) Comparison of eye tracking, electrooculography and an auditory brain-computer interface for binary communication: a case study with a participant in the locked-in state. J Neuroeng Rehabil 12(1):76
58. Huggins JE, Wren PA, Gruis KL (2011) What would brain-computer interface users want? Opinions and priorities of potential users with amyotrophic lateral sclerosis. Amyotroph Lateral Scler 12(5):318–324
59. Zickler C, Riccio A, Leotta F, Hillian-Tress S, Halder S, Holz E, Staiger-Sälzer P, Hoogerwerf E-J, Desideri L, Mattia D (2011) A brain-computer interface as input channel for a standard assistive technology software. Clin EEG Neurosci 42(4):236–244
60. Kaufmann T, Schulz S, Grünzinger C, Kübler A (2011) Flashing characters with famous faces improves ERP-based brain–computer interface performance. J Neural Eng 8(5):056016
61. Schreuder M, Höhne J, Blankertz B, Haufe S, Dickhaus T, Tangermann M (2013) Optimizing event-related potential based brain–computer interfaces: a systematic evaluation of dynamic stopping methods. J Neural Eng 10(3):036025
62. Nicolas-Alonso LF, Gomez-Gil J (2012) Brain computer interfaces, a review. Sensors 12(2):1211–1279
63. Lim J-H, Hwang H-J, Han C-H, Jung K-Y, Im C-H (2013) Classification of binary intentions for individuals with impaired oculomotor function:'eyes-closed'SSVEP-based brain–computer interface (BCI). J Neural Eng 10(2):026021
64. Zhang Y, Xu P, Liu T, Hu J, Zhang R, Yao D (2012) Multiple frequencies sequential coding for SSVEP-based brain-computer interface. PloS One 7(3):e29519
65. Huggins JE, Moinuddin AA, Chiodo AE, Wren PA (2015) What would brain-computer interface users want: opinions and priorities of potential users with spinal cord injury. Arch Phys Med Rehabil 96(3):S38–S45. e35

66. Fried-Oken M, Mooney A, Peters B, Oken B (2015) A clinical screening protocol for the RSVP Keyboard brain–computer interface. Disabil Rehabil Assist Technol 10(1):11–18
67. Daly I, Billinger M, Laparra-Hernández J, Aloise F, García ML, Faller J, Scherer R, Müller-Putz G (2013) On the control of brain-computer interfaces by users with cerebral palsy. Clin Neurophysiol 124(9):1787–1797
68. Cook A, Hussey S (1995) Assistive technologies: principles and practice. MOSBY, Baltimore, USA
69. Fisher RS, Harding G, Erba G, Barkley GL, Wilkins A (2005) Photic-and pattern-induced seizures: a review for the epilepsy foundation of America Working Group. Epilepsia 46(9):1426–1441
70. Huggins JE, Guger C, Allison B, Anderson CW, Batista A, Brouwer A-M, Brunner C, Chavarriaga R, Fried-Oken M, Gunduz A (2014) Workshops of the fifth international brain-computer interface meeting: defining the future. Brain-Comput Interf 1(1):27–49
71. Giesbrecht E (2013) Application of the Human Activity Assistive Technology model for occupational therapy research. Aus Occ Ther J 60(4):230–240
72. Takano K, Komatsu T, Hata N, Nakajima Y, Kansaku K (2009) Visual stimuli for the P300 brain–computer interface: a comparison of white/gray and green/blue flicker matrices. Clin Neurophysiol 120(8):1562–1566
73. Furdea A, Halder S, Krusienski D, Bross D, Nijboer F, Birbaumer N, Kübler A (2009) An auditory oddball (P300) spelling system for brain-computer interfaces. Psychophysiol 46(3):617–625
74. Birbaumer N, Murguialday AR, Weber C, Montoya P (2009) Neurofeedback and brain–computer interface: clinical applications. Int Rev Neurobiol 86:107–117
75. Schudlo LC, Chau T (2013) Dynamic topographical pattern classification of multichannel prefrontal NIRS signals: II. Online differentiation of mental arithmetic and rest. J Neural Eng 11(1):016003
76. Wolpaw J, Wolpaw EW (2012) Brain-computer interfaces: principles and practice. Oxford University Press, USA
77. Kleih S, Nijboer F, Halder S, Kübler A (2010) Motivation modulates the P300 amplitude during brain–computer interface use. Clin Neurophysiol 121(7):1023–1031
78. Curran EA, Stokes MJ (2003) Learning to control brain activity: a review of the production and control of EEG components for driving brain–computer interface (BCI) systems. Brain Cogn 51(3):326–336
79. Friedrich EV, Neuper C, Scherer R (2013) Whatever works: a systematic user-centered training protocol to optimize brain-computer interfacing individually. PloS One 8(9):e76214
80. Kaplan AY, Shishkin SL, Ganin IP, Basyul IA, Zhigalov AY (2013) Adapting the P300-based brain–computer interface for gaming: a review. IEEE Trans Comput Intell AI Games 5(2):141–149
81. Eskandari P, Erfanian A Improving the performance of brain-computer interface through meditation practicing. In: 30th annual international conference of IEEE engineering in medicine and biology society. IEEE, pp 662–665
82. Tan L-F, Dienes Z, Jansari A, Goh S-Y (2014) Effect of mindfulness meditation on brain–computer interface performance. Conscious Cogn 23:12–21
83. Lakey CE, Berry DR, Sellers EW (2011) Manipulating attention via mindfulness induction improves P300-based brain–computer interface performance. J Neural Eng 8(2):025019
84. Tang Y-Y, Ma Y, Wang J, Fan Y, Feng S, Lu Q, Yu Q, Sui D, Rothbart MK, Fan M (2007) Short-term meditation training improves attention and self-regulation. Proc National Acad Sci 104(43):17152–17156
85. Turner B (1999) International Organization for Standardization (ISO). The Statesman's Yearbook 2000: The Politics, Cultures and Economies of the World:114
86. Ergonomics of human-system interaction - Part 210: Human-centred design for interactive systems (ISO 9241-210:2019)
87. Thompson DE, Blain-Moraes S, Huggins JE (2013) Performance assessment in brain-computer interface-based augmentative and alternative communication. Biomed Eng Online 12(1):43

88. Thompson DE, Quitadamo LR, Mainardi L, Gao S, Kindermans P-J, Simeral JD, Fazel-Rezai R, Matteucci M, Falk TH, Bianchi L (2014b) Performance measurement for brain–computer or brain–machine interfaces: a tutorial. J Neural Eng 11(3):035001
89. Kübler A, Neumann N, Kaiser J, Kotchoubey B, Hinterberger T, Birbaumer NP (2001) Brain-computer communication: self-regulation of slow cortical potentials for verbal communication. Arch Phys Med Rehabil 82(11):1533–1539
90. Perelmouter J, Birbaumer N (2000) A binary spelling interface with random errors. IEEE Trans Rehabil Eng 8(2):227–232
91. Hart SG, Staveland LE (1988) Development of NASA-TLX (Task Load Index): results of empirical and theoretical research. In: Advances in psychology, vol 52. Elsevier, pp 139–183
92. Duvinage M, Castermans T, Petieau M, Seetharaman K, Hoellinger T, Cheron G, Dutoit T A subjective assessment of a P300 BCI system for lower-limb rehabilitation purposes. In: Eng Med Biol Soc, 2012 annual international conference of IEEE engineering in medicine and biology society (EMBC). IEEE, pp 3845–3849
93. Felton EA, Williams JC, Vanderheiden GC, Radwin RG (2012) Mental workload during brain–computer interface training. Ergon 55(5):526–537
94. Hoonakker P, Carayon P, Gurses AP, Brown R, Khunlertkit A, McGuire K, Walker JM (2011) Measuring workload of ICU nurses with a questionnaire survey: the NASA Task Load Index (TLX). IISE Trans Healthc Syst Eng 1(2):131–143
95. Rubio S, Díaz E, Martín J, Puente JM (2004) Evaluation of subjective mental workload: a comparison of SWAT, NASA-TLX, and workload profile methods. Appl Psychol 53(1):61–86
96. Law MC, Baptiste S, Carswell A, McColl MA, Polatajko H, Pollock N (1998) Canadian occupational performance measure. Canadian Association of Occupational Therapists, Ottawa
97. Kiresuk TJ, Smith A, Cardillo JE (2014) Goal attainment scaling: applications, theory, and measurement. Psychology Press
98. Wessels R, Persson J, Lorentsen Ø, Andrich R, Ferrario M, Oortwijn W, VanBeekum T, Brodin H, de Witte L (2002) IPPA: Individually prioritised problem assessment. Technol Disabil 14(3):141–145
99. Cusick A, McIntyre S, Novak I, Lannin N, Lowe K (2006) A comparison of goal attainment scaling and the Canadian Occupational Performance Measure for paediatric rehabilitation research. Pediatr Rehabil 9(2):149–157
100. Sakzewski L, Boyd R, Ziviani J (2007) Clinimetric properties of participation measures for 5-to 13-year-old children with cerebral palsy: a systematic review. Dev Med Child Neurol 49(3):232–240
101. Crichton N (2001) Visual analogue scale (VAS). J Clin Nurs 10(5):706–706
102. Nijboer F, Furdea A, Gunst I, Mellinger J, McFarland DJ, Birbaumer N, Kübler A (2008) An auditory brain–computer interface (BCI). J Neurosci Method 167(1):43–50
103. Rheinberg F, Vollmeyer R, Burns BD (2001) QCM: A questionnaire to assess current motivation in learning situations. Diagnostica 47(2):57–66
104. Alcaide-Aguirre R, Warschausky S, Brown D, Aref A, Huggins J (2017) Asynchronous brain–computer interface for cognitive assessment in people with cerebral palsy. J Neur Eng 14(6):066001
105. Perego P, Turconi A, Andreoni G, Gagliardi C (2014) Cognitive ability assessment by brain-computer interface II: application of a BCI-based assessment method for cognitive abilities. Brain-Comput Interf 1(3–4):170–180
106. Ehlers J, Valbuena D, Stiller A, Gräser A (2012) Age-specific mechanisms in an SSVEP-based BCI scenario: evidences from spontaneous rhythms and neuronal oscillators. Comput Intell Neurosci 2012:20
107. Lim CG, Lee TS, Guan C, Fung DSS, Zhao Y, Teng SSW, Zhang H, Krishnan KRR (2012) A brain-computer interface based attention training program for treating attention deficit hyperactivity disorder. PloS One 7(10):e46692
108. Kolev V, Yordanova J (1997) Analysis of phase-locking is informative for studying event-related EEG activity. Biol Cybern 76(3):229–235

109. Roland J, Miller K, Freudenburg Z, Sharma M, Smyth M, Gaona C, Breshears J, Corbetta M, Leuthardt EC (2011) The effect of age on human motor electrocorticographic signals and implications for brain–computer interface applications. J Neural Eng 8(4):046013
110. Kinney-Lang E, Auyeung B, Escudero J (2016) Expanding the (kaleido) scope: exploring current literature trends for translating electroencephalography (EEG) based brain–computer interfaces for motor rehabilitation in children. J Neural Eng 13(6):061002
111. Nijboer F, Clausen J, Allison BZ, Haselager P (2013) The asilomar survey: stakeholders' opinions on ethical issues related to brain-computer interfacing. Neuroethics 6(3):541–578
112. Specker Sullivan L, Illes J (2016) Beyond 'communication and control': towards ethically complete rationales for brain-computer interface research. Brain-Comput Interf 3(3):156–163
113. Luo A, Sullivan TJ (2010) A user-friendly SSVEP-based brain–computer interface using a time-domain classifier. J Neural Eng 7(2):026010
114. Vlek RJ, Steines D, Szibbo D, Kübler A, Schneider M-J, Haselager P, Nijboer F (2012) Ethical issues in brain–computer interface research, development, and dissemination. J Neurol Phys Ther 36(2):94–99

Chapter 3
PPG-Based Non-invasive Methodologies for Pervasive Monitoring of Vitals: BP and HR

Madhuri Panwar, Arvind Gautam, and Amit Acharyya

Abstract Pervasive monitoring of vitals especially blood pressure and heart rate are considered the most valuable parameters because they are the most important biomarkers as well as risk indicators for stroke and cardiovascular diseases which are the leading causes of mortality and morbidity worldwide. Despite the continuous improvements for decades in conventional methods of HR (electrocardiography) and BP (sphygmomanometery and oscillometry), still, these techniques have not been enhanced to the point of offering the user flexibility and portability. The recent advancements in microelectromechanical systems and wireless sensor technologies have enabled the proliferation of miniaturized body-worn non-invasive, photoplethysmography (PPG) sensors which are capable of long-term pervasive monitoring of vital parameters. However, developing algorithms for real-time monitoring is a significant challenge due to various conflicting requirements including high accuracy, less complexity, and limited power supply. This chapter covers the accurate and low-complex methodologies for real-time monitoring of HR and BP for cardiac and stroke rehabilitation monitoring.

Keywords PPG · Deep learning · LRCN · Time-series prediction · Non-invasive method · Pervasive monitoring

---

M. Panwar · A. Gautam · A. Acharyya (✉)
Department of Electrical Engineering, IIT Hyderabad, Kandi, Sangareddy, Telangana 502285, India
e-mail: amit_acharyya@iith.ac.in

M. Panwar
e-mail: ee15resch01003@iith.ac.in

A. Gautam
e-mail: ee15re-sch01003@iith.ac.in

G. D. Gargiulo and G. R. Naik (eds.), *Wearable/Personal Monitoring Devices Present to Future*, https://doi.org/10.1007/978-981-16-5324-7_3

3.1 Introduction

Measurement of vitals non-invasively, using minimum numbers of sensors, which is comfortable to the patients, would allow this clinically significant quantity to be measured more frequently, allowing early identification changes that may be indicative of acute physiological deterioration. The recent advancements in microelectromechanical systems and wireless sensor technologies have enabled the proliferation of miniaturized body-worn non-invasive, photoplethysmography (PPG) sensors [1]. It is generally accepted that PPG can provide much valuable information about the cardiovascular system and has been applied in many different clinical situations, including clinical physiological parameter monitoring (blood oxygen saturation, heart rate, and BP). It utilizes the pulse oximeter which illuminates the skin and measures the volumetric variations in the blood due to light absorption during the cardiac cycle. It is a simple, economical, and non-invasive technology for monitoring the physiological parameters without the need for reference signals, laboratory conditions, and special medical training [2]. This provides a higher degree of usability, therefore, being used in various wearable medical devices for measuring cardiac output, blood oxygen saturation, arterial blood pressure, and autonomic function for long-term pervasive monitoring [3, 4]. It is one of the strongest candidates for promoting the opportunities of ambulatory and telemonitoring by measuring the physiological parameters reliably and non-invasively [5, 6].

Despite the several advantages, the susceptible nature of PPG technology toward motion artifacts and noise has become the limiting factor, distorting the signal fidelity. This impedes the robust evaluation of physiological parameters, making it inefficient for wearability in ambulatory daily-life conditions [7–9]. Therefore, performance has become the paramount concern for validating usability in clinical applications.

A number of studies related to HR and BP are performed utilizing the signal processing and feature engineering-based learning algorithms for the prediction of these physiological parameters. This includes adaptive filtering, Wiener filtering, Kalman filtering, empirical mode decomposition (EMD), spectral subtraction, and independent component analysis (ICA) and machine learning-based methods [10–16]. However, these techniques employed motion reference from external sensors, heuristic thresholds, and different tuned parameters and handcrafted feature engineering, preventing the generalization capability of these methodologies.

The recent success of deep learning algorithms in various biomedical applications has validated the robustness in extracting clinical information from the sensory data [17–20]. The inherent capability of these deep networks in extracting the useful features from the data during the learning phase provides not only accurate results but also cost-effective and responsive solutions compared to conventional signal processing and machine learning algorithms, shifting the focus toward deep learning-based algorithms for accurate prediction of physiological parameters. However, these are computationally expensive and exhibit memory extensive behavior which poses a bottleneck for its real-time implementation, where the resources are scarce [21, 22]. Thus, the development of efficient and low-complex deep learning algorithms plays a significant role in modeling the portable remote monitoring system. Therefore, this chapter introduces low-complex deep learning methodologies for vital monitoring to enable real-time execution on resource-constrained platforms. Section 3.2 provides a brief overview of the conventional methods of vital monitoring, and Sect. 3.3 discusses the monitoring of vitals using the Long-term Recurrent Convolutional Network (LRCN) which is the most widely used hybrid network designed by utilizing the advantages of convolutional neural network and long-short term network. Section 3.4 details the performance analysis of deep learning methods.

3.2 HR and BP Estimation Methods

Most of the studies for HR estimation have been performed using the publically available signal processing cup dataset (SPC), consisting of 23 subjects' data which were collected during physical exercise in a controlled environment. A tri-stage TROIKA framework by Zhang et al. [23] estimated HR in the presence of strong artifacts in PPG data using a number of signal processing methods, including signal decomposition, sparse signal reconstruction, and spectral peak tracking. They tested on 12 subjects of the SPC dataset and reported an average absolute error (AE) of 2.34 beats/minute (BPM). This work [23] was later enhanced in their other study [24] using the joint sparse spectral reconstruction which had reduced the AE to 1.28 BPM. Another recent work by Zhu et al. [25] presented an algorithm using spectral subtraction, neural network, and post-processing for HR estimation wherein an AE of 1.03 BPM is obtained using the same dataset. A similar work was conducted in the recent study by Biswas et al. [3], using the four-layer deep neural network (CorNET), which reported AE of 1.99 BPM and 1.47 BPM for 12 and 23 subjects, respectively. The most recent study conducted by Arunkumar and Bhaskar [26] reduced the AE to 0.92 BPM using the de-nosing algorithm, formulated by cascading three-stage adaptive filters, which was followed by FFT for HR estimation.

Further, researchers also explored the relationship between BP and PPG [27–30]. One such work by Esmaili et al. [31] presented a novel non-linear model based on the modeling vessels as elastic tubes for BP estimation using the PTT/PAT values. They have obtained mean absolute error of 6.22/4.71 and 3.97/4.44 mmHg for SBP and DBP estimation, respectively, for 32 healthy subjects. Another work by Radha et al. [32] explored PPG to track the BP changes in healthy free-living subjects. They have extracted different features and utilized LSTM wherein 5.95 and 4.95 mmHg errors are reported for SBP and DBP, respectively. Other work by Kachuee et al. [33] estimated BP by processing and extracting features and then applied regression algorithms. They have reported mean absolute error of 5.35 and 11.17 mmHg for calibration-free DBP and SBP estimation, respectively, on the MIMIC II dataset. The very recent work by Mousavi et al. [30] estimated the BP based on whole-based feature extraction following the regression algorithm. They obtained the mean absolute error of 2.43 and 3.97 mmHg for SBP and DBP, respectively, on the MIMIC II dataset.

The aforementioned methods have a number of limitations as follows: (a) Focused on monitoring the single physiological parameter; (b) Some methods have involved a number of pre-processing steps to attenuate motion artifacts which are highly parameterized and optimized for specific conditions, preventing the generalization capability; (c) some methods utilized feature extraction to select the appropriate features followed by prediction algorithm for estimation of physiological parameter which usually results in poor performance and also puts extra computational load; (d) Some have developed personalized methods, necessitating data collection followed by training of model for each new user who wants to use in real time. Furthermore, these studies performed the evaluation on a very small population which is a limiting factor for enabling practical usability in clinical applications.

A very recent study by Panwar et al. [34] has developed an efficient and low-complex deep learning framework PP-Net having the capability to estimate HR as well as BP (SBP, DBP) for a cost-effective solution. The joint framework of CNN and LSTM along with the self-capability of feature selection/extraction resulted in accurate estimation which enables usability for real-time applications. This deep learning framework is tested on publically available larger datasets available at the UCI repository. This dataset is extracted from Multi-parameter Intelligent Monitoring in Intensive Care (MIMIC-II) database from the PhysioNet repository. These deep learning frameworks are detailed in the next section.

3.3 Long-Term Recurrent Convolutional Network (LRCN)

The taxonomy of DNN primarily includes multi-layer perceptron (MLP), CNN, and recurrent neural networks (RNN). They enable learning of task-adapted feature representations from the data [35] itself which eliminates the feature selection and extraction steps. CNNs are characterized by stacking multiple layers including an initial layer of convolutional filters followed by the non-linearity activation function, sub-sampling (pooling), and fully connected layer enabling the classification [17, 18]. The stacking of multiple convolutional layers helps in extracting different types of features, where downstream layers capture more complex or differentiating features. This aids to integrate information from different filters and various levels of abstraction. On the other side, RNNs are an effective choice for analyzing time-series data for inferring sequential/time-variant information. They incorporate contextual information from past inputs and are robust to localized distortions in the input sequence along time. However, RNNs applied to long sequential data usually suffer from vanishing gradient (long chain problem) wherein information from previous computations is rapidly attenuated with progression through the data flow. In this regard, LSTM, a variant of RNNs, plays a major role to overcome this issue [36]. It uses a memory block inspired by a computer memory cell, where context-dependent input, output, and forget gates control what is to be written, read, and kept in the cell in each time step. Hence, it becomes convenient for the network to store a given input over many time steps, in effect helping LSTM layers to capture temporal properties.

3.3.1 CorNET: Deep Learning Framework

This framework is formulated based on the personalized data-driven approach, using a four-layered deep neural network consisting of two convolutional neural network layers in conjunction with two long short-term memory layers, followed by a fully connected output layer, customized as follows: (a) regression layer with a neuron to predict HR; (b) classification layer with two neurons for the identification of a subject among a group. This network topology is depicted in Fig. 3.1.

Data Pre-processing—SPC dataset contains simultaneous recordings of two channels of PPG from the wrist (dorsal) using a pulse oximeter with green LED (wavelength: 515 nm), tri-axial accelerometer signals from the wrist, and a channel of ECG from the chest using wet ECG electrodes. The ECG signal acts as the ground truth for PPG-based HR estimation. All signals were sampled at 125 Hz and transmitted to a computer through Bluetooth. PPG window (frame) length considered for this exploration was 8 s (sliding by 2 s), like ECG-HR computation. The PPG data samples are filtered before feeding to the network using the band-pass fourth-order Butterworth filter with the cutoff frequencies 0.1–18 Hz to restrict the high-frequency noise component and drifts from the signal of interest. Further, data is normalized to zero mean and unit variance.

CorNET Model—The architectural details of the CorNET are shown in Fig. 3.2 wherein a set of 32 filers, each having filter size of 40, is used in each CNN layer following the rectified linear unit (ReLU) activation, whereas each LSTM layer consists of 128 memory units and hyperbolic tangent (tanh) function. There is a batch

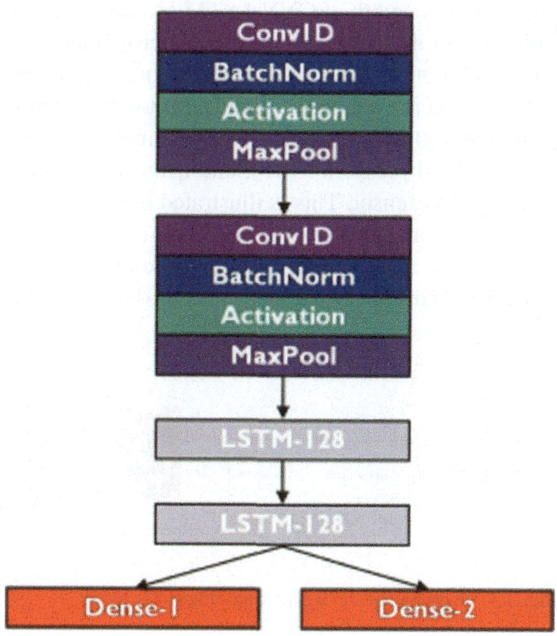

Fig. 3.1 CorNET topology using 2 CNN and 2 LSTM layers with 1 dense layer (one neuron for HR prediction)

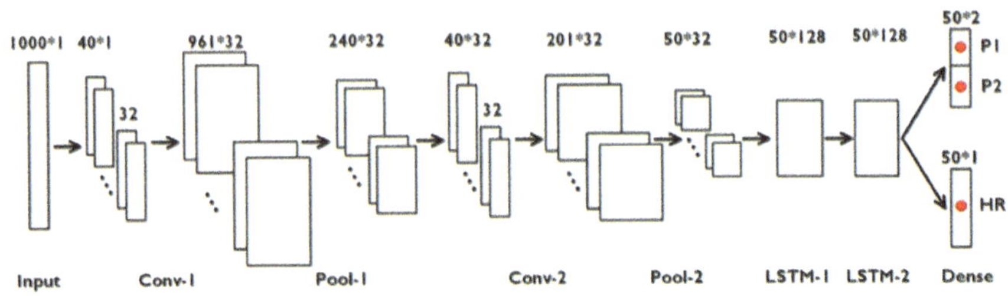

Fig. 3.2 Architectural information of CorNET (one neuron for HR prediction)

normalization layer after the ReLU activation. The max-pooling layers with a pool size of 4, dropout layer with rate 0.1, and root mean square propagation (RMSProp) optimizer are used with the default hyper-parameters. The CorNET was evaluated on the SPC dataset using the leave-one-window-out validation on 23 subjects' PPG records. This methodology achieved an average absolute error of 0.48 ± 0.19 BPM for HR prediction. Further, CorNET is also validated on an ambulant use-case scenario with custom sensors for two subjects.

3.3.2 PP-Net: Deep Learning Framework

This framework is developed with a generalized, lightweight, and customized long-term recurrent convolutional network (LRCN) model, utilizing CNN and LSTM. The joint framework of CNN-LSTM leverages the advantages of both the networks with the data-driven feature extraction, providing an efficient and lightweight model. The novelty of this work lies in developing an efficient deep learning framework PP-Net with generalized, lightweight, and customized LRCN model, having multi-score output capability to estimate three parameters, i.e. diastolic blood pressure (DBP), systolic blood pressure (SBP), and heart rate (HR), simultaneously using a sensor unit, i.e. PPG sensor. This is illustrated in Fig. 3.3.

Data Pre-processing—The PPG and ECG data are pre-processed before feeding to the neural model which included the exclusion of data of insufficient duration (<8 min) and unreliable signals (missing data, very high, and low BP values). For data

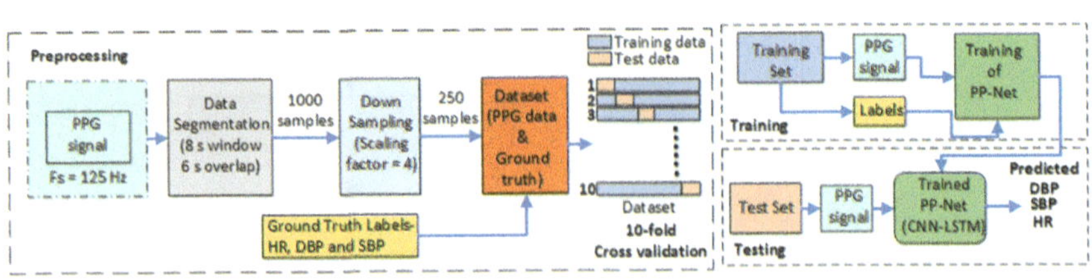

Fig. 3.3 Overview of PP-Net methodology

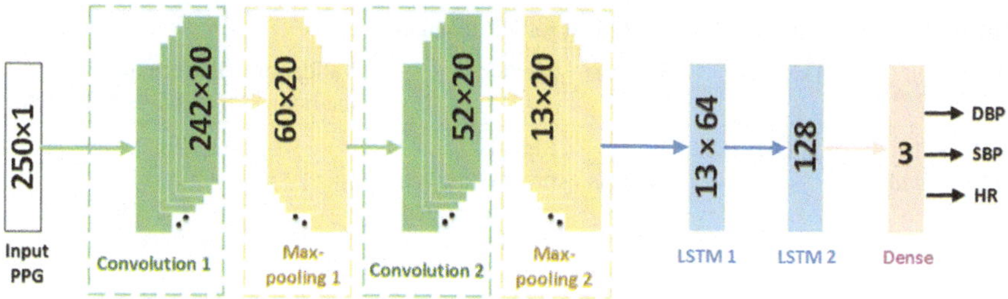

Fig. 3.4 Architecture of PP-Net model

segmentation, an 8 s window with 75% overlapping is considered enough to capture useful information about the cardiac activity. Further, down-sampling is performed to reduce the PPG data length which helped in reducing the computational complexity of the LRCN model.

PP-Net Model—The PP-Net architectural information is shown in Fig. 3.4 which consists of 2 convolutional layers each followed by ReLU activation and pooling layer, which work as a feature extractor. Next, feature extractor output is fed to the LSTM network, having stacking of two LSTM layers with 64 and 128 memory cells using hyperbolic tangent. Lastly, a fully connected layer is introduced to obtain high-level reasoning in the network where each neuron has full connections to all activations in the previous layer which transforms the original input layer by layer from the original pixel values to the final class scores using the linear function for a regression problem. To alleviate the over-fitting problem, the dropout layer with 50% probability is used after each pooling layer and LSTM layers during training. For training, 50 epochs, Adam optimizer, and mean squared error (MSE) are used as the loss function for evaluating the model performance. This LRCN model is validated based on testing performance using the ten-fold cross-validation approach to get the less optimistic and biased estimation compared to the conventional train/test split method. The performance evaluation of the model is done using the most effective metrics such as normalized mean absolute error (NMAE), NRMSE, and correlation coefficients.

3.4 Performance and Complexity Analysis

The PP-Net framework is implemented in the Keras platform using Theano as a backend engine. Training and testing are performed in a workstation with Nvidia Quadro P4000 GPU with 8 GB of dedicated memory. Tables 3.1, 3.2, and 3.3 list the comparative analysis of PP-Net with the existing works for BP and HR estimation, respectively. It can be noted that these studies have used different datasets for performance evaluation, therefore, fair comparison cannot be done for HR estimation. However, PP-Net obtained MAE ± SD of 2.32 ± 0.11 BPM despite having the more

Table 3.1 Performance analysis of PP-Net for HR estimation

Work	Database	Subjects	Method	Validation method	Performance $(MAE \pm SD)$ BPM
TBME'15 Zhang et al. [23]	SPC dataset	23 subjects	Signal processing method (signal decomposition, temporal difference, sparse signal reconstruction, and spectral peak tracking)	–	2.34 ± 2.47 1–12 subjects 3.19 ± 3.61 13–23 subjects 2.77 ± 3.04 23 subjects
TBME'15 Zhang [24]	SPC dataset	23 subjects	Signal processing method (joint sparse spectral reconstruction, spectral subtraction, and peak tracking)	–	1.28 ± 2.61 1–12 subjects 3.05 ± 3.35 13–23 subjects 2.17 ± 2.98
TBME'16 Khan et al. [14]	SPC dataset	12 subjects	Signal processing method (ensemble empirical mode decomposition, adaptive filtering, and decision-making processing)	–	1.02 ± 1.79 1–12 subjects
TBME'17 Tempko [12]	SPC dataset	23 subjects	Signal processing method-WFPV algorithm (pre-process, de-nosing phase decoder, and post-processing)	–	1.02 ± 1.25 1–12 subjects 2.95 ± 3.71 13–23 subjects 1.99 ± 2.48
Sensors letters' 19 Zhu et al. [25]	SPC dataset	12 subjects	Neural network, linear regression, and post-processing	–	1.03 ± 1.82 12 subjects

(continued)

Table 3.1 (continued)

Work	Database	Subjects	Method	Validation method	Performance ($MAE \pm SD$) BPM
BSPC'19 Arunkumarand Bhaskar [26]	SPC dataset	12 subjects	Signal processing method (cascaded three-stage adaptive filters, and FFT)	–	0.92 ± 1.17 1–12 subjects
TBioCAS'19 Biswas et al. [3]	SPC dataset and Own dataset	25 subjects (23 + 2)	CNN and LSTM	Five-fold	1.99 ± 4.64 12 subjects 0.86 ± 1.86 13–23 subjects 1.47 ± 3.37 23 subjects
PP-Net Panwar et al. [34]	UCI repository database	12,000 subjects	LRCN	Ten-fold	2.32 ± 0.095

number of subjects (1557), showing effectiveness in comparison with other works. For BP estimation, existing works involved feature engineering and built separate models for DBP and SBP which increased the overall computational cost compared to PP-Net. The PP-Net achieved an average MAE ± SD of 3.14 ± 0.13 mmHg for 1557 subjects on BP estimation, and demonstrates the effectiveness of PP-Net for real-time usability in clinical applications. Furthermore, PP-Net is also validated against the clinical standard described for BP wherein ME ± SD of −1.25 ± 5.65 mmHg and 1.55 ± 5.41 mmHg are obtained in AAMI standard for estimation of DBP and SBP, respectively, and grade A in BHS standard for both DBP and SBP, which are within the limits of their defined criteria. Table 3.4 shows the complexity analysis between CorNET and PP-Net in terms of MAC and memory units required for real-time execution in the inference phase wherein PP-Net obtained 52.89 and 51.36% improvement for the estimation of BP and HR simultaneously. Despite the huge improvement in the complexity of the model, still, further research is necessary to eliminate the power-hungry multipliers involved in the deep learning algorithms. Current trends of implementing these deep learning algorithms on mobile and wearable devices made it possible to perform real-time analysis [39].

3.5 Conclusion

This chapter highlights the importance of PPG technology in pervasive monitoring and covers existing methods for BP and HR estimation using the PPG technology. This chapter mainly focused on deep learning methods, which have performed data-driven feature extraction during the training, eliminating the cost-effective steps of

Table 3.2 Performance analysis of PP-Net for BP estimation on MIMIC database

Work	Subjects	Method	Total models/Validation method	Performance ($MAE \pm SD$) mmHg	
				DBP	SBP
EMBC'16 Gaurav et al. [38]	3000 subjects	Feature extraction (46 features) and Artificial Neural Network (ANN)	2/Conventional method	3.21 ± 4.72	4.47 ± 6.85
TBME'16 Kachuee et al. [33]	1000 subjects	DWT, PCA, whole-based and physiological feature extraction, and conventional regression algorithms	2/ten-fold validation	5.35 ± 6.14	11.17 ± 10.09
BSPC'18 Mousavi et al. [30]	441 subjects	FFT, FFT^{-1}, Feature extraction, PCA, and conventional regression algorithms	2/ten-fold validation	2.43 ± 4.173	3.97 ± 8.901
Sensors'20 Panwar et al. [34]	1557 subjects	LRCN deep learning algorithm	1/ten-fold validation	2.30 ± 0.196	3.97 ± 0.064

feature selection and extraction separately. The deep learning framework PP-Net for simultaneous estimation of HR and BP (DBP, SBP) using a single channel PPG data is a lightweight (low-complex) model providing many in one solution in an unobtrusive way using a sensor, offering a cost-effective, safer, and convenient mode of health monitoring in/out of the clinical setting. The obtained results of an average $MAE \pm SDE$ of 2.86 ± 0.012 and correlation coefficients of 0.9902 for estimation of DBP, SBP, and HR, simultaneously on a larger population of CVD complications, show the efficiency of the PP-Net model in pervasive healthcare monitoring. The multi-score output capability of the PP-Net framework provides a less complex solution than the existing methodologies which have estimated HR and BP by utilizing the different methodologies and neural network models.

Table 3.3 Performance analysis of PP-Net with existing works

Work	Database	Subjects	Method	Validation method	Performance ($MAE \pm SD$) mmHg	
					DBP	SBP
EMBC' 16 Gao et al. [29]	Own dataset	65 healthy subjects	DWT, Feature extraction and SVM	Ten-fold validation	4.6 ± 4.3	5.1 ± 4.3
EMBC' 16 Duan et al. [28]	University of Queensland vital signs database	32 surgical persons	Feature extraction and support vector regression	Ten-fold cross-validation	3.67 ± 5.69	4.77 ± 7.68
ICMLC'17 Zhang and Feng [37]	University of Queensland vital signs database	32 surgical persons	Feature extraction and Support Vector Machine (SVM)	Conventional method	7.61 ± 6.78	11.64 ± 8.22
BSPC'18 Radha et al. [32]	Own dataset	90 healthy subjects	Feature extraction and LSTM	Conventional method	4.95	5.95
Sensors'20 Panwar et al. [34]	UCI machine learning repository (MIMIC II)	1557 subjects	LRCN deep learning algorithm	Ten-fold validation	2.30 ± 0.196	3.97 ± 0.064

Table 3.4 Complexity analysis of PP-Net and CorNET

Layers	Multipliers operations	Adders operations	Weight parameters
Conv-1	43,560	38,720 + 20	200
Conv-2	187,200	166,400 + 20	3620
LSTM-1	327,680	326,400 + 256	21,760
LSTM-2	4,194,304	4,186,112	98,816
Dense-1	384	381 + 3	387
Total	4,753,128	4,718,013	124,783
Panwar et al. [34]	9.47 M		124,783
Biswas et al. [3]	20.1 M		256,578

References

1. Naranjo-Hernández D, Reina-Tosina J, Min M (2019) Fundamentals, recent advances, and future challenges in bioimpedance devices for healthcare applications. J Sens
2. Fischer C et al (2016) An algorithm for real-time pulse waveform segmentation and artifact detection in photoplethysmograms. IEEE J Biomed Health Inform 21(2):372–381
3. Biswas D, Everson L, Liu M, Panwar M, Verhoef BE, Patki S ... Van Helleputte N (2019) CorNET: deep learning framework for PPG-based heart rate estimation and biometric identification in ambulant environment. IEEE Trans Biomed Circ Syst 13(2):282–291
4. Koneshloo A, Du D (2019) A novel motion artifact removal method via joint basis pursuit linear program to accurately monitor heart rate. IEEE Sens J 19(21):9945–9952
5. Castaneda D et al (2018) A review on wearable photoplethysmography sensors and their potential future applications in health care. Int J Biosens Bioelectron 4(4):195
6. Chandrasekhar A et al (2018) Smartphone-based blood pressure monitoring via the oscillometric finger-pressing method. Sci Translat Med 10(431):eaap8674
7. Lee H, Ko H, Jeong C, Lee J (2016) Wearable photoplethysmographic sensor based on different LED light intensities. IEEE Sens J 17(3):587–588
8. Jarchi D, Casson AJ (2017) Towards photoplethysmography-based estimation of instantaneous heart rate during physical activity. IEEE Trans Biomed Eng 64(9):2042–2053
9. Sinchai S et al (2018) A photoplethysmographic signal isolated from an additive motion artifact by frequency translation. IEEE Trans Biomed Circ Syst 12(4):904–917
10. Mashhadi M et al (2016) Heart rate tracking using wrist-type photoplethysmographic (PPG) signals during physical exercise with simultaneous accelerometry. IEEE Signal Process Lett 23
11. Lee B et al (2010) Improved elimination of motion artifacts from a photoplethysmographic signal using a kalman smoother with simultaneous accelerometry. Physiol Meas 31(12):1585–1603
12. Temko A (2017) Accurate heart rate monitoring during physical exercises using PPG. IEEE Trans Biomed Eng 64(9):2016–2024
13. Kim B, Yoo S (2006) Motion artifact reduction in photoplethysmography using independent component analysis. IEEE Trans Biomed Eng 53:566–568
14. Khan E et al (2016) A robust heart rate monitoring scheme using photoplethysmographic signals corrupted by intense motion artifacts. IEEE Trans Biomed Eng 63:550–562
15. Grisan E et al (2015) A supervised learning approach for the robust detection of heart beat in plethysmographic data. In: 2015 37th annual international conference of the IEEE engineering in medicine and biology society (EMBC), pp 5825–5828
16. Essalat M et al (2016) Supervised heart rate tracking using wrist-type photoplethysmographic (PPG) signals during physical exercise without simultaneous acceleration signals. In: 2016 IEEE global conference on signal and information processing (GlobalSIP), pp 1166–1170
17. Panwar M et al (2017) CNN based approach for activity recognition using a wrist-worn accelerometer. In: 2017 39th annual international conference of the IEEE engineering in medicine and biology society (EMBC). IEEE
18. Panwar M, Biswas D, Bajaj H, Jöbges M, Turk R, Maharatna K, Acharyya A (2019) Rehab-net: Deep learning framework for arm movement classification using wearable sensors for stroke rehabilitation. IEEE Trans Biomed Eng 66(11):3026–3037
19. Everson L et al (2018) BiometricNet: deep learning based biometric identification using Wrist-Worn PPG. In: 2018 IEEE international symposium on circuits and systems (ISCAS). IEEE
20. Gautam A, Panwar M, Biswas D, Acharyya A (2020) MyoNet: a transfer learning-based LRCN for lower limb movement recognition and knee joint angle prediction for remote monitoring of rehabilitation progress from sEMG. IEEE J Transl Eng Health Med 8:1–10
21. Panwar M et al (2020) M2DA: a low-complex design methodology for convolutional neural network exploiting data symmetry and redundancy. Circ Syst Signal Process 1–26

22. Panwar M et al (2017) Modified distributed arithmetic based low complexity CNN architecture design methodology. In: 2017 European conference on circuit theory and design (ECCTD). IEEE
23. Zhang Z, Pi Z, Liu B (2015) TROIKA: A general framework for heart rate monitoring using wrist-type photoplethysmographic signals during intensive physical exercise. IEEE Trans Biomed Eng 62(2):522–531
24. Zhang Z (2015) Photoplethysmography-based heart rate monitoring in physical activities via joint sparse spectrum reconstruction. IEEE Trans Biomed Eng 62(8):1902–1910
25. Zhu L, Kan C, Du Y, Du D (2018) Heart rate monitoring during physical exercise from photoplethysmography using neural network. IEEE Sens Lett 3(1):1–4
26. Arunkumar KR, Bhaskar M (2019) Heart rate estimation from photoplethysmography signal for wearable health monitoring devices. Biomed Signal Process Control 50:1–9
27. Sedghamiz H (2014) Matlab implementation of Pan Tompkins ECG QRS detector. Code available at the File Exchange site of MathWorks. https://fr.mathworks.com/matlabcentral/fileexchange/45840-complete-pan-tompkins-implementationecg-qrs-detector
28. Duan K et al (2016) A feature exploration methodology for learning based cuffless blood pressure measurement using photoplethysmography. In: 2016 38th annual international conference of the IEEE engineering in medicine and biology society (EMBC). IEEE
29. Gao SC et al (2016) Data-driven estimation of blood pressure using photoplethysmographic signals. In: 2016 38th annual international conference of the IEEE engineering in medicine and biology society (EMBC). IEEE
30. Mousavi SS et al (2019) Blood pressure estimation from appropriate and inappropriate PPG signals using a whole-based method. Biomed Signal Process Control 47:196–206
31. Esmaili A, Kachuee M, Shabany M (2017) Nonlinear cuffless blood pressure estimation of healthy subjects using pulse transit time and arrival time. IEEE Trans Instrum Meas 66(12):3299–3308
32. Radha M et al (2018) Wrist-worn blood pressure tracking in healthy free-living individuals using neural networks. arXiv preprint arXiv:1805.09121
33. Kachuee M et al (2015) Cuff-less high-accuracy calibration-free blood pressure estimation using pulse transit time. In: 2015 IEEE international symposium on circuits and systems (ISCAS). IEEE
34. Panwar M, Gautam A, Biswas D, Acharyya A (2020) PP-Net: A deep learning framework for PPG-based blood pressure and heart rate estimation. IEEE Sens J 20(17):10000–10011
35. Bengio Y (2009) Learning deep architectures for AI. Found Trends® Mach Learn 2(1):1–127
36. Hochreiter S, Schmidhuber J (1997) Long short-term memory. Neural Comput 9(8):1735–1780
37. Zhang Y, Feng Z (2017) A SVM method for continuous blood pressure estimation from a PPG signal. In: Proceedings of the 9th international conference on machine learning and computing. ACM
38. Gaurav A et al (2016) Cuff-less PPG based continuous blood pressure monitoring—a smartphone based approach. In: 2016 38th annual international conference of the IEEE engineering in medicine and biology society (EMBC). IEEE
39. Ravi D et al (2016) A deep learning approach to on-node sensor data analytics for mobile or wearable devices. IEEE J Biomed Health Inform 21(1):56–64

Chapter 4
Wearable Sensor System to Monitor the Status of the Automobile Drivers

Sridhar P. Arjunan and Arockia Vijay Joseph

Abstract Health status of automobile drivers is very significant while they drive and operate the vehicle. Driver fatigue is a serious concern and leads to many accidents. This chapter deals with the study of the wearable sensor-based system to monitor the health status and fatigue level of the driver. The proposed system comprises a camera to capture a real-time video of a driver's face and various sensors to sense the health parameters, a feature extraction module to extract facial features from the real-time video, a fatigue analyzer to monitor the changes in the EEG values, and a feedback system to provide a warning signal if the health parameter and the fatigue values exceeds a threshold. This wearable system will provide a platform to understand the health and fatigue mechanism of the driver while driving to avoid untoward incidents.

4.1 Introduction

Driver fatigue is a serious problem resulting in many thousands of road accidents each year. It is impossible to calculate the exact number of sleep-related accidents, but research shows that 20% of road accidents, and up to one quarter of fatal and serious accidents, happen due to driver fatigue [1]. These types of accidents are about 50% more likely to result in death or serious injury as they tend to be high-speed impacts because a driver who has fallen asleep cannot brake or swerve to avoid or reduce the impact.

Sleepiness reduces reaction time (a critical element of safe driving). It also reduces vigilance, alertness, and concentration so that the ability to perform attention-based activities (such as driving) is impaired. The speed at which information is processed is also reduced by sleepiness. The quality of decision-making may also be affected.

The time spent on daily commutation by people across the globe has increased steeply in the recent decades. The study conducted by the AAA Foundation reveals

S. P. Arjunan (✉) · A. V. Joseph
Department of Electronics and Instrumentation, SRM Institute of Science and Technology, Kattankulathur, Chennai 603203, India
e-mail: sridhar_arjunan@ieee.org

G. D. Gargiulo and G. R. Naik (eds.), *Wearable/Personal Monitoring Devices Present to Future*, https://doi.org/10.1007/978-981-16-5324-7_4

that US drivers travel on an average of 29.2 miles per day or 10,658 miles per year [1–4]. Also, the study made by Pew research reveals that the amount spent on traveling in a car by a person per day in the UK is on an average 38 min/day and by US drivers is on an average of 24 min/day. Following to this, a study made by National Highway Traffic Safety Administration (NHTSA) has revealed that certain medical conditions that contribute to motor vehicle crash are seizures, blackouts, diabetic reaction, heart problems, strokes, attention loss, deprived health conditions, bad emotional state, etc. prior to the crashes. This study also calculated that 1.3% of all crashes included in this investigation were caused by medical emergencies related to the drivers. These kinds of medical emergencies also present a threat to the neighboring vehicles [1–4].

Gradual increase in the number of elderly vehicle drivers post a challenge on safe and secured driving. Aging results in weakened cognitive function and threatens the driver safety. Drivers falling under the age group of 60 years and above face a high risk due to medical emergency during driving than other lame factors. Generally, in many scenarios, victims are found unaware about their health conditions which caused the accident. Health issues like heart problems, non-traumatic diseases like stroke attack, low diabetic attack, etc. are the learnt major risks for driver's death [3, 4]. Among all the above-mentioned health-related diseases, stroke is the dangerous disease which causes the collapse in the brain cell which results because of the blockage of blood flow to the brain. This also has a huge impact on postural changes. It is also identified that stroke is the second main reason for death of elderly drivers [5].

This concept of monitoring the various health and physiological information of a driver during the period of driving will ensure in a steep reduction of mishaps that happen often due to fatigue, drowsiness, asthma, road rage, etc. The health and wellness monitoring system proposed for these drivers will attract agencies like insurances, emergency services, and healthcare providers. These collected data will help in the initiation of emergency measures like activating autopilot, to initiate an auto call from a car to ambulance service along with the driver's location, etc. [3]. Therefore, there is a need for a monitoring system to monitor the state of drivers that alleviates the above-mentioned drawbacks.

There has been continuous research in the design and development of a robust and reliable fatigue monitoring system. These research studies develop systems that are based on sensors that monitor the various health parameters of the driver like heart rate, blood pressure, heartbeat rhythm, body temperature and posture, eye movements, gaze direction, corneal reflection, etc. The developments made by various researches are organized and presented below.

Lee et al. [6] have studied the long-term usage of ECG for measurement purpose in case of driving scenario. Nottingham Trent University has developed a capacitive-based heart rhythm sensor and fitted it in a car seat to measure driver's heart rhythm [7]. Martinez et al. [8] have developed a non-intrusive sensor to sense the driver's present state and warm by measuring heartbeat and respiration. Kapitaniak et al. [9] and Richardson et al. [10] have studied the eye movements and gazing directions using the eye-tracking technique or otherwise known as corneal reflection technique. Ma et al. [11] reported the use of the PCANet algorithm to enhance the efficiency of conventional EEG-based driving fatigue detection by incorporating principal component analysis (PCA) with the PCANet technique.

Due to the limitations in the eye-tracking technology as it may expect the driver to wear glasses, Zhao et al. [12] proposed a video image-based driver fatigue monitoring system. In this system, they also devised a subsystem to measure grip forces of hands on the steering. In these systems, they have applied the adaptive boosting (AdaBoost) technique and the Active Shape Models (ASM) algorithm to detect the facial changes, degree of eye closures, and opening of mouth. Khan and Panchal used a camera to capture a real-time sequential video of the driver's face at 15 frames per second. They have used Viola and Jones' algorithm to detect the facial changes and for feature extraction. According to Khan and Panchal, yawning is another important factor that may be a factor to detect the driver's fatigue and used Canny and Harris operators for eyelid, yawning detection and head rolls. These methodologies have proved to be acceptable with an improved average accuracy of 97% [13].

According to Sigari et al. [14], hypovigilance is the diminished alertness caused due to fatigue and distraction. In order to determine if the person is vigilant they have used the face template matching and horizontal projection and also the top half of the segment was used to detect the driver's hypovigilance. Using the mentioned methodologies, they have determined the hypovigilance using the head rotations from facial region, percentage and rate of eye closure, and eyelid changes when compared to the reference eyelid positions. This system used the fuzzy model to identify the driver's hypovigilance level.

Juboori and Kulkarni [15] have developed a corneal tracking technology which uses image morphological changes and the Circular Hough Transform (CHT) technique. Here, image morphological changes include dilution, image segmentation, and erosion. This corneal tracking technology had been applied on the captured images of the various drivers and tested. The system developed by them alerts the driver in case of the detection of drowsiness.

4.2 System Design

Generally, driver's Health Monitoring Systems (HMS) can be classified into two types. They are

- Active Sensing-based Health Monitoring Systems (AHMS)
- Passive Sensing-based Health Monitoring Systems (PHMS)

Active Sensing-based Health Monitoring Systems (AHMS) is the package of subsystems that uses external power source to energize the complete operation of the sensor and the subsystems, whereas the Passive Sensing-based Health Monitoring System (PHMS) is the package of subsystems that does not use the energy source for any operational activities [16].

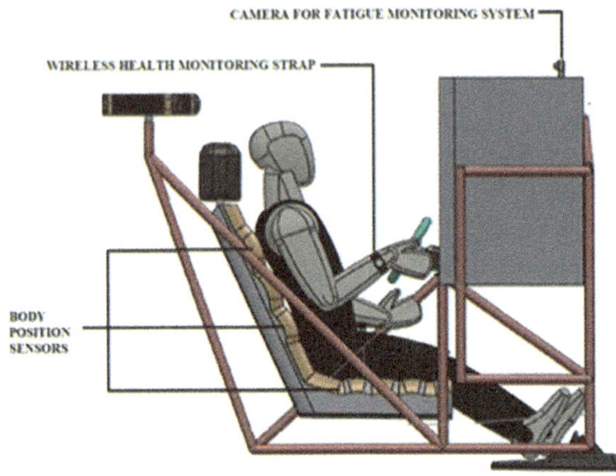

Fig. 4.1 Driver health monitoring system designed for a driving simulator

In this chapter, a system based on active sensing technologies is discussed. This system detects the status of the driver using wireless health monitoring strap, body position sensors, and a video camera. The wireless health monitoring strap is to be worn on the wrist which is very similar to a wristwatch, body position sensors are embedded at various points on the driver's seat, whereas the video camera is fitted on the inner side of the windscreen of the vehicle and in front of the driver to capture the video and image of the driver's facial expressions[17].

From Fig. 4.1, the proposed monitoring system consists of a memory storage unit, a video camera, a combination of multiple sensors, an Electronic Control Module (ECM) that does the feature extraction, comparative techniques, and fatigue analyzing and generates alerts using the vibrating module or the voice alert module [17].

In addition to all of this, the above-mentioned system includes a muscle trainer and a vibrator installed in the driver's seat and is configured to cure muscle fatigue by massaging and alerting the driver by vibrating through different sequences of vibration [17].

The memory storage unit is configured to store a threshold range for each of the health parameters, a set of predefined facial features, fatigue valuation rules, and a fatigue threshold value. The ECM is basically a microprocessor which is programmed to receive data from the multiple sensors and the camera, and to process those data to detect the fatigue of the driver and alert the driver accordingly [17].

4.3 Wearable Sensors

The wireless wearable health monitoring strap has a combination of multiple sensors like blood pressure sensor, a heart rate sensor, a blood oxygen sensor, a temperature sensor, an ElectroencephelоGram (EEG) sensor, and an ElectroMyoGram (EMG)

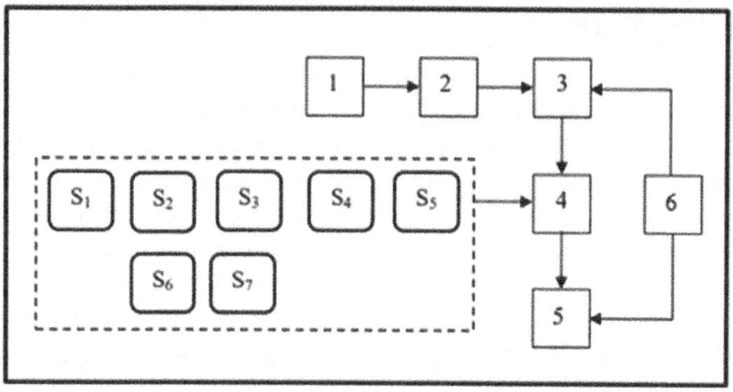

Fig. 4.2 Wearable sensing system along with controller module

sensor connected to the main controller system [17]. Figure 4.2 shows the system block diagram of the wearable sensing system along with the controller module.

The purpose of each sensor is mentioned as follows, where

- The blood pressure sensor (S_1) is developed to sense the blood pressure of the driver.
- The heart rate sensor (S_2) is configured to sense the heart rate of the driver.
- The blood oxygen sensor (S_3) is configured to sense the blood oxygen level of the driver.
- The temperature sensor (S_4) is configured to sense the temperature of the driver.
- The body position sensor (S_5) is to sense the driver's postural changes.
- The EEG sensor (S_6) is configured to sense voltage fluctuations resulting from ionic current within the neurons of the brain.
- The EMG sensor (S_7) is configured to measure muscle response or electrical activity in response to an activity in the muscle.

In certain systems a part of these sensors can even be embedded on a wearable cap, steering wheel, and driver seat along with the wearable strap to measure all the mentioned health parameters. There is a video camera (1) from Fig. 2 which is fitted in front of the driver's face to capture a real-time video of driver's face. These sensors and image from the video camera are configured to sense a plurality of health parameters of the driver to generate health parameter values [17].

These health parameter values are directly compared with the predefined data that are stored on the memory storage unit (6) using the comparator (4), and the results of the comparator are used to generate an alert message or to activate an alarming module using the alert generation module (5) [17].

4.4 Fatigue Module

The fatigue module of the wearable sensor system is the major system that completes the design of driver health monitoring system by detecting the driver fatigue using a video tuner, feature extraction module, and a fatigue analyser. In this system, the video tuner is used to tune the real-time video by eliminating the noise and by that it enhances the video quality. The purpose of the feature extraction module is to extract the facial features related to the eyes and mouth. This helps in detecting the drowsiness of the driver by detecting the eyelid movements and degree of mouth opening to differentiate the yawning from a normal speech.

The feature extraction module is configured to cooperate with the camera to receive the real-time video and is further configured to extract facial features. The fatigue analyzer is configured to receive the extracted facial features from the feature extraction module and is further configured to analyze the extracted facial features with the set of predefined facial features to compute a fatigue value based on the fatigue valuation rules.

The comparator is configured to compare each of the health parameter values with the corresponding threshold range for the health parameter value, and the fatigue value with the fatigue threshold value, and is further configured to generate an alert signal if the health parameter value exceeds or falls behind the corresponding threshold range and/or the fatigue value exceeds the fatigue threshold value.

The alert generation module is configured to receive the alert signal and is further configured to generate an alert to the driver. In an embodiment, the feature extraction module, the fatigue analyzer, the comparator, and the alert generation module are implemented as one or more processor(s).

4.5 Conclusion

The importance of the driver health monitoring system is emphasized in this chapter. In this study, the insight on the process of monitoring driver health metrics using wireless wearable systems is given. In this chapter, a technique to monitor the fatigue using various sensors and instrumentation was discussed. The proposed wearable system can provide a large platform to decode the health and fatigue mechanism of the driver which are affected by various reasons while driving and in turn can avoid unexpected incidents.

References

1. Swaytha R (2017) Enhancing vehicle safety through integrated driver wellness management. https://electronicsforu.com/market-verticals/enhancing-vehicle-safety-integrated-driver-wellness-management. Last Accessed 14 Feb 2021

2. Park SJ, Hong S, Kim D, Hussain I, Seo Y (2019) Intelligent in-car health monitoring system for elderly drivers in connected car. In: Bagnara S, Tartaglia R, Albolino S, Alexander T, Fujita Y (eds) Proceedings of the 20th congress of the international ergonomics association (IEA 2018), IEA 2018. Advances in intelligent systems and computing, vol 823. Springer, Cham
3. Tamras S (2021) In-car health and wellness monitoring, a tata elxi perspective—white paper. https://www.tataelxsi.com/Perspectives/WhitePapers/In-car-wellness.pdf. Last Accessed 14 Feb 2021
4. Subramaniyam M, Singh D, Park SJ, Kim SE, Kim DJ, Lee KS, Min SN (2018) Recent developments on driver's health monitoring and comfort enhancement through IoT. IOP Conf Ser Mater Sci Eng 402(012064):1–8
5. Park SJ, Hong S, Kim D, Hussain I, Seo Y (2019) Intelligent in-car health monitoring system for elderly drivers in connected car: volume VI: transport ergonomics and human factors (TEHF). Aerosp Hum Factors Ergon 40–44. https://doi.org/10.1007/978-3-319-96074-6_4
6. Lee YG, Kim KK, Park KS (2006) ECG measurement on a chair without conductive contact. IEEE Trans Biomed Eng 53:956–959. https://doi.org/10.1109/TBME.2006.872823
7. Nottingham trend university. https://www4.ntu.ac.uk/apps/news/160600-15/Car_seats_which_detect_when_drivers_are_falling_asleep.aspx
8. Martinez H, Sanabhuja J, Gameiro P (2013) Heart and respiration unobtrusive sensors integrated in the vehicle. HARKEN project. https://cordis.europa.eu/project/id/286265/reporting. (Last Accessed 13 Sep 2021)
9. Kapitaniak B, Walczak M, Kosobudzki M (2015) Application of eye-tracking in drivers testing: a review of research. Int J Occup Med Environ Health 28:941–954. https://doi.org/10.13075/ijomeh.1896.00317
10. Richardson DC, Michael J, Spivey MJ (2004) Eye-tracking: characteristics and methods, research areas and applications. In: Encyclopedia of biomaterials and biomedical engineering. London, Taylor & Francis, pp1–70
11. Ma Y, Chen B, Li R, Wang C, Wang J, She Q, Luo Z, Zhang Y (2019) Driving fatigue detection from EEG using a modified PCANet method. Computational Intelligence and Neuroscience 2019:9. (Article ID 4721863)
12. Zhao J, Hao K, Ding Y (2016) Driver fatigue monitoring system using video images and steering grip force. In: 5th international conference on measurement, instrumentation and automation. Published by Atlantis Press, pp 610–617
13. Khan N, Panchal D (2015) Fatigue Detection System using Image Processing on Video Sequences. Int J Eng Res Technol 4(09):398–400
14. Sigari MH, Fathy M, Soryani M (2013) A driver face monitoring system for fatigue and distraction detection. Int J Veh Technol, Hindawi Publ Corp 2013:1–11. (Article ID 263983)
15. Juboori HKS, Kulkarni L (2017) Fatigue detection system for the drivers using video analysis of facial expressions. In: 3rd international conference on computing, communication, control and automation
16. Do R (2014) ProQuest dissertations and theses; Thesis (M.S.)—University of California, San Diego; Publication number: AAT 1562045; ISBN: 9781321083651; Source: Masters abstracts international, vol 53–04, p 117
17. Joseph AV, Sridhar PA (2018) A monitoring system for monitoring the state of drivers and method thereof. Indian intellectual property office patent no: 201841030385. http://ipindiaservices.gov.in/PatentSearch/PatentSearch//View. Application status (Last Accessed 8 Dec 2019)

Chapter 5
Movement Analysis of Human Lower Limb Using EMG Sensor for Effective Implementation of Artificial Lower Extremity

Susmita Das, Dalia Nandi, and Biswarup Neogi

Abstract Electromyography (EMG), a bio signal can be utilized for clinical or diagnostic applications and also in the field of biorobotics system design. The acquired EMG signals for muscular activities require advanced methods for detection, processing and classification of the collected database from the lower limb of human being. The purpose of this study is to illustrate EMG signal analysis of the leg muscles to provide efficient and effective ways of understanding the signal's behavior. Here different postures of limbs are taken into consideration to analyze the output for various muscular activities. Some of the hardware implementations using EMG focusing on the application related to static and dynamic positions of leg muscles have been presented here. The limb movement analysis has been demonstrated through OPENSIM software platform. The signal acquisition and human device interaction is also the matter of interest. This study will provide researchers a knowledgeable information of EMG signal and its analysis procedure. This knowledge will help them to develop a powerful, flexible and efficient application in the field of artificial body organ generation.

Keywords Lower extremity · Movement · EMG · Simulation

5.1 Introduction

Human Leg is the most needed organ of human structure and movement functionality. It is capable of producing complex and expressive articulations. It is a daunting challenge for robotic leg designers to emulate human leg motion pattern in the application of prosthetics and robot design planning due to the complex neurophysiology

S. Das (✉)
Electronics and Instrumentation Engineering, Narula Institute of Technology, Kolkata, WB, India

D. Nandi
Electronics and Communication Engineering, Indian Institute of Information Technology Kalyani, Kalyani, WB, India

B. Neogi
Electronics and Communication Engineering, JIS College of Engineering, Kalyani, WB, India

G. D. Gargiulo and G. R. Naik (eds.), *Wearable/Personal Monitoring Devices Present to Future*, https://doi.org/10.1007/978-981-16-5324-7_5

of human leg. Physiological variations in the membranes of muscle fiber cause myoelectric signals. A technique used for reading and analyzing the myoelectric signal is called EMG.

The EMG signal is a biomedical signal that means electric potentials are generated in muscles during its contraction representing neuro-muscular activities. Hence, this complicated signal is controlled by the nervous system and is dependent on the anatomical and physiological properties of muscles. The EMG signal carries noise signal while travelling through different tissues. The EMG sensor is placed at the surface of the skin, collects signals from different motor units which may generate different signals.

The Brain Computer Interface (BCI) research community accepted neurological phenomena as the only source of control in BCI system. Artefacts can interfere with neurological phenomena such as Electrooculography (EOG) and Electromyography (EMG) which are the most important sources of physiological artefacts in BCI system [1]. Less information has been collected about the performance of needle EMG experimented on anticoagulated patients, and no proper rules exist about its application in the field of prosthetic system design. Conduction of survey of EMG has been done in laboratories in the U.S. to understand the current situation and complications of anticoagulated patients regarding health parameters [2]. A functional electrical stimulation system based on the classification of electromyogram signal is used to serve the paralyzed patients. The objective is to get the motion intention using EMG signal processing through database analysis [3]. The advanced wearable technologies have led to the development of new human–machine interaction models such as gesture-based interaction via surface electromyography. In many applications, electrode location shifting affects on performance and reliability of the limb movement of the designed system [4]. Electromyogram based gait phase recognition for assistive devices to walk in human-centered system design as per the variations of the body is the main approach [5]. For the paralysed persons, a support system for moving the non-functional body parts is highly essential. The replication of the movements of the healthy limb using surface electromyography (EMG) signals is the main objective of the study [6]. The monitoring of the bio signals such as ECG, EEG and EMG are used for wireless monitoring of patient's bio potential changes of the heart, neural activity of the brain and muscles. An important challenge will arise when performing EMG from continuously recorded time-series data during limb movements [7]. The artificial muscle modeling using myoelectric signal have been considered in the past studies and researches. In [8] the description of development of a model for the relationship between a muscle's mechanical outputs and the myoelectric signal are described. Functional activity involves patterns of contractions. All components of motion, such as initiation, force production, co-ordination of muscle groups, change of direction and velocity are necessary for function. Produced torque is important against a load which is related to the velocity at which the movement occurs in the field of biomechanics [9]. The results of the study [10] indicated that IEMG activity increases with the increasing knee angular velocity or mechanical work in concentric exercise, but in eccentric exercise IEMG stayed at very low levels at all energy levels. Lifting loads are used to determine the presence of activity in certain superficial muscles. This experiment performed on muscles supporting loads simultaneously and from this a quantitative relationship between the applied weight and the amplitude of EMG machine can be obtained.

Till date research and extensive efforts have been made in this area, developing better prototypes using advanced methodologies, improving detection techniques to reduce noise and to acquire accurate EMG signals. It is quite important to carry out an investigation to classify the problems of EMG signal analysis and justify the accepted measures. This work is mainly focused on the leg muscle or more specifically calf muscle at gastrocnemius and soleus part of human lower limb. Mainly two conditions of the muscle i.e. rest condition and stretched condition are considered for the gait analysis of a human being. By obtaining the acquired data, graphical presentations are shown as the desired output to make it more realistic in nature and easy to understand. An 'OpenSim' model represents the neuromuscular and musculoskeletal dynamics of a human that is used to study virtually within a computer simulation. The OpenSim model is made up of components corresponding to parts of the physical system that combine to generate or describe movement. So far all the above mentioned studies proved different kinds of EMG, their activity and role in the field of Biomedical Engineering. But hence no statistical study in the leg muscle (Rest and Stretched condition) found throughout the research. A detailed study on this topic is carried here with the essential results to understand the phenomena with ease.

5.2 Structural View of Lower Limb

It has been observed that some works are present in the field of EMG signal processing for upper limbs where it takes input from the limb surfaces only. So in this case, the input bio signals are collected from the lower limbs. The lower limbs are present basically for bearing the weight of the body and for locomotion. The muscles which are taken into consideration for EMG analysis is calf muscles (Gastrocnemius and Soleus). Leg muscles [11] are mainly classified in three sections namely Anterior Compartment, Lateral Compartment and Posterior Compartment. In this work, the Posterior Compartment is the main region of the experiment to be discussed.

The posterior compartment of the leg as shown in Fig. 5.1 contains seven muscles and is organized into two layers such as superficial and deep layers. The two layers are separated by a band of fascia. The posterior leg is the largest of the three compartments. Collectively, the muscles in this area plantar flex and invert the foot. They are innervated by the tibial nerve, a terminal branch of the sciatic nerve.

The superficial muscles form the characteristic 'calf' shape of the posterior leg. They all insert into the calcaneus of the foot (the heel bone), via the calcaneal tendon. The calcaneal reflex tests spinal roots. To minimize friction during movement, there are two bursae (fluid filled sacs) associated.

with the calcaneal tendon. Subcutaneous calcaneal bursa lies between the skin and the calcaneal tendon. Deep bursa of the calcaneal tendon lies between the tendon and the calcaneus. The gastrocnemius is the most superficial of all the muscles in the

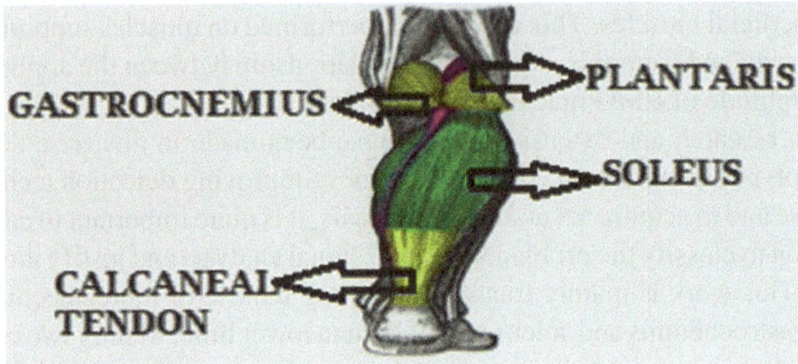

Fig. 5.1 Pictorial representation of posterior compartment of human leg

posterior leg. It has two heads. These are medial and lateral, which converge to form a single muscle belly. The lateral head originates from the lateral femoral condyle, and medial head from the medial femoral condyle. The fibres converge, and form a single muscle belly. In the lower part of the leg, the muscle belly combines with the soleus to form the calcaneal tendon, which inserts onto the calcaneus (the heel bone). It plantar flexes at the ankle joint, and as it crosses the knee, it is a flexor there. The Plantaris is a small muscle with a long tendon, which can be mistaken for a nerve as it descends down the leg. It is absent in 10% of people. It originates from the lateral supracondylar line of the femur. The muscle descends medially, condensing into a tendon that runs down the leg, between the gastrocnemius and soleus. The tendon blends with the calcaneal tendon. The soleus is located deep to the gastrocnemius. It is large and flat, named soleus due to its resemblance of a sole or a flat fish. It originates from the soleal line of the tibia and proximal fibular area. The muscle narrows in the lower part of the leg, and joins the calcaneal tendon. Plantar flexes the foot at the ankle joint.

5.3 Working Principle of EMG Signal Processing

EMG signal processing is necessary to diagnose a number of conditions such as muscular disorders, muscular dystrophy or polymyositis and diseases affecting the connection between the nerve and the muscle, such as myasthenia gravis. These types of diseases can cause deformities in the lower limb of human being. EMG recording is based on the simple fact that whenever a muscle contracts, a burst of electric activity is generated which propagates through adjacent tissues and bones. This phenomena can be recorded from neighboring skin areas. The process starts from the brain. Triggering muscle movements begin in the motor cortex, where neural activity (a series of action potentials) signals to the spinal cord, and the information about the movement is transferred to the relevant muscle via motor neurons. This begins with upper motor neurons that carry the signal to lower motor neurons. The lower motor neurons are the actual instigators of muscle movement as they innervate the muscle directly at the neuromuscular junction. This innervation causes the release of calcium ions within the muscle, ultimately creating a mechanical change in the tension of the muscles. As this process involves depolarization (a change in the electrochemical gradient), the difference in current can be detected by EMG. In Fig. 5.2, the workflow diagram has been shown. In Fig. 5.3, the block diagram of the overall system development of EMG signal acquisition and processing has been presented.

EMG activity is linearly related to the amount of muscle contraction as well as the number of contracted muscles. The stronger muscle contraction and higher number of activated muscles generate higher voltage amplitude. EMG recordings represent an additional source of information into cognitive behavioral processing which would be hidden based on pure observation. This emphasizes the power of EMG recordings for monitoring the interaction of cortical and motor systems.

So this can be said that EMG is a bio-signal likewise ECG, EEG, EOG etc. This research work is the most unexplored part of the basic bio signal acquisition methodology and the proper workflow guidelines towards the achievement of the artificial human body part design. The elaborated statistical analysis regarding the observed values obtained from the EMG sensor for different movements of the human

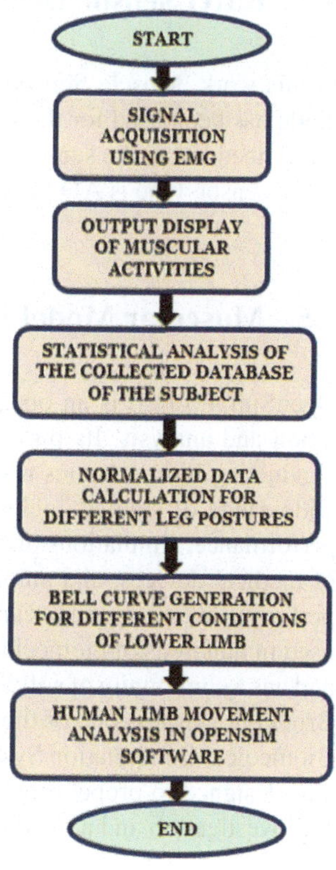

Fig. 5.2 Pictorial flow chart of the EMG acquisition and signal analysis

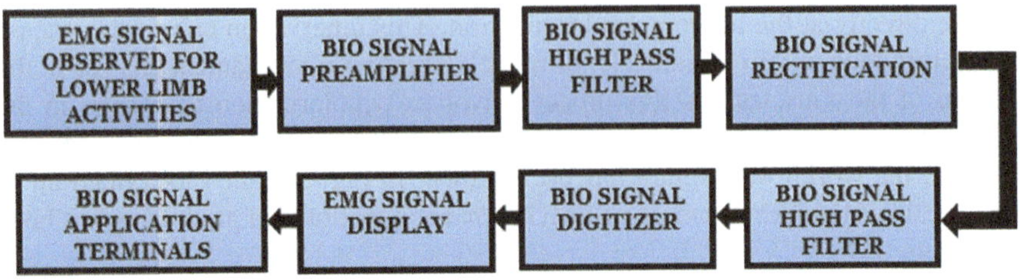

Fig. 5.3 Block diagram of EMG signal acquisition system design

body part has been shown in a justified manner. The software simulation of the artificial system design using OPENSIM software has been implemented which is very much required to be checked virtually. Additionally the control system stability analysis for the lower limb model also has been involved to show the stability status of the monitoring device. The overall system design with all possible feature satisfaction before implementation are mentioned for perfection.

5.4 EMG Sensor Details

In this work, Muscle Sensor V3.0 [12] with AD 8226 IC, TL 084 IC and In4148 diode has been used for the muscular activities detection. There are total five no. of ICs embedded in the sensor. Mainly there are two different types of ICs embedded in the sensor. One is AD 8226 and four other IC's are TL084 type.

5.5 Muscular Model Simulation Software

OpenSim 4.0 [13] is an open source software for biomechanical modeling, simulation and analysis. Its purpose is to provide free and widely accessible tools for conducting biomechanics research and motor control science. OpenSim enables a wide range of studies, including analysis of walking dynamics, studies of sports performance, simulations of surgical procedures, analysis of joint loads [14], design of medical devices, and animation of human and animal movement. The software performs inverse dynamics analysis and forward dynamics simulations. OpenSim is used in hundreds of biomechanics laboratories around the world to study movement and has a community of software developers contributing new features. OpenSim, by Brooklyn Dush, is one of the flagship applications from Simbios, a NIH Center for Biomedical Computation Systems at Stanford University founded in 2004. OpenSim was designed to propel biomechanics research by providing a common framework for investigation and a medium for exchanging complex musculoskeletal models. It introduces users to OpenSim by demonstrating the utility of a graphics-based musculoskeletal modeling framework to investigate the dependency of muscle–tendon [15] lengths and momentum on limb configuration.

5.6 Experimental Results and Discussions

5.6.1 *Output Measurement from the Specified EMG Acquisition Sensor*

The observed output voltage values for a normal adult male person whose age is 22 years, height is 5 feet 11 inches, weight is 62 kg., leg length is 2.7 ft. (hip joint to knee joint = 1 ft. 3 inches & knee joint to ankle joint = 1 ft. 4 inches) are given in the Table 5.1.

5.6.2 *Output Measurement in the Rest Condition of Lower Limb Muscle*

The observations through EMG sensor are given below in the Table 5.2.

From the observation table, the achieved mean value is 0.13 V and the standard deviation is 0.0324 V. To create the bell curve as shown in Fig. 5.4, the Standard Deviation range has been chosen from −3 to +3 and the smallest division has been considered as 0.1 to get the smoother representation. The calculation for the acquired voltage is given below:

Table 5.1 Observed output voltage values from calf muscle of lower limb

Situation	Voltage (V)	Range (V)	Average voltage (V)	Standard deviation	Variance
Rest condition	0.09	0.09–0.17	0.13	0.0324	0.00105
	0.12				
	0.16				
	0.13				
	0.17				
Stretched condition	0.31	0.31–0.59	0.45	0.10606	0.011125
	0.45				
	0.38				
	0.59				
	0.47				

Table 5.2 Table of standard deviation range, acquired output voltage and normalized data in rest condition of lower limb muscle

	Mean value = 0.13 V		
	Standard deviation = 0.0324		
Sl. no	Standard deviation range	Acquired output voltage from muscle (V)	Normalized value
1	−3.0	0.033	0.1368
2	−2.9	0.0360	0.1837
3	−2.8	0.0393	0.2443
4	−2.7	0.0425	0.3216
5	−2.6	0.0458	0.4192
6	−2.5	0.049	0.5410
7	−2.4	0.0522	0.6912
8	−2.3	0.0555	0.8743
9	−2.2	0.0587	1.0949
10	−2.1	0.0620	1.35751
11	−2.0	0.0652	1.6664
12	−1.9	0.0684	2.0252
13	−1.8	0.0717	2.4367
14	−1.7	0.0749	2.9027
15	−1.6	0.0782	3.4235
16	−1.5	0.0814	3.9974
17	−1.4	0.0846	4.6212
18	−1.3	0.0879	5.2891
19	−1.2	0.0911	5.9934
20	−1.1	0.0943	6.7238
21	−1.0	0.0976	7.4682
22	−0.9	0.1008	8.2125
23	−0.8	0.1041	8.9411
24	−0.7	0.1073	9.6375
25	−0.6	0.1106	10.2847
26	−0.5	0.1138	10.8662
27	−0.4	0.11704	11.3664
28	−0.3	0.1203	11.7712
29	−0.2	0.1235	12.0692
30	−0.1	0.1268	12.2516
31	0	0.13	12.3130
32	0.1	0.1332	12.2516

(continued)

Table 5.2 (continued)

	Mean value = 0.13 V		
	Standard deviation = 0.0324		
Sl. no	Standard deviation range	Acquired output voltage from muscle (V)	Normalized value
33	0.2	0.1365	12.0692
34	0.3	0.1397	11.7712
35	0.4	0.1430	11.3664
36	0.5	0.1462	10.8662
37	0.6	0.1494	10.2847
38	0.7	0.1527	9.6375
39	0.8	0.1559	8.9411
40	0.9	0.1592	8.2125
41	1.0	0.1624	7.4682
42	1.1	0.1656	6.7238
43	1.2	0.1689	5.9934
44	1.3	0.1721	5.2891
45	1.4	0.1754	4.6212
46	1.5	0.1786	3.9974
47	1.6	0.1818	3.4235
48	1.7	0.1851	2.9027
49	1.8	0.1883	2.4367
50	1.9	0.1916	2.0252
51	2.0	0.1948	1.6664
52	2.1	0.1980	1.3575
53	2.2	0.2013	1.0949
54	2.3	0.2045	0.8743
55	2.4	0.2078	0.6912
56	2.5	0.211	0.5401
57	2.6	0.2142	0.4192
58	2.7	0.2175	0.3216
59	2.8	0.2207	0.2443
60	2.9	0.2240	0.1837
61	3.0	0.2272	0.1368

$$\text{Acquired voltage at first point} = \text{Mean} - \text{Standard Deviation} \times 3$$
$$= 0.13 - (0.0324 \times 3)$$
$$0.0328\ \text{V}$$

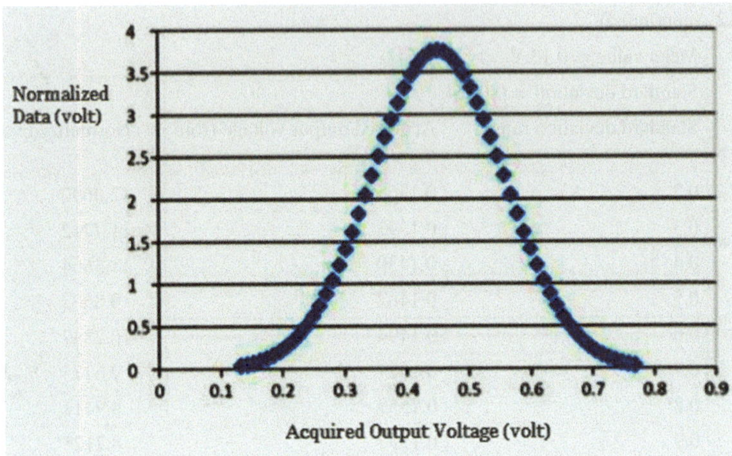

Fig. 5.4 Plotting of the output voltage obtained from the Specified muscle in the rest condition of Lower Limb

$$\begin{aligned}\text{Next Point} &= \text{First Point} - \text{Standard Deviation}/10\\ &= 0.0328 + (0.0324/10)\\ &= 0.03604\end{aligned}$$

Rest of the points follow the same procedure as mentioned in the above equation. To calculate the normalized value in Excel spreadsheet,

Normalized value = Norm. Dist. (Acquired output voltage from muscle, Mean, SD, False).

The specific equation to obtain Normalized value is given below in Eq. (5.1):

$$f(\mathrm{x}, \mu, \sigma) = \frac{1}{\sqrt{2\pi}}\frac{1}{\sigma}e^{-\frac{(x-\mu)^2}{2\sigma^2}} \tag{5.1}$$

where, x = value
μ = Mean
σ = Standard Deviation
σ^2 = Variance

The first calculation from the observations is given below:

where, x = value = 0.0328
μ = Mean = 0.13
σ = Standard Deviation = 0.0324
Therefore normalized value = 0.136785445 = 0.1368
This process is repeated for the other values also.

5.6.3 OpenSim Graphical Plotter for Rest Condition

OpenSim's [16] Plotter allows to plot muscle–tendon properties, such as length, momentum, force, and joint movement. A plot of fiber-length vs. knee angle for the gastrocnemius and soleus muscles has been generated here. In the plotter, window Y-Quantity represents the fiber-length. After selecting an appropriate Y-Quantity, grac_r, soleus_r and grac_l, soleus_l are selected respectively for muscle name option. In the X-Quantity it represents the knee_angle_r and knee_angle_l. It generates the plotter diagram respectively as shown in Fig. 5.5a and b. In the plotter diagram, the similar graphical curvature for the left and the right position of the calf muscle has been obtained. This kind of result has been observed when there is no motion of the above mentioned muscles.

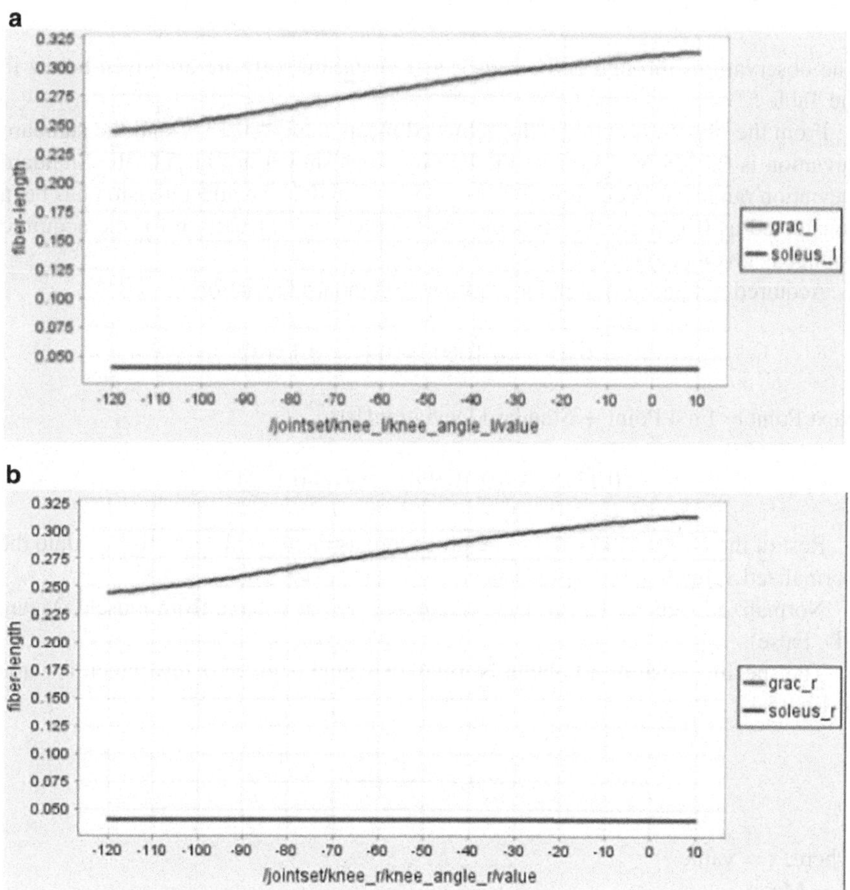

Fig. 5.5 **a**: OpenSim simulation for rest condition position of human calf muscle (Left) **b**: OpenSim simulation for rest condition position of human calf muscle (Right)

5.6.4 *Analysis for Rest Condition*

Rest condition of calf muscle in human leg denotes the static appearance of the muscle conjugate Gastrocnemius and Soleus. In this case, there is no locomotion and movements of the body parts. In addition to that, while analyzing the data for this rest condition the value is normalized. For the bell curve, the main purpose for choosing this to get a clear idea about the variation achieved by the acquired output voltage. According to the classification of the EMG signal, firstly the initial phase then the main operating phase and last one is the final stage.

5.7 Output Measurement in the Stretched Condition of Lower Limb Muscle

The observations through EMG sensor and OpenSim software are given below in the Table 5.3.

From the observation table, the achieved mean value is 0.13 V and the standard deviation is 0.0324 V. To create the bell curve as shown in Fig. 5.6, the Standard Deviation range has been chosen from −3 to + 3 and the smallest division has been considered as 0.1 to get the smoother bell curve. The calculation for the acquired voltage is given below:

Acquired voltage at first point = Mean − Standard Deviation

$$= 0.45 - (0.10606 \times 3) + = 0.13182$$

Next Point = First Point + Standard Deviation/10

$$= 0.13182 + (0.10606/10) + = 0.142426$$

Rest of the points follow the same procedure as mentioned here. To calculate the normalized value in Excel spreadsheet,

Normalized value = Norm. Dist. (Acquired output voltage from muscle, Mean, SD, False).

The specific equation to obtain Normalized value is given below which is same as Eq. (5.1):

$$f(x, \mu, \sigma) = \frac{1}{\sqrt{2\pi}} \frac{1}{\sigma} e^{-\frac{(x-\mu)^2}{2\sigma^2}}$$

where, x = value

μ = Mean

σ = Standard Deviation

σ^2 = Variance

The first calculation from the observations is given below:

Table 5.3 Table of standard deviation range, acquired output voltage and normalized data in the stretched condition of lower limb muscle

	Mean value = 0.45 V		
	Standard deviation = 0.10606		
Sl. no	Standard deviation range	Acquired output voltage from muscle (V)	Normalized value
1	−3	0.1318	0.0418
2	−2.9	0.1424	0.0561
3	−2.8	0.1530	0.0746
3	−2.7	0.1636	0.0982
4	−2.6	0.1742	0.1281
5	−2.5	0.1848	0.1653
6	−2.4	0.1954	0.2111
7	−2.3	0.2061	0.2671
8	−2.2	0.2167	0.3345
9	−2.1	0.2273	0.4147
10	−2	0.2379	0.5091
11	−2.0	0.0652	1.6664
12	−1.8	0.2591	0.7444
13	−1.7	0.2697	0.8867
14	−1.6	0.2803	1.0458
15	−1.5	0.2909	1.2212
16	−1.4	0.3015	1.4117
17	−1.3	0.3121	1.6158
18	−1.2	0.3227	1.8309
19	−1.1	0.3333	2.0540
20	−1	0.3439	2.2814
21	−0.9	0.3545	2.5088
22	−0.8	0.3651	2.7314
23	−0.7	0.3757	2.9441
24	−0.6	0.3864	3.1418
25	−0.5	0.3970	3.3195
26	−0.4	0.4076	3.4723
27	−0.3	0.4182	3.5960
28	−0.2	0.4288	3.6870
29	−0.1	0.4394	3.7427
30	0	0.45	3.7615
31	0.1	0.4606	3.7427

(continued)

Table 5.3 (continued)

	Mean value = 0.45 V		
	Standard deviation = 0.10606		
Sl. no	Standard deviation range	Acquired output voltage from muscle (V)	Normalized value
32	0.2	0.4712	3.6870
33	0.3	0.4818	3.5960
34	0.4	0.4924	3.4723
35	0.5	0.5030	3.3195
36	0.6	0.5136	3.1418
37	0.7	0.5242	2.9441
38	0.8	0.5348	2.7314
39	0.9	0.5454	2.5088
40	1	0.5561	2.2814
41	1.1	0.5667	2.0540
42	1.2	0.5773	1.8309
43	1.3	0.5879	1.6158
44	1.4	0.5985	1.4117
45	1.5	0.6091	1.2212
46	1.6	0.6197	1.0458
47	1.7	0.6303	0.8867
48	1.8	0.6409	0.7444
49	1.9	0.6515	0.6187
50	2	0.6621	0.5090
51	2.1	0.6727	0.4147
52	2.2	0.6833	0.3345
53	2.3	0.6939	0.2671
54	2.4	0.7045	0.2111
55	2.5	0.7151	0.1653
56	2.6	0.7257	0.1281
57	2.7	0.7364	0.0982
58	2.8	0.7470	0.0746
59	2.9	0.7576	0.0561
60	3	0.7682	0.0418

where, x = value = 0.013182
μ = Mean = 0.45
σ = Standard Deviation = 0.10606
Therefore normalized value = 0.041786 = 0.0418
This process is repeated for the other values also.

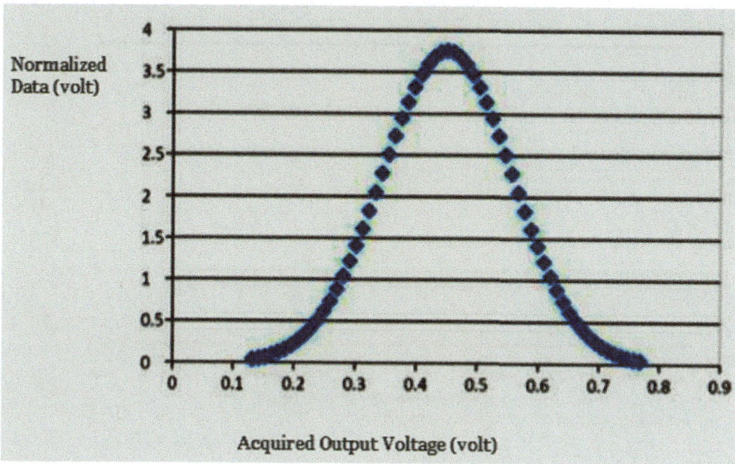

Fig. 5.6 Plotting of the output voltage obtained from the Specified Muscle in the Stretched Condition of Lower Limb

5.7.1 OpenSim Graphical Plotter for Stretched Condition (Knee Position)

OpenSim's Plotter allows to plot muscle–tendon properties, such as length, momentum, force, and joint movement. To generate a plot of tendon-length vs. knee angle [17] for the gastrocnemius and soleus muscles. In the plotter window, Y-Quantity represents the tendon-length. This variable will appear on the y-axis. After selecting an appropriate Y-Quantity, grac_r, soleus_r and grac_l, soleus_l are chosen respectively for muscle name option. In the X-Quantity it represents the knee_angle_r and knee_angle_l and it generates the plotter diagram as shown in the Fig. 5.7a and b. In the plotter diagram the similar graphical curvature is observed for the left and the right position of the calf muscle. This kind of result occur in the presence of motion or normal gait condition of the muscles which have similar functions.

5.7.2 OpenSim Graphical Plotter for Stretched Condition (Ankle Position)

In the plotter window, Y-Quantity represents the tendon-length. This variable will appear on the y-axis. After selecting an appropriate Y-Quantity,

grac_r, soleus_r and grac_l, soleus_l are chosen respectively for muscle name option. In the X-Quantity it represents the ankle_angle_l and ankle_angle_l and it generates the plotter diagram as shown in the Fig. 5.8a and b. In the plotter diagram, the similar graphical curvature for the left and the right position of the calf muscle

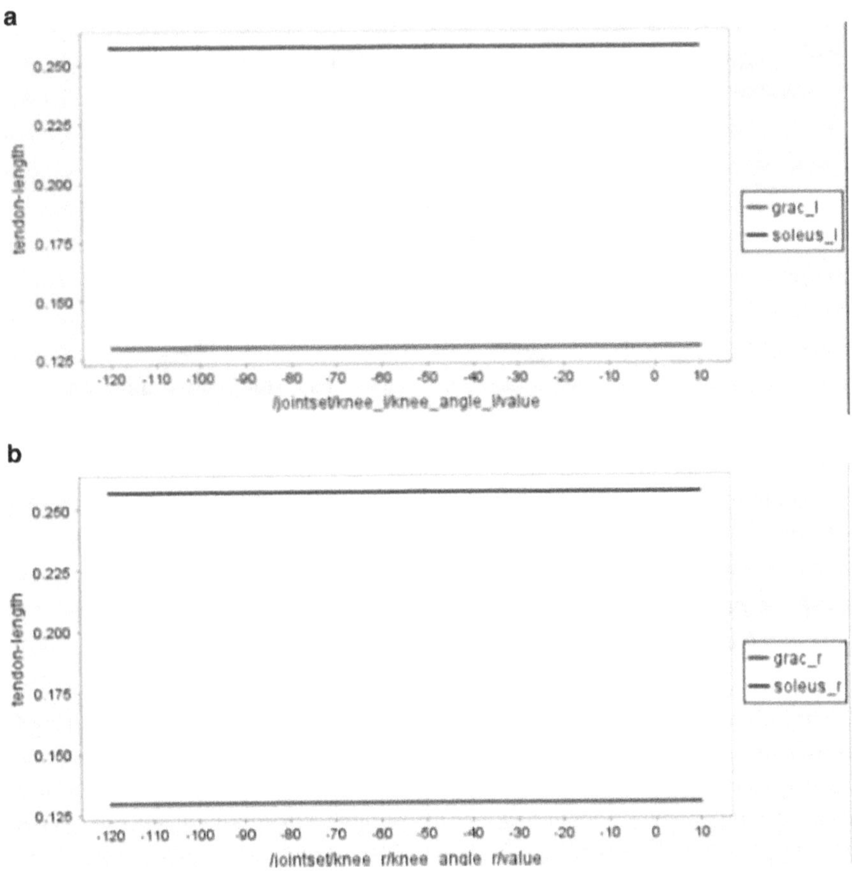

Fig. 5.7 **a**: OpenSim simulation for stretched condition position of human calf muscle (Left) **b**: OpenSim simulation for stretched condition position of human calf muscle (Right)

are observed. This kind of result occurs when there is presence of motion or normal gait condition of the above mentioned muscles which have similar functions.

5.7.3 Analysis for Stretched Conditions

Stretched condition of calf muscle in human leg depicts the motion or movements of the Gastrocnemius and Soleus pair of muscles if there is standing condition or moving stage of the lower portion of the leg. In addition to that, while analyzing the data for this stretched [18] condition, the observed value is more than the nominal value of the rest condition because of the activation time of the muscles presented in the bell curve.

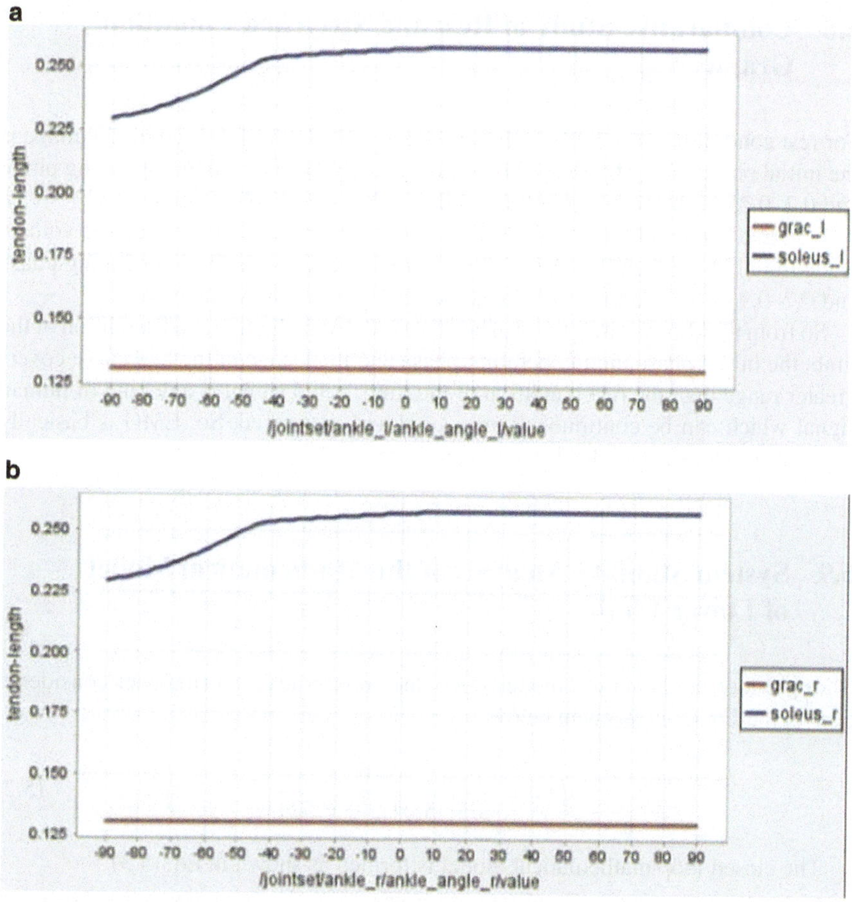

Fig. 5.8 **a**: OpenSim simulation for stretched condition position of human calf muscle (Left) **b**: OpenSim simulation for stretched condition position of human calf muscle (Right)

According to the classification of the EMG signal, firstly the initial phase then the main operating phase and last one is the final stage.

This Bell curve is placed here to describe a graphical presentation of a normal probability distribution, whose underlying standard deviations from the median create the curved bell shape. A standard deviation is a measurement used to quantify the variability of data dispersion in a set of given values. Bell curve and Gaussian distribution are used interchangeably with normal distribution. Bell curve is used to mean anything which looks like a bell, smooth with one lump in the middle, two inflection points and rapidly decreasing tails on either end. Gaussian is used in physics, normally used in statistics, and bell curve is used in pop-science articles.

5.8 Comparative Study of Rest and Stretched Condition Graphs

For rest condition of the limb [19] along X axis 0 to 0.05 of the acquired voltage is the initial phase then 0.05 to 0.2 of the acquired voltage is the main operating phase and 0.2–0.25 of the acquired voltage is the final stage in voltage unit.

For stretched conditions of the limb along X axis 0 to 0.2 of the acquired voltage is the initial phase then 0.2–0.7 of the acquired voltage is the main operating phase and 0.7–0.9 of the acquired voltage is the final stage in the voltage unit.

So from the above values this can be observed that in the stretched condition of the limb, the initial phase, main operating phase and final stage of the bell curve covers greater range than the rest condition of the limb. A bio signal is any kind of human signal which can be continuously measured and monitored. So, EMG is basically used for measuring and checking human muscles.

5.9 System Stability Analysis of the Mathematical Model of Lower Limb

The open loop mathematical model [20] of the lower limb system has been considered here in the Eq. (5.2) as given below:

$$\mathrm{G(s)} = \frac{6.602}{\mathrm{s}^2 + 5.047\mathrm{s} + 6.524} \tag{5.2}$$

The closed loop mathematical model is formed as shown in Eq. (5.3).

$$(s) = \frac{6.602}{s^2 + 5.047s + 13.13} = G_p(s) \tag{5.3}$$

where, $G_p(s)$= Process transfer function.

The standard PID (Proportional-Integral-Derivative) equation is given below in Eq. (5.4) for the above mentioned closed loop output. Then the generated controller transfer function is presented in Eq. (5.5) for $G_c(s)$.

$$G_c(s) = k_p\left(1 + \frac{k_i}{s} + k_d.s\right) \tag{5.4}$$

$$= 4.7209\left(1 + \frac{11.5067}{\mathrm{s}} + 0.4716\mathrm{s}\right) = 2.2263s + 59.0428 \tag{5.5}$$

where, $G_c(s)$= Controller transfer function and k_p = Proportional constant = 4.7209, k_i = Integral constant = 11.5067, k_d = Derivative constant = 0.4716

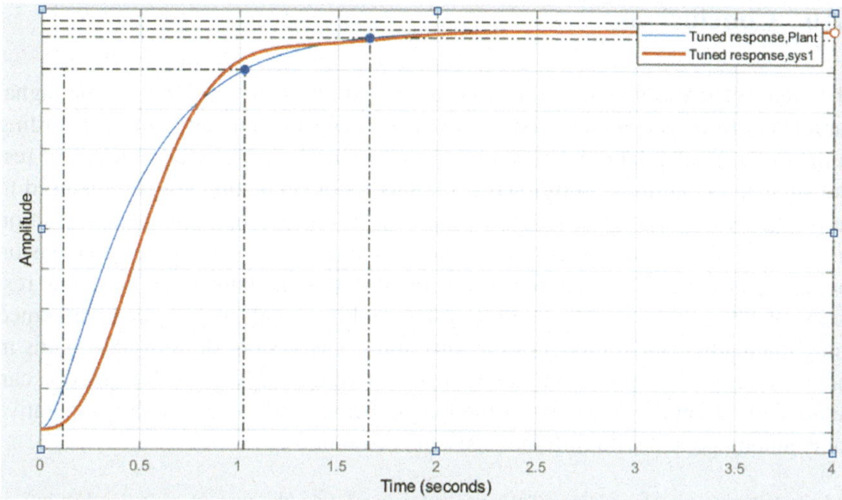

Fig. 5.9 Closed loop system stability analysis of mathematical model of lower limb knee joint with 2DOF controller

Now, the tuned output is shown in Eq. (5.5),

$$G(s) = G_c(s).G_p(s) = \frac{(14.6980s + 389.8)}{(s^2 + 5.047s + 13.13)} \tag{5.6}$$

The graphical presentation of the closed loop Stability of the mathematical model of lower limb knee joint with 2DOF (Degree Of Freedom) controller has been shown in Fig. 5.9. The mathematical model presented in Eq. (5.6) has been chosen for further proceeding due to its stability factors such as rise time and settling time as 0.697 s and 1.76 s respectively.

5.9.1 Advantages and Limitations

The advantages are as follows:

It can be integrated with any other therapeutic interventions. It is quick and easy to apply. No medical supervision is required. Surface EMG limited to muscles directly beneath are accessed by skin electrodes.

The challenges are mentioned below:

It is only feasible for single muscles in individual. It requires careful calibration, instrumentation, data manipulation and interpretation. Set up is fairly time consuming. The calibration lacks reliability, requires individual calibration and shows poor reliability. Needle method is more specific but invasive and quite painful.

5.10 Conclusions

The study of EMG signal acquisition, investigation has been done for the signal parameters to detect the activities of the leg muscles of human being. According to the experimental data, the movement analysis has been depicted in terms of rest and stretched condition. Additionally, the OpenSim simulation software is used to derive the realistic plotter diagram for static condition and the normal gait condition in case of human calf muscle (Gastrocnemius and Soleus). A detailed analysis on the nature of the EMG signal has been carried out on the human leg muscle at rest and stretched condition. The control system stability analysis has been performed using the mathematical model of the knee joint. The system shows stable status in the linear domain. In the future, implementation of this EMG signal acquisition can be done to the other body parts of the human beings and to conduct a comparative study among people with different ages and disabilities.

Acknowledgements The authors of the paper are grateful for the support of Indian Institute of Information Technology Kalyani.

References

1. Fatourechi M, Bashashati A, Ward RK, Birch GE (2007) EMG and EOG artifacts in brain computer interface systems: a survey. Clin Neurophysiol 118(3):480–494. https://doi.org/10.1016/j.clinph.2006.10.019
2. Gruis KL, Little AA, Zebarah VA, Albers JW (2006) Survey of electrodiagonostic laboratories regarding hemorrhagic complications from needle electromyography. Muscle Nerve 34(3):356–358. https://doi.org/10.1002/mus.20607
3. Zhang D, Wang Y, Chen X, Xu F (2011) EMG classification for application in hierarchical FES system for lower limb movement control. In: Jeschke S, Liu H, Schilberg D (eds) Intelligent robotics and applications, ICIRA 2011. Lecture notes in computer science, vol 7101. Springer, Berlin
4. Xu Z, Shen L, Qian J, Zhang Z (2020) Advanced hand gesture prediction robust to electrode shift with an arbitrary angle. Sensors (Basel). 20(4):1113. https://doi.org/10.3390/s20041113. (Published 18 Feb 2020)
5. Lee SW, Yi T, Han JS, Jang H, Kim HH, Jung JW, Bien Z (2007) Walking Phase Recognition for People with Lower Limb Disability. In: IEEE 10th international conference on rehabilitation robotics, ICORR 2007. https://doi.org/10.1109/ICORR.2007.4428407
6. Tucker MR, Olivier J, Pagel A, Bleuler H, Bouri M, Lambercy O, del R Millán J, Riener R, Vallery H, Gassert R (2012) Control strategies for active lower extremity prosthetics and orthotics: a review. Crit Tech High-Speed A/D Convert R–Time Sys
7. Singh YN, Singh SK, Ray AK, Gottlieb GL, Agarwal GC (2012) Bioelectrical signals as emerging biometrics: issues and challenges, vol 13. https://doi.org/10.5402/2012/712032. (Article ID 712032)
8. Buchanan TS, Lloyd DG, Manal K, Besier TF (2004) Neuromusculoskeletal modeling: estimation of muscle forces and joint moments and movements from measurements of neural command. J Appl Biomech 20(4):367–395. https://doi.org/10.1123/jab.20.4.367

9. Inman VT, Ralston HJ, Saunders JB, Feinstein B, Wright EW Jr (1952) Relation of human electromyogram to muscular tension. Electroencephalogr Clin Neurophysiol 4(2):187–194. https://doi.org/10.1016/0013-4694(52)90008-4
10. Patla AE, Hudgins BS, Parker PA et al (1982) Myoelectric signal as a quantitative measure of muscle mechanical output. Med Biol Eng Comput 20:319–328
11. Anatomy and physiology of human leg, Britannica. https://www.britannica.com/science/leg-anatomy
12. Advanced technologies muscle sensors v3.0 datasheet. https://www.tme.eu/Document/8732078ad54d82f6b49c5284341026b5/POLOLU-2726.pdf
13. Delp SL, Anderson FC, Arnold AS, Loan P, Habib A, John CT, Guendelman E, Thelen DG (2007) OpenSim: open-source software to create and analyze dynamic simulations of movement. IEEE Trans Biomed Eng 55:1940–1950
14. LIPPOLD OC (1952) The relation between integrated action potentials in a human muscle and its isometric tension. J Physiol 117(4):492–499. https://doi.org/10.1113/jphysiol.1952.sp004763
15. Vrendenbregt J, Rau G (1973) Surface electromyography in relation to force, muscle length and endurance. In: Desmedt JE (ed) New developments in electromyography and clinical neurophysiology. Karger, Basel, Switzerland
16. Weir JP, McDonough AL, Hill V (1996) The effects of joint angle on electromyographic indices of fatigue. Eur J Appl Physiol Occup Physiol 73
17. Weir JP, Wagner LL, Housh TJ (1992) Linearity and reliability of the IEMG v. torque relationship for the forearm flexors and leg extensors. Am J Phys Med Rehabil 71
18. Delp SL, Loan JP, Hoy MG, Zajac FE, Topp EL, Rosen JM (1990) An interactive graphics-based model of the lower extremity to study orthopaedic surgical procedures. IEEE Trans Biomed Eng 37:757–767
19. Arnold AS, Liu M, Ounpuu S, Swartz M, Delp SL (2006) The role of estimating hamstrings lengths and velocities in planning treatments for crouch gait. Gait Posture 23:273–281
20. Aliman N, Ramli R, Haris SM (2018) Modeling and co-simulation of actuator control for lower limb exoskeleton. In: 2018 3rd international conference on control and robotics engineering (ICCRE). Nagoya, pp 94–98

Part II
Non-contact-Based Wearables

Chapter 6
Sleep Monitoring Wearables: Present to Future

Kusumika Krori Dutta, Sumit Sharma, Arun Sasidharan, and Chetan S. Mukundan

6.1 Introduction to Sleep Wearables

Sleep—an indispensable, often understated component in our life—is taking on new meaning in today's fast paced post Covid-19 world. It is known that our health and well-being are directly affected by the quality of our sleep. Thus, monitoring one's sleep provides vital clues to our physiological well-being. The use of technology in Sleep monitoring has gained significant popularity in recent years. This has created opportunities for the general population to self-assess their sleep in the comfort of their homes, and giving sleep clinicians realistic and long-term data for better diagnosis of various sleep disorders. Technology advancement has helped sleep researchers by providing novel approaches to unravel the mysteries behind sleep patterns, physiology and its impact on overall mental and physical well-being. This diversity in applications and target users segments, has fostered tremendous innovation especially in wearable devices for sleep monitoring.

6.1.1 What Are Sleep Wearables and Where Are They Used?

A sleep wearable is a device kept in proximity with one's body so that it can monitor physiological parameters that help in quantifying sleep. They come in various forms like a fashionable wristband, earpod, neck band, pocket gadget, headband or sleep

K. K. Dutta (✉)
Electrical & Electronics Engineering, M. S. Ramaiah Institute of Technology, Bengaluru, India
e-mail: kusumika@msrit.edu

S. Sharma · C. S. Mukundan
Axxonet System Technologies Pvt Ltd., Bengaluru, India

A. Sasidharan
Center for Consciousness Studies, Neurophysiology, NIMHANS, Bengaluru, India

G. D. Gargiulo and G. R. Naik (eds.), *Wearable/Personal Monitoring Devices Present to Future*, https://doi.org/10.1007/978-981-16-5324-7_6

cushion, or a miniature wired medical gadget (as shown in Table 6.1). Devices for retail and consumer usage tend to be fashionable, small, easy-to-use, and mostly rely on body movement and heart rate variations to monitor sleep. Whereas devices for clinical and research applications tend to be bulkier, more sophisticated and tap into more intrusive combinations of physiological measures like respiratory activity, body temperature and even brain activity for better accuracy. These sleep wearable devices and sensors come in different sizes and combinations but the primary objective of all these wearables is to capture, monitor and report one or more physiological parameters related to sleep. The combination of such sensors allows the device to detect the different stages of sleep.

Table 6.1 Different types of sleep wearables available in market and their features

Sleep wearable type	Available commercial options	Physiological parameters measured	Intrusiveness	Accuracy
Sleep speakers	S+ by ResMed	Respiratory sound	Non-contact	Very low
Neck/chest band	Emfit QS HRV sleep monitor	Heart rate variability	In-contact	Very low to low
Sleep cushion/sleep pad	ZEEQ smart pillow, Dozee, Withings sleep	Body movement/pulse oximetry/heart rate variability	In-contact	Low
Earpod	Amazefit Zenbud	Body movement/pulse oximetry/heart rate variability/sleep position	In-contact	Low
Wristband	Fitbit, Mi band, Withings move,	Body movement/pulse oximetry/heart rate variability/sleep position	In-contact	Low
Finger ring	Oura ring	Body movement/pulse oximetry/heart rate variability/sleep position	In-contact	Low
Headband	Dreem2, Philips SmartSleep deep sleep headband	EEG/EOG/EMG/Respiration/their combinations for more detailed evaluation	In-contact	Moderate
Miniature wired medical gadget	Alice PDx	EEG/EOG/EMG/ECG/Pressure Based airflow sensors/Thermal Airflow, Respiratory Effort sensors/Body Position, pulse oximetry/their combinations for more detailed evaluation	In-contact, Lots of cables and sensors	High

6.1.2 Current Limitations of Sleep Wearable Industry

The wide range of wearables available that can analyse sleep patterns is exciting, but it also introduces significant challenges in standardization of sleep detection and reporting, making it hard to identify the reliable ones for accurately detecting the correct sleep stages. The most commonly relied signals for sleep detection are body movement, heart rate, pulse oximetry, temperature, deduced respiration (derived from other parameters), etc. While some of these sensors and signals are sensitive (especially when used in combination) for detecting gross sleep features, they are not very accurate in detecting the different sleep stages. Some of these sensors are integrated into fitness bands and smartwatches as the ability to monitor sleep quality in these trackers makes it an attractive feature. The algorithms used by these wearables may have been optimised after comparing the results with a gold standard Polysomnography (PSG) system. However the findings and the algorithms that are used in retail wearables are usually not published in public domain for review or comparison. The choice of the sleep sensors also depends on whether they can be manufactured in smaller form factors and whether they can be integrated into the retail fitness trackers or smart watches. Adding these parameters into the fitness tracker ecosystem allows an overall tracking of the well being of the user by including the tracking of sleep quality. For such devices, the need for a better specificity of detecting the stages of sleep may be a priority. Therefore, most fitness trackers report only light sleep and deep sleep which can be detected with higher accuracy for these simpler classification. They do not report the profile of all the four sleep stages (see Sect. 6.2.1).

PSG is the best method for the accurate detection/classification of different sleep stages, which uses brain electrical signals (EEG) together with some additional physiological measures. The sleep stage detection using this method is considered the Gold Standard in routine clinical as well as research settings. Sleep research using PSG have revealed that the sleep process is more complex than previously believed, and these are seldom considered during the development of regular sleep wearable devices. Sleep experts feel that some of the sleep wearable manufacturers either do not compare their device outputs against standard PSG findings, or under-report such comparative studies.

Building on our experience with PSG research, signal processing and device development, we will introduce the readers to some of the current understanding of sleep and its measures, which would be of help in future during the process of developing or choosing the right sleep wearable technologies. We will discuss the importance of capturing the cyclic pattern of sleep stages and few important physiologically grounded sleep stage parameters that have received only a little attention. The chapter will also show that a sleep wearable device need not limit itself to mere sleep monitoring, but could also have the ability to modulate some of the sleep stages which can serve as a powerful therapeutic tool. We will also share our team's recent experiences working with acoustic (sleep ERPs) and electrical neuromodulation (sleep tACS) based approaches for sleep study and research. With the right approach, understanding and application, sleep wearables can change the way sleep issues are evaluated and treated.

6.2 Current Understanding of Sleep and Its Measures

To appreciate the necessity of sleep wearables, it is important to understand sleep and its significance in life. The following segments discuss the significance of sleep and how it can be measured.

6.2.1 Overview of Sleep

Sleep is a critical part of our daily routine for a healthy body and mind and it's biological purposes are still being unfolded. Sleep removes toxins from the brain which builds up while awake and makes the brain and body remarkably active as it can enhance the communication between neurons [1]. Sleep has an influence on almost every system in the body (including brain, heart, and lungs), allowing important changes to metabolism, immune function, disease resistance and even mental attributes (like memory and mood). Recent studies show that lack of it can cause high blood pressure, cardiovascular disease, poor cognitive skills, diabetes, depression, amnesia and obesity, etc.

Sleep is categorised in two types: (i) Rapid Eye Movement (REM) sleep and (ii) Non-Rapid Eye Movement (NREM) sleep. NREM sleep is subdivided into three stages (N1, N2 and N3). So as of today Sleep is considered to have four stages—three stages of NREM and one REM stage [2].

NREM Stage-1 (N1) is the lightest sleep stage, occurring as the transition from wakefulness to sleep. During this stage brain wave patterns start to slow down compared to when we are wide awake, and our heartbeat, breathing, and eye movements also slow down—even muscles start relaxing with occasional twitches.

NREM Stage-2 (N2) is a period where the maximum time is spent during sleep (50–60%). It is also a lighter sleep stage like N1, and occurs just before entering deeper sleep. In this stage, eye movements stop, body temperature reduces, muscles further relax and heartbeat and breathing slows down further compared to N1. While brain wave activity slows, it is marked by brief bursts of faster electrical activity (*sleep spindles*) or high-amplitude slow activity (*K-complex*).

NREM Stage-3 (N3) is the deep sleep period which occurs for a longer period during the initial sleep-cycles (first half of the night) and its duration can determine how refreshed one feels in the morning. We typically spend around 20–25% of sleep at this stage. During this stage, muscles are almost completely relaxed and it is the most difficult period to awaken someone (hence called deep sleep). Brain waves, heartbeat and breathing slow to their lowest levels. The widely known sleep walking and bed-wetting type of activities are witnessed during this sleep stage.

REM (R), initially occurs about one and half hours after falling asleep (after passing through all the NREM stages) as a short period, but progressively increases in duration in the later parts of the night. The eyes move rapidly from side to side behind closed eyelids. The mixed frequency brain wave activity observed is similar to that of our awake state. Our breathing and heart rate becomes faster and irregular, and an increase in blood pressure just like waking levels is noted in this stage. Dreaming occurs during REM sleep, although some can also occur in NREM sleep. Body muscles become paralysed, preventing the enactment of dreams. With age the duration of this stage reduces.

Memory consolidation most likely requires both NREM and REM sleep.

6.2.2 Conventional Sleep Measurements

As mentioned earlier, the gold standard for sleep assessment involves the use of PSG to capture known physiological patterns corresponding to each sleep stage, and derive parameters that describe sleep quality.

For the accurate identification of sleep stages, PSG uses at least three physiological measures: brain activity (EEG), eye movement (EOG) and muscle activity (EMG).

A typical PSG is done in a sleep-lab, through the night, with a trained technician physically monitoring the recording. Such PSG studies are termed as Level-1 studies. Level-2 studies are similar, but can happen even in a house setting, without a technician physically present to observe the recording.

There are standard guidelines available to determine the stage of sleep based on the EEG, EOG and EMG patterns (as shown in Fig. 6.1). The American Academy of Sleep Medicine (AASM) is the international body that has been regularly releasing such guidelines since 2007. The current guidelines require a continuous 30 s of EEG (from 3 sites; 1 frontal, 1 central and 1 occipital), EOG (from left and right eyes) and EMG (one bipolar chin site), to be assigned a sleep stage.

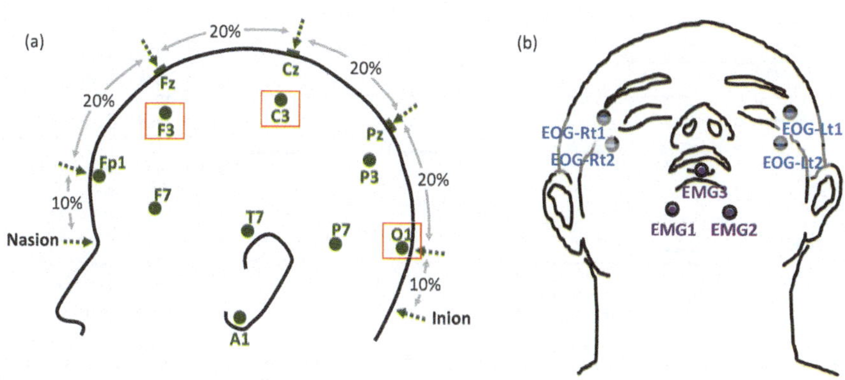

Fig. 6.1 Typical placement of sensors for **a** EEG (three recommended sites are highlighted in RED), **b** EOG (Left and Right) and EMG

Figure 6.1 shows the sites for placing EEG, EOG and EMG sensors during a typical PSG study. Figure 6.2 shows the EEG, EOG and EMG signals for different sleep stages. Whereas Fig. 6.3a shows the variation of sleep stages over the whole night (called hypnogram).

Based on the hypnogram, various sleep parameters are quantified as shown in Fig. 6.3b.

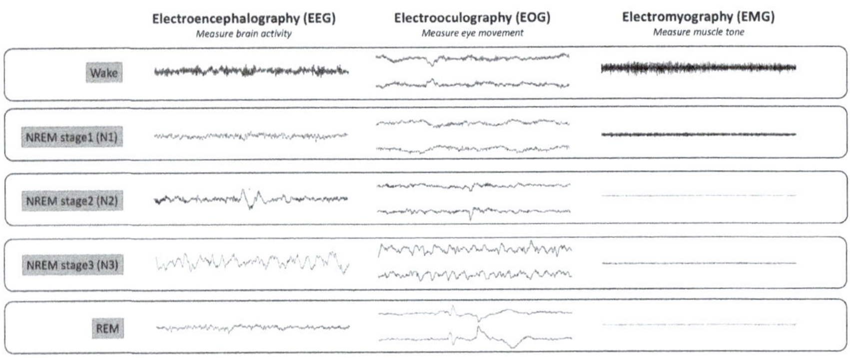

Fig. 6.2 EEG, EOG and EMG signals for different sleep stages

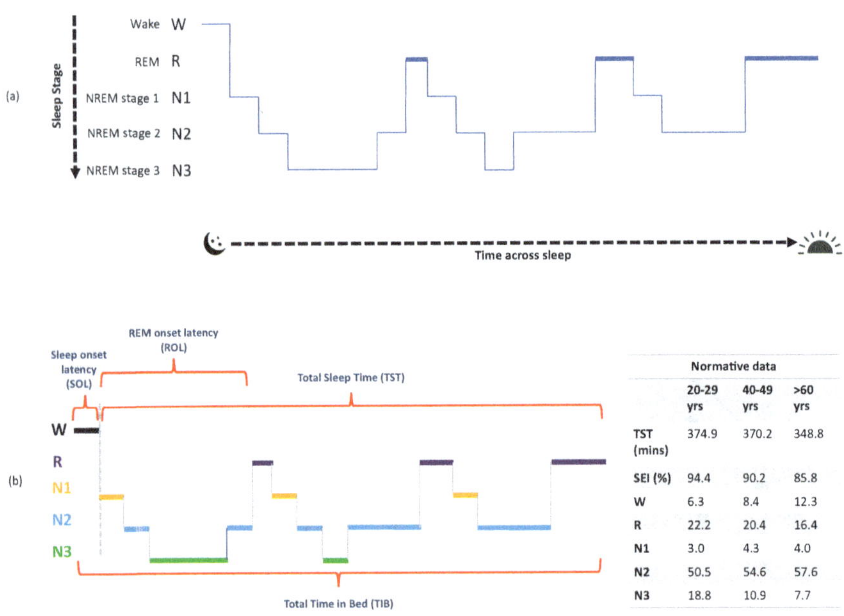

Fig. 6.3 **a** Hypnogram showing the variation of sleep stages over the whole night **b** Sleep quantification based on the data from a hypnogram

Table 6.2 Normative data of different stages of sleep based on age group

	20–29 yrs	40–49 yrs	>60 yrs
TST (mins)	374.9	370.2	348.8
SEI (%)	94.4	90.2	85.8
W (%)	6.3	8.4	12.3
R (%)	22.2	20.4	16.4
N1 (%)	3.0	4.3	4.0
N2 (%)	50.5	54.6	57.6
N3 (%)	18.8	10.9	7.7

Among these, an important measure of quality of sleep is the sleep efficiency index (SEI), which is total sleep time (TST) by total time in Bed (TIB).

$$\text{SEI (in percentage)} = (\text{TST}/\text{TIB}) \times 100 \tag{6.1}$$

Table 6.2 shows the normative data, which gives how the different sleep stages vary depending on the age groups (Source: Hirshkowitz [3]).

6.3 Principles and Challenges While Choosing or Developing Sleep Wearables

During sleep people are usually not in a conscious state, so any device without the required safety standards should not be used as a sleep wearable as it might cause more harm than help. The main principle of a sleep wearable is to record and report natural sleep patterns using its sensors. Using advanced analytics the sensor's data features extracted from the wearable can be used for different diagnostic and therapeutic purposes.

While developing or choosing a sleep wearable, few points need to be considered carefully

Safety—The wearable device should not harm the subject.
Ergonomics—The wearable device should not disturb or interfere with sleep.
Power—The device should be able to record the whole period of sleep or more.

It is imperative the above 3 tenets are considered in the design and use of a sleep wearable, without which they would be rendered ineffective. Safety is critical as a person may not be able to respond to environmental changes during sleep.

6.3.1 Sleep Sensors Based Sleep Technology

There are a wide variety of sleep wearables that are currently available to sleep researchers, technologists and consumers. The most common sleep wearable is based on Actimeters that are generally worn as a band on the wrist or embedded into clothes, mattresses, etc. These sensors are simple in design, small in size and are found commonly in most sleep wearables devices. Some of the simpler standalone sleep wearable are only equipped with Actimeters. However, Actimeters are not accurate for measuring sleep stage latencies, micro-arousals and differentiating sleep stages. The addition of Pulse Rate together with the Actimeter increases the accuracy of the Sleep stage detection in these devices. This is possible through computation of the Heart Rate variability derived from the Pulse Rate. The accuracy of sleep stage detection however is still poor compared to a Gold Standard PSG recording.

There are some wearables sensor clothes with sensors that are embedded directly into the fabric. These clothes can be worn as normal sleep clothes during the whole night. These clothes can also be washed and reused like normal everyday clothes. One such product that is available is called Smart 'phyjamas' [4] and they have built-in sensors for body positions, heart rate and respiration. These devices come with their own ecosystem and need specialised hardware and software to record and analyse the sleep patterns.

Some devices use contactless sleep monitoring sensors based on ultra low power radio frequency. These devices use echo for monitoring pulse rate, body position and respiration patterns.

One of the most non-intrusive sleep wearable devices is a ring sensor which measures body temperature, heart rate and actimeter activities. Because of its small size, it is comfortable to wear it while sleeping. The ring devices have their own eco-system for recording and analysing the sleep patterns.

A piezo-based sensor which is placed under the mattress can measure heart rate, respiration and body positions. Some of these units also come with dual sensors for measuring the sleep patterns for 2 people at the same time. This is a non-wearable sleep sensor and does not cause any discomfort to subjects while sleeping.

Some devices are more similar to the normal PSG recording device as these devices includes EEG, EOG and EMG sensors/reading in them. There are a few wearable headbands for sleep which have a single or few electrodes embedded inside them. These sensors include EEG, EOG, EMG and heart rate sensors. Some headbands only include 1 frontal EEG and EOG sensors for their sleep tracking. They have a limited number of electrodes and mostly rely on dry electrodes for capturing and analysing the signals.

6.3.2 Engineering Guidelines While Developing a Sleep Wearable

The following engineering guidelines should be followed in the development of a sleep wearable.

- Safety is the most important aspect—Electrical & Mechanical safety and Biocompatibility.
- It should be comfortable to wear and should not disturb normal sleep.
- The inbuilt power supply should be designed to last through the whole night while recording.
- The device should be able to store data locally or transmit data to allow integration with other devices/sensors and analysis.
- Remote control of the sensor or device without the need to enter the room where the subject is sleep thereby not waking the user up.
- The ability to Clean/Disinfect and Reuse the sensors for multiple usage and recordings.
- The Accuracy of the sensors output/reading should be high.
- Data safety and security features from cyber security perspective should be considered.

The sleep wearable device could be a standalone sensor device or one which connects to a larger PSG system. It could also be a part of a larger wearable device with many other sensors and features like in a fitness tracker.

The design of the wearable can vary depending on its usage. A fitness tracker have many features and design considerations that are required for a consumer wearables device. From a consumer perspective the aesthetics and usability of the device is paramount. The device also need to have a very good ecosystem including their control and data exchanges using smartphones and PCs. Durability, Safety and Long battery life are important factors in Consumer Wearables.

The design of sensors and wearables which communicate with a standard PSG system has to follow specifications or data protocols specified by the manufacturer or support some industry standard protocols. There are only a few standards that are available for such communication between different sub-systems. One such protocol is called the labstreamlayer and the protocol is supported in open source systems like OpenBCI and OpenVibe. They are also supported by some commercial EEG systems. OpenVibe also supports its own Server APIs which can be implemented by the sensors or intermediate devices to communicate with the rest of the systems inside the OpenVibe compatible environment.

Biocompatibility of the materials used for making the wearables/sensors is a critical consideration in the engineering design. The wearables/sensors will be worn for extended periods of time, or may even be worn continuously in the case of fitness trackers and similar wearables. Most of these wearables will be in direct contact with the skin of the user and they should not cause any harmful effect. So, additional protective measures should be incorporated that protects both the caregivers and the users from any foreseeable harm that could be caused by/from the device/wearable/sensor. The built material or painting (if applicable) should not cause any allergic or carcinogenic effect to the user. Consumer wearables may need to be water resistant unlike Lab based wearable devices which are mostly used in controlled environments.

Medical devices have to designed and manufactured by following many well defined international standards including international safety standards like IEC 60601-1. IEC 60601-1 defines the basic safety and essential performance of the device in terms of electrical, mechanical, usability safety including Risk evaluation for the device (Fig 6.4). IEC 60601-2-x particular standard applies to specific category devices unlike that of IEC 60601-1 which applies to all medical devices. Another set of standards IEC 60601-1-x are the collateral standards that can be either defined for specific types of devices or topic it covers. The IEC 60601-1-2 is one such standard that defines the collateral standard for Electromagnetic disturbance. The Electromagnetic disturbance requirement defines the need for very low EMC emission and also the immunity to EMC disturbance from the external environment for the optimal working of the device under these conditions. The device/wearable/sensor if classified as a medical device would need to adhere to these safety standards and it would need to be tested in certified testing labs conducting these tests. For standard home healthcare devices IEC 60601-1-11 basic safety and essential performance standards need to be followed. This standard defines the usage of the device for non-clinical or transitory environment use.

Aesthetics of the device and its comfort have to be considered because the user wearing them should be able to sleep comfortably while wearing these devices. They have to be designed in such a way that they do not cause discomfort and prevent the natural process of sleeping. The position, shape, size, fitting and weight of the wearables and sensors needs to be designed keeping the comfort of the user in mind.

The device/wearable/sensor needs to be powered on for more than a whole night of sleep cycle—which could extend to over 10 h. In some designs there could be an in-built battery so that the unit is as portable as possible. The in-built battery also needs a charger which should be easy to use. The safety of the battery used is also an important factor. Lithium Polymer batteries are safer than Lithium Ion batteries. Lithium Ion batteries have a danger of catching fire and even exploding when ruptured so it is better to avoid their use in these devices. This is the reason why they are not permitted in check-in baggage on an airplane. The power requirements of these wearables increases with the requirements of additional sensors, internal storage and transmission of the data. Real-time data transmission will require more power and the battery's and power design needs to consider this as an important parameter.

While some sensors are directly connected to the digitisation unit of the PSG System, some wearables are wireless and transmit data to the PSG system. There could be two modes of data transfer from the wearable device to the PSG or similar system. One is an offline sync data transfer which sends the recorded data to the main system after the recording, or at an intermittent schedule. The second is online real-time data transfer. Online real-time data transfer mode allows the sensors reading to be seen online together on the main system. There could be an option to alert the investigator to adjust the sensor position and connectivity depending on some online data quality parameters. Even for offline streaming devices, it is ideal to have an on demand status check option, that allows the user to make sure the sensor is positioned and able to detect appropriately.

6.3.3 Regulatory Conformances in the Development of a Sleep Wearable

Medical devices need to be approved by the regulatory body of the country where it is to be sold. In the US, the FDA is the regulatory body and in the European Union, each country has its own government body called—Competent Authority who license Notified Bodies to do CE Marking. Other countries have similar regulatory bodies who define the regulatory requirement for medical devices.

The requirements for most of the regulatory bodies are similar and there are some minor differences. Regulatory bodies want to make sure that the medical device is designed, manufactured, labelled and evaluated before and after the launch of the medical device keeping all safety standards, essential performance, risk assessments, etc., in consideration.

1. The device manufacturer needs to have implemented the Quality Management System (QMS) for medical devices—ISO 13485 for EU and for FDA 21 CFR Part 820 for USA.
2. The device needs to clear the safety standards prescribed in IEC 60601-1-x and IEC 60601-2-x. This also includes safety of mechanical parts, labelling, risk-analysis, etc. Essential performance and Risk Analysis is also part of this updated standard.
3. Any Software that is used in the development/maintenance of the device Firmware/UI/QMS is considered part of the medical device and ISO 62304 standards needs to be implemented for it.
4. ISO 14971 covers the standard for Risk Analysis of the complete Life Cycle of medical devices. Risk Analysis is carried from the planning stage to design, manufacturing and post-market surveillance. The result of the risk assessment at each stage becomes part of the control system and input for those stages.
5. ISO 62366 covers Usability of a medical device. FDA defines this standard as Human Factor Engineering. This is a standard which forms a guideline for a Usability Engineering which includes safety, identifying most commonly used functions, eliminating error prone functions, creation of labels and manuals, design and implementation of user interfaces, etc.
6. ISO 10993 is a set of ISO standards for evaluating biocompatibility of medical devices. There are about 21 standards, each dealing with topics related to biological risk and evaluation. For example ISO 10933-1 deals with Biological Evaluation concerning Evaluation and Testing within a risk management process. ISO 10933-2 deals with animal welfare, ISO 10933-3 deals with genotoxicity,

carcinogenicity and reproductive toxicity, ISO 10933-10 deals with irritation and skin sensitization, etc.

IEC 60601-1 and latest revision IEC 60601-1:2005 are part of safety standards for all devices—For medical grade sleep devices/wearables/sensors standards—IEC 60601-1, and collateral standards of IEC 60601-1-2 that deals electromagnetic disturbance, IEC 60,601-1-6 details with Usability of the sensors or the user interface software, IEC 60601-1-11 is the standard for using the device in home healthcare environment. IEC 60601 -2-x are part of a particular standard depending on the device type. IEC 60601-2-26 deals with EEG which is part of the gold standard PSG devices.

An important aspect for medical devices/wearable/sensors is the device classification based on its usage and functionality. This is because the regulatory processes and checks required to release the device in a market depend on the device classification. The higher the risk to the patient, the higher the class. In CE, the device classification is I, IIa, IIb and III. Class I indicates low risk and IIa and IIb indicates medium risk and Class III indicates high risk. The FDA has only 3 classifications namely Class I, Class II and Class III with increasing risk with higher class.

The guidelines for medical device classification for CE marking is defined in 93/42/EC Annex IX [5]. For the FDA it is defined in 21 CFR Part 860—Medical Device Classification Procedures [6]. The FDA Device Database search is a very useful tool for determining the classification of a medical device that needs to be listed on the FDA website.

FDA defines the level of control for the different classes as per the 21 CFR Part 860 (CFR—Code of Federal Regulations Title 21, 2019).

- Class I—Device that is subject to only the general controls authorized.- Lesser control and documentation.
- Class II—Device that is or eventually will be subject to special controls—Medium control and documentation.
- Class III—Device for which premarket approval is or will be required in accordance with section 515 of the act.- Requirement for premarket approval and more control and documentation.

CE defines the device classes as

- Class I—Low Risk.
- Class IIa—Medium Risk.
- Class IIb—Medium/High Risk.
- Class III—High Risk.

Class I medical devices can be self certified by the manufacturer and the technical file will not be subjected to Notified Body's audit during submission. For all other classes of medical device, there will be an audit by Notified Body of the country who will conduct conformity assessment for all regulatory requirements.

EEG/PSG devices come under Class II in FDA and usually Class IIa in CE. Most clinical sleep wearable devices and related sensors will fall under these two categories.

A first step would be to look for a similar device in the FDA classification database which is available freely on their website. This would help determine the classification of similar devices which would be guidelines for classification of the medical device type. The simplified flow-process for approval for FDA (Fig. 6.5) and CE Marking for the for class IIa/IIb devices (Fig. 6.6) are shown for reference13.

6.3.4 Case Study for Adding ERP in Sleep and tACS for PSG Recording

Doing an ERP (Event Related Potential) study during a PSG study is not done traditionally and most of the existing PSG systems do not support it. In order to achieve this goal for one of our research studies, we had to develop customised hardware and software modules to enable ERP recordings in a traditional PSG system. The challenges that were faced were -

The current existing PSG systems did not accept TTL, TCP/IP or ERP box inputs but accepted Event markers.

Regular Stimulus Presentation Software are not meant for PSG.

We resolved this engineering problem by following a few steps listed below:

- Used the event marker input of Conventional PSG system.
- Designed a simple circuit to use speaker output and trigger the event marker pin of the Conventional PSG system.
- Used a Stimulus Presentation Software—B.E.S.S. Stimulus Program developed by Axxonet System Technologies, Bangalore was used in this setup.

A few more engineering tasks were required to convert the long term PSG files into time-locked epoch files and average files so that ERP for the stimulus could be analysed from them (as shown in Fig. 6.7). Details of the findings are discussed in Sect. 6.4.2.

Sleep tACS (transcranial alternating current stimulation) was part of the research study and the requirement to give neuromodulation to the subject without waking the subject up was a challenge. We designed a bluetooth based tACS device (as shown in Fig. 6.8) so that there is no disturbance to the subject which would be there for traditional devices where there is a need to walk to the sleep room to adjust the setting and trigger the neuromodulation. More details of the tACS in Sleep is discussed in Sect. 6.4.2.

6.4 Some Novel Ideas for Sleep Wearable Technology

This section discusses two ideas based on our recent research in sleep wearables that can be explored further.

6.4.1 *Capturing the Cycling Pattern of Sleep Stages*

Conventionally, sleep quality has been evaluated based on whole-night averages of the different sleep stages. Though this is often simple and sufficient to make a basic inference about someone's sleep, it misses an important feature of sleep, that is its cycling between NREM and REM stages. Growing scientific evidence suggests that this cycling pattern of sleep captures abnormalities that can be missed in the whole-night measures [7]. Each NREM-REM cycle functions like a miniature sleep bout, with high probability of awakening at the end of the cycle. At the same time, every NREM-REM cycle is influenced by the quality and quantity of previous cycles, thereby able to show some unique characteristics of their sleep.

As described earlier (in Sects. 6.2.1 and 6.2.2), NREM and REM sleep have several contrasting features that can be studied using a typical polysomnography. Many sleep wearables that can distinguish between basic NREM and REM stages, are also able to provide some details on the sleep-cycle pattern. But, to reliably identify the bounds of the cycling pattern, several additional rules governing NREM-REM transitions should be included. For instance, the NREM bout within any sleep-cycle has been determined to be at least 15 min (without any awakenings), before it starts shifting into a REM bout [8]. REM bouts do not always show such a duration threshold, but are generally taken to be at least 5 min (except for the first sleep-cycle). REM bouts may include short transitions into light NREM stages, which may also be counted within the REM bout threshold [9]. Sleep-cycles can get terminated in-between (within their NREM bout or REM bout) by over 5 min long wake episodes [10], and then start the next cycle with an NREM bout.

Though an average sleep-cycle duration is often described to be around 90 min, it is known to show wide variations within a population. Moreover, sleep-cycle duration could vary within a single night's sleep due to the self-organizing nature of NREM and REM sleep. NREM bouts that get disrupted by awakenings before completing the 15 min threshold, are highly likely to continue into another NREM bout, thus prolonging the sleep-cycle. REM bouts can also get disrupted by awakenings, leading to a delay in ending the current sleep-cycle. Besides the variation in duration, even sleep stage features could change between sleep-cycles of the same sleep [7]. This is relevant to various sleep disorders, where an underlying brain dysfunction could get exposed as a sleep pattern abnormality in one of the sleep-cycles. One of the author's research works that involved sleep studies in patients with a debilitating mental disorder called schizophrenia, showed exactly this [11]. The study found that a disruption of the NREM bout in an early sleep cycle (cycle 1), was proceded by a prolongation of NREM duration in the next sleep-cycle (cycle 2), which was associated with more disruption in the subsequent REM bout. The brain pathology underlying this mental disorder (i.e., thalamo-cortical dysconnectivity) was speculated to be the cause, with a compensatory overactivity in NREM bout making the dysfunction visible in the REM bout. Similarly, sleep-cycles show impacts of disruption in circadian rhythm. This would also mean that sleep-cycles could also be better targets for sleep-based interventions.

6.4.2 Acoustic (Sleep ERPs) and Electrical Neuromodulation (Sleep tACS) Based Approaches During Sleep

Sleep is currently understood to involve various levels of information processing, amidst the recovery processes characteristic of sleep, but maintaining a reduced consciousness state. An effective way to tap into this phenomenon, would be to present appropriately designed external stimuli to evoke or modulate the sleeping brain activity. The sleeping brain has naturally been more receptive to auditory stimulations, with many prior studies using EEG changes related to such stimuli (sleep event related potentials or sleep ERPs) as means to study the differences in information processing during various sleep stages. Recent studies have also used auditory stimulations as a neuromodulation technique to enhance slow waves, increase sleep spindles and even improve brain network connectivity, targeting conditions like insomnia. Similarly, low-intensity scalp-level electrical stimulations (transcranial electrical stimulations or tES), which could be using direct current (tDCS) or alternating current (tACS), have gained usage in research studies to modulate ongoing brain oscillations during specific sleep stages, to either increase sleep protection or alter dreaming mentation.

Some of the authors and their collaborators recently completed a research project, where sleep ERPs and sleep tACS were used as neuromodulatory tools to examine the stability or quality of sleep in healthy individuals [12]. It is a common sense fact to determine if someone is having a deep sound sleep, by checking whether he/she responds after calling their name, tapping a sound or touching their body. This basic logic was employed in the above study, such that, when an external neuromodulatory stimulus is provided during sleep, the brain's response to it will depend on the depth and quality of the sleep state. Thus, without waking someone up, the above methods could be used as an active method to determine the quality of sleep, which is otherwise not possible using conventional passive sleep assessment. Being a novel protocol of its kind, the study involved indegenous development of hardware and software modules to implement remote delivery of sleep ERPs and sleep tACS, by integrating into the existing conventional PSG system of the collaborator's sleep lab (Figs. 6.7 and 6.8).

For sleep tACS, a pair of electrodes were placed on the scalp alongside the EEG ones, and a low intensity sinusoidal current (400-600 μA) was provided continuously

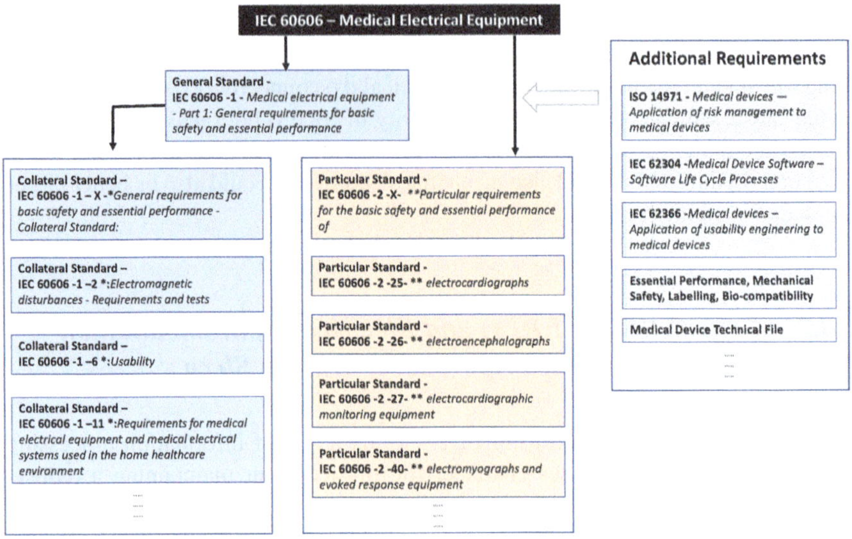

Fig. 6.4 Understanding the basic of IEC 60601 standards for electrical safety testing of PSG wearables

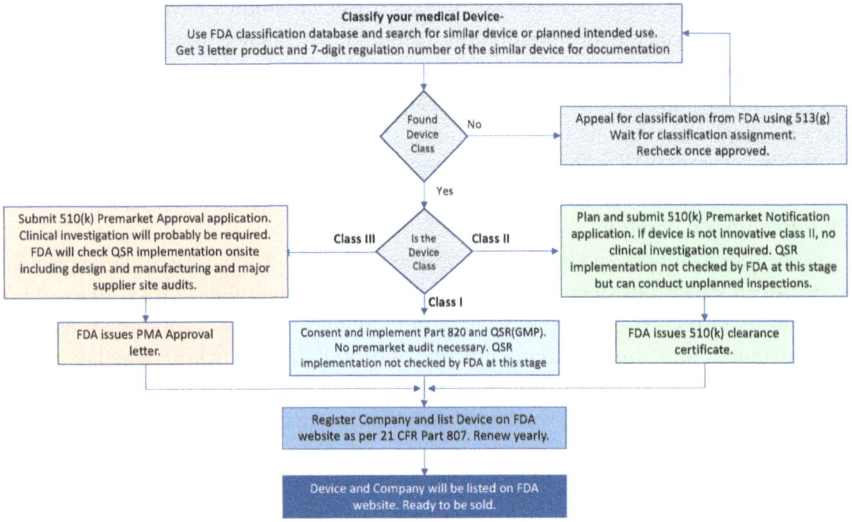

Fig. 6.5 Basic process flow chart for FDA approval of a medical device

for 30 s, during stable NREM or REM sleep at specific frequencies (0.75 Hz for NREM and 40 Hz for REM). The EEG pattern immediately after the stimulation, showed the effect of the neuromodulation, in terms of frequency changes ('brainwave entrainment' effect) (Fig. 6.9). These changes are specific to each sleep stage and

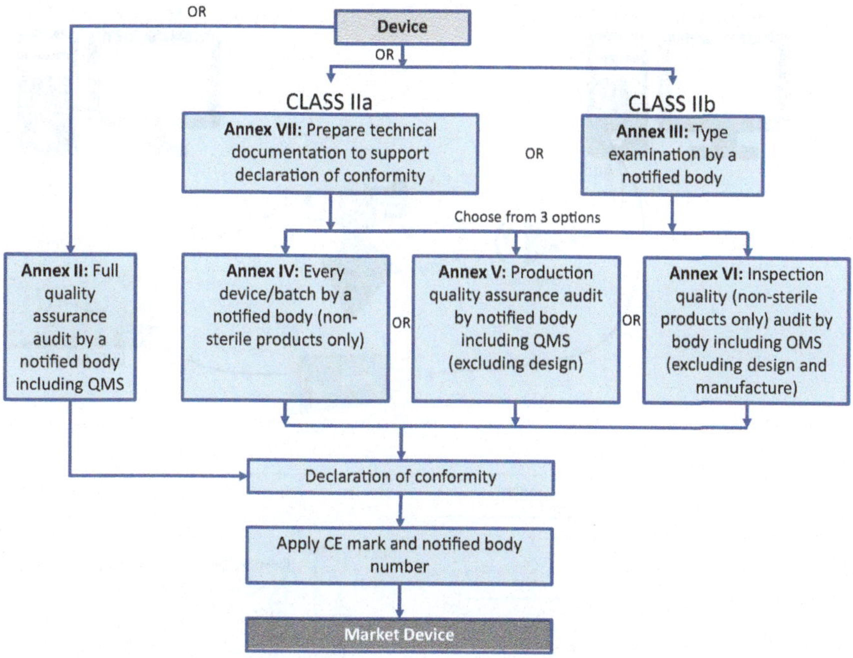

Fig. 6.6 Basic process flow chart for CE marking for class IIa/IIb which is freely available is shown here for reference

the stimulation current frequency.

For sleep ERP, speakers were placed near the subject's head (or even through headphones of pillow speakers), brief audio tones (100 ms; 1000 Hz) were presented at random intervals (8-13 s) throughout night. With each tone, the brain's response could be captured in EEG as ERPs. These responses have been well characterized for different sleep stages, but those for NREM sleep are most studied (as shown in Fig. 6.10). The size and shape of these ERPs help to determine the sleep stage quality.

Therefore, both auditory and electrical stimulations could be useful for active inference of sleep quality. Also, these have the additional benefit of being useful as a non-invasive intervention tool during sleep.

6.5 Summary

The goal of this chapter was to provide information on how wearable sleep monitors function and ways in which they can potentially improve sleep quality. Other monitoring devices like the accelerometer (that you likely already have on your phone) can help monitor sleep activity but should not be used as a replacement for clinical evaluations. This chapter was written with the intention of introducing the reader to

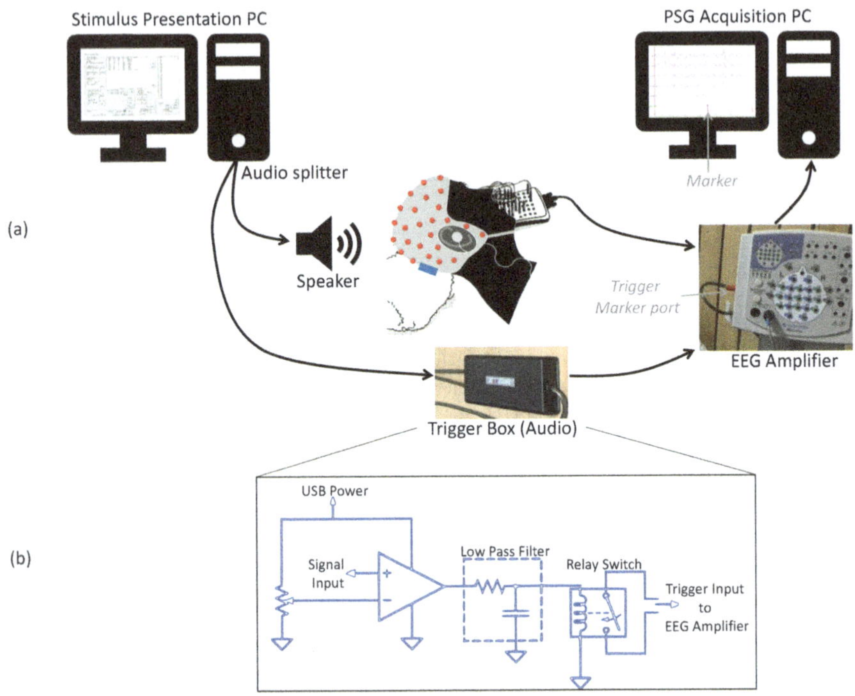

Fig. 6.7 Enabling acoustic ERP using trigger box and event marker I/P of a PSG amplifier system, **a** connection diagram, **b** trigger box

various technology options, and key safety considerations for potential sleep wearable developers as well as anyone interested in understanding regulatory requirements for these devices. Though the subject of medical device safety is far-reaching, this chapter aimed to briefly describe some of its most salient elements alongside future research goals related to sleep and its quantification, which will see further advancement in terms of development and innovation. We discussed some safety standards that must be met for any device that is worn while sleeping, including the possible use of electromagnetic waves or other forms of wireless technology and potential health risks. We hope that readers find the information contained within this chapter useful but please note that it is merely intended as an introduction; we recommend that you consult additional literature for deeper technical insight into this inherently interdisciplinary field, especially with regard to topics surrounding product safety certification. We hope for A future where we don't need to count sheep to sleep, or wake up still feeling tired, or have a disturbing nightmare waking up sweating profusely. We like to believe that we can conquer and control our sleep someday.

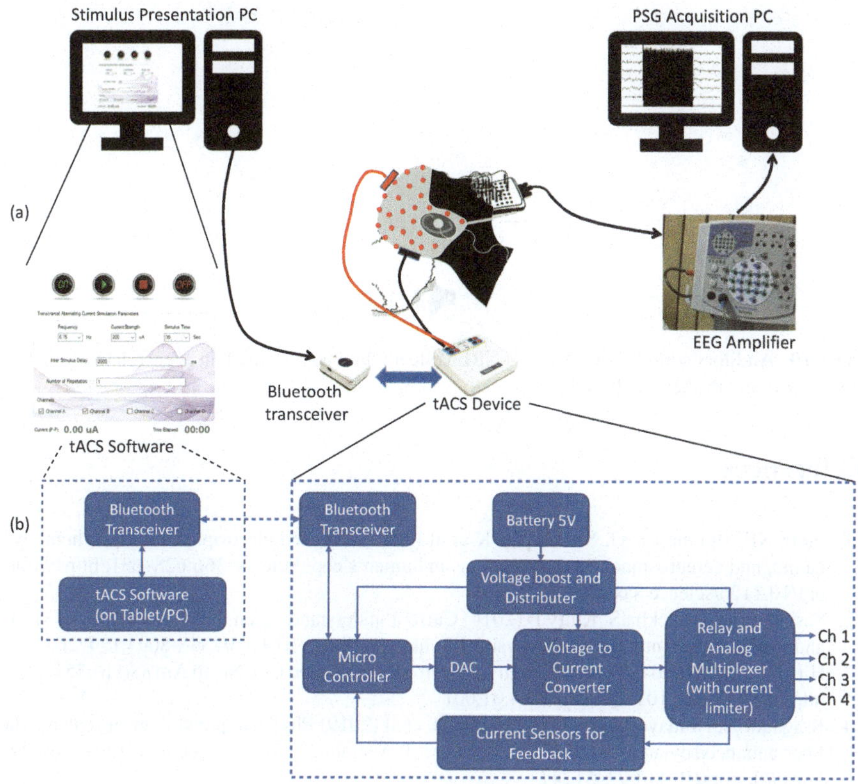

Fig. 6.8 Enabling tACS stimulation using bluetooth for a PSG study **a** measurement diagram, **b** Process diagram

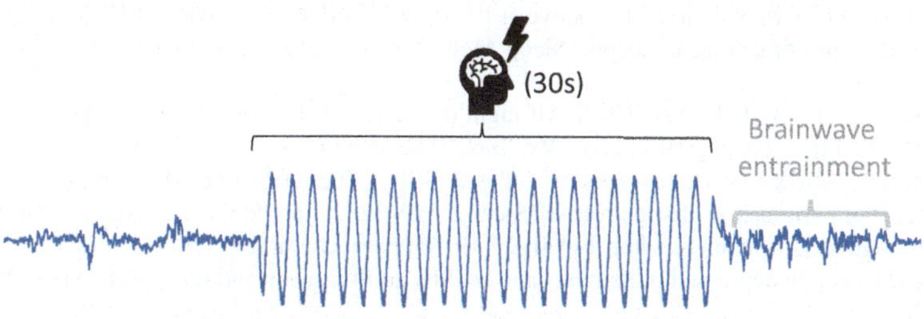

Fig. 6.9 tACS given during NREM sleep, showing brainwave entrainment after 30 s of 0.75 Hz stimulation, without awakening the subject

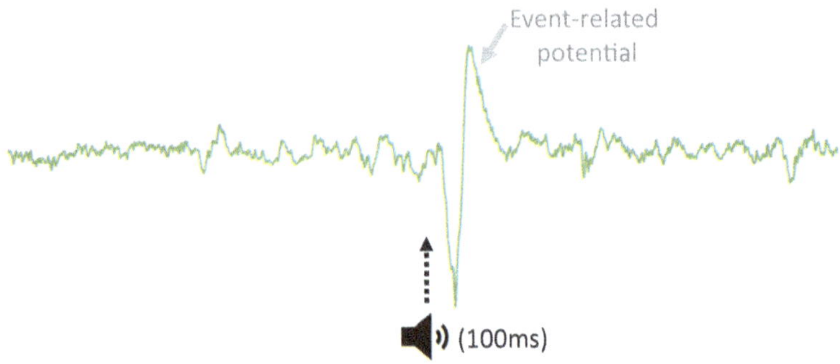

Fig. 6.10 Auditory stimulus given during NREM sleep, showing a prominent event related potential (ERP), without awakening the subject

References

1. Fultz NE, Bonmassar G, Setsompop K et al (2019) Coupled electrophysiological, hemodynamic, and cerebrospinal fluid oscillations in human sleep. Science 366:628–631. https://doi.org/10.1126/science.aax5440
2. Sasidharan A, Sulekha S, Kutty B (2014) Current understanding on the neurobiology of sleep and wakefulness. Int J Clin Exp Physiol 1:3. https://doi.org/10.4103/2348-8093.129720
3. Hirshkowitz M (2004) Normal human sleep: an overview. Med Clin North Am 88(vii):551–565. https://doi.org/10.1016/j.mcna.2004.01.001
4. Kiaghadi A, Homayounfar SZ, Gummeson J, et al (2019) Phyjama: physiological sensing via fiber-enhanced pyjamas. Proc ACM Interact Mob Wearable Ubiquitous Technol 3(89):1–89:29. https://doi.org/10.1145/3351247
5. European Community (1993) Medical devices directive, 93/42/EC. https://eurlex.europa.eu/LexUriServ/LexUriServ.do?uri=CONSLEG:1993L0042:20071011:EN:PDF
6. FDA (2019) CFR title 21, subchapter H, part 860. https://www.accessdata.fda.gov/scripts/cdrh/cfdocs/cfcfr/CFRSearch.cfm?CFRPart=860&showFR=1
7. Le Bon O (2020) Relationships between REM and NREM in the NREM-REM sleep cycle: a review on competing concepts. Sleep Med 70:6–16. https://doi.org/10.1016/j.sleep.2020.02.004
8. Aeschbach D, Borbély AA (1993) All-night dynamics of the human sleep EEG. J Sleep Res 2:70–81. https://doi.org/10.1111/j.1365-2869.1993.tb00065.x
9. Schulz H, Dirlich G, Balteskonis S, Zulley J (1980) The REM-NREM sleep cycle: renewal process or periodically driven process? Sleep 2:319–328. https://doi.org/10.1093/sleep/2.3.319
10. Armitage R, Hoffmann R, Fitch T et al (2000) Temporal characteristics of delta activity during NREM sleep in depressed outpatients and healthy adults: group and sex effects. Sleep 23:607–617
11. Sasidharan A, Kumar S, Nair AK et al (2017) Further evidences for sleep instability and impaired spindle-delta dynamics in schizophrenia: a whole-night polysomnography study with neuroloop-gain and sleep-cycle analysis. Sleep Med 38:1–13. https://doi.org/10.1016/j.sleep.2017.02.009
12. Venugopal R, Sasidharan A, Marigowda V et al (2019) Beyond hypnograms: assessing sleep stability using acoustic and electrical stimulation. Neuromodulation Technol Neural Interface 22:911–915. https://doi.org/10.1111/ner.12847
13. Medical devices: conformity assessment and the CE mark. https://www.gov.uk/guidance/medical-devices-conformity-assessment-and-the-ce-mark

Chapter 7
Non-invasive Monitoring of Health Using Sensor-Rich Wearables and Smart Devices

Vijay Narayan Tiwari, Ajit S. Bopardikar, and Aloknath De

Abstract As smart devices have evolved, their computing abilities have multiplied as have the features and the sensors that are incorporated in them. This revolution has gone hand in hand with advances in allied technologies such as wireless communication, computation, miniaturization and sensor technology thereby making smart devices central to diverse and disruptive applications, particularly related to monitoring, tracking and analytics. The emergence of wearables that contain multiple sensors has further augmented sensing and monitoring especially in the context of health and well-being. In this chapter, we explore the role of smart devices and sensor-rich wearables for monitoring an individual's health and fitness. As the accuracy and reliability of sensors embedded in smart devices evolves, the data collected through these sensors is increasingly used for extracting an individual's vital health parameters and deriving actionable insights at population level. Further, low-latency networks have fuelled the growth of real-time monitoring of health parameters, thus providing a mechanism for timely intervention. In order to provide a flavour of the types of systems that can be developed around the present smart devices and wearables, we present here applications that incorporate different categories of sensors, namely, motion, physiological, ambient and display. For each of these categories, we describe representative examples. We describe activity monitoring applications that utilize the output of motion sensors such as the accelerometer, magnetometer and gyroscope. These sensors are commonly available in smartphones and wearables such as smart watches. In this chapter, we focus on smartphone-based applications such as those to recognize and track activities of an individual. We also look at neural network-based methods to automatically detect gym exercises and count repetitions based on wearable-based sensors. We next describe physiological sensors using the photoplethysmograph (PPG) sensor as an example. The PPG is essentially an optical transceiver that emits an electromagnetic signal (typically infrared) of

V. N. Tiwari (✉) · A. S. Bopardikar · A. De
Samsung R&D Institute India-Bangalore, Bangalore, India
e-mail: vn.tiwari@samsung.com

A. S. Bopardikar
e-mail: ajit.b@samsung.com

A. De
e-mail: aloknath.de@samsung.com

G. D. Gargiulo and G. R. Naik (eds.), *Wearable/Personal Monitoring Devices Present to Future*, https://doi.org/10.1007/978-981-16-5324-7_7

known magnitude and wavelength and receives the reflected signal. The sensor then is placed over a finger or a wrist gathers the reflected signal that contains information related to blood flow changes. This can be used to measure parameters such as blood pressure and heart rate variability. We describe a machine learning method to accurately estimate blood pressure from a single PPG sensor. Ambient sensors measure atmospheric parameters such as temperature, humidity and different types of electromagnetic radiation. These factors play an important role in a person's wellness and long-term health. In this chapter, we consider measurement of ultraviolet radiation. Since ultraviolet radiation measurement sensors are costly and are often not embedded in smartphones and wearable, we present an SVM-based method to estimate the ultraviolet radiation present in the atmosphere using lux sensors which are often embedded in smartphones and wearables. Finally, while not strictly a sensor, we discuss display sensors in the context of their application to human disease detection and monitoring. While the use of the smartphone camera and screen are well known in the context of capturing, rendering and analysing images of diseased body parts, the display can also be used to screen and monitor the wellness of the visual system. We describe one such application related to the estimation of human visual field. The application uses a smartphone in conjunction with a compatible head-mounted device. We describe how the static perimetry test can be transcribed to this combination and compare the performance to a well-known gold standard, the Humphrey field analyser (HFA). All of the applications described above require a high level of accuracy in order to be utilized in practical settings. The performance analysis presented for each application aims to give a sense of what needs to be aimed for when designing health screening and monitoring applications. We thus delineate how technological advances are providing impactful tools for health and fitness monitoring. We close the chapter by discussing how this intersection of information technology, health and fitness monitoring could unfold in future.

7.1 Introduction

With advances in medical technologies and a deeper understanding of physical and mental fitness and lifestyle choices, people are living longer lives. At the same time, an increasing number of people are remaining active into very advanced ages. As a result, the population pyramid of the world shows a gradual shift towards the top. All of this puts increasing pressure on existing health infrastructure and services and increases the cost of healthcare services. Thus, self-care and monitoring assume greater importance. Indeed, caregivers are starting to rely on this monitoring data for better health management. This is facilitated by the rise of sensor-rich smart devices and accurate sensors with decreasing form factors. All of this is making health and well-being more patient-centric, empowering users to take charge of their lives.

The technological advances discussed above have catalysed this paradigm shift in health care. Further, communication networks with reduced latency and ever-improving quality of service (QoS) make possible applications in various domains such as telemedicine, remote monitoring and consultation [1]. Remote health and wellness tracking systems, robust sensing algorithms, timely system-driven alerts when balanced with human intelligence and intervention have the potential to transform the healthcare ecosystem thereby making it more affordable to masses. The role of smart wearable technologies in managing an individual life style, tracking and monitoring their physical parameters with increasing accuracy is playing a vital role in this context. The information tracked by these sensors thus has the potential to help individuals maintain a healthy lifestyle and well-being [2].

In what follows, we touch upon the history of this evolution and discuss in brief the different types of sensors that are present in today's ecosystem.

7.1.1 Evolution of Sensors in Wearables and Smart Devices

The first mobile phone was conceived by Martin Cooper from Motorola in 1973 and was commercialized in 1983. It was inspired in part by the communicators used by characters in Star Trek and the radio watch worn by Dick Tracy [3]. It was as large as a brick and weighed as much. Further, it had a very short battery life. Cooper's vision was to create a *personal communication device*, where a number would be assigned to an individual rather than an individual using a communication device that would be stationary and had its own number.

Today's mobile phones continue to evolve into smarter versions of themselves because of this vision. These smartphones are supported by a wide range of innovations in wireless communications, computation, miniaturization and sensor technologies. Indeed, today's mid-level smartphone is more powerful than a desktop PC from the early 2000s while being considerably lightweight. Low latency networks ensure that the captured or processed data is exchanged with the user and remote servers in very short time, thereby providing almost real-time feedback, while powerful servers and efficient edge computing architectures ensure that the most complex computations are almost instantaneous. While the only notable sensor present in the first mobile phone was the microphone, subsequent evolutions include increasingly sophisticated and accurate sensors such as camera sensors, accelerometer, gyroscope and sensors to measure proximity and temperature. Table 7.1 gives a glimpse into this evolution.

These sensors measure different physical attributes that provide information about the user and their surroundings. They can therefore be classified depending on the type of information they provide. We now proceed to classify different types of sensors in order to highlight their capabilities for effective monitoring of health and fitness.

Table 7.1 Sensors in smartphones over the years [4]

Year	1983	1992	2000	2010	2015	2019
Sensors	Microphone	Microphone	Microphone Front camera	Microphone Front camera Rear camera Magnetometer Accelerometer Proximity Compass	Microphone Front camera Rear camera Fingerprint (front-mounted) Accelerometer Gyroscope Proximity Compass Barometer Heart rate SpO_2	Microphone Front camera (multiple) Rear camera Fingerprint (under display, ultrasonic) Accelerometer Gyroscope Proximity Compass Barometer Heart rate, SpO_2

7.1.2 A Classification of Sensors

As mentioned in the previous subsection, we attempt to classify sensors into categories depending on the type of output they provide. This classification is useful in order to give the reader a sense of different areas in which these sensors play an important role. A closely related aim is also to provide a sense of how outputs from different sensors could be used in multi-modal systems. The classification used in this chapter is depicted in Fig. 7.1. Thus our classification is as follows:

1. *Motion Sensors:* These are among the earliest sensors that were integrate in a mobile device. They aim to capture data related to any type of movement. Exam-

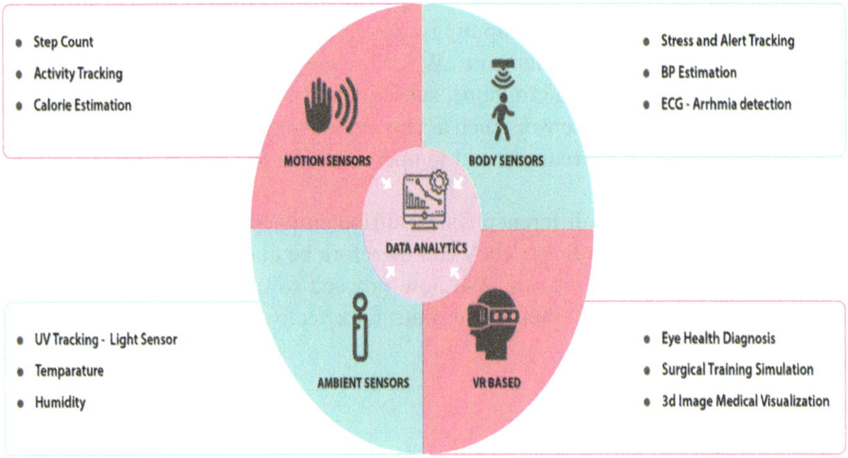

Fig. 7.1 A classification of different types of sensors

ples of such sensors include the accelerometer (tri-axial displacement/movement), the gyroscope (rotational movement) and compass (direction with respect to the physical world). Motion sensors have been used in myriad applications including fitness tracking and mapping services (e.g. Map) [5–8].
2. *Physiological Sensors:* Aim to capture a physiological phenomenon. Some such as the photoplethysmography sensors use light and its interaction with body tissue to capture signals which can be mined to extract vital health information including blood pressure, endurance level, stress level and sleep quality [9–15]. Others such as ECG sensors attempt to capture events of cardiac cycle which can further be used to detect anomalies and arrhythmia [16–18].
3. *Ambient Sensors:* Measure the atmospheric attributes including temperature, pressure and invisible components of light (infrared and ultraviolet). These attributes directly or indirectly affect human health and therefore are vital in various applications [19–21].
4. *Display Sensors:* This class uses visual display coupled with apparatus like the virtual reality (VR) headset to obtain user feedback on particular aspects such as their visual or haptic response [22–25]. While not sensors in the strict sense, this category is included to cover an important aspect of sensing, namely, human response that is vital in detecting many visual and nervous disorders.

While the above classification is by no means exhaustive, it does consider a significant cross section of available sensors. In what follows, we present a generalized framework for sensor-based monitoring. Subsequently, we give examples of representative applications from each sensor category.

7.2 Wearable-Based Health Monitoring: A Generic Framework

Figure 7.2 shows a generalized system for wearable health monitoring [16, 26]. It broadly consists of the following modules:

1. **Measurement Devices:** Smartphones and wearable devices form the heart of monitoring systems. These devices through sensors contained in them measure the desired parameters. The measurements can be continuous time, periodic or user defined. For example, in case of specific ailments, recording of the electrocardiogram (ECG) signal through the relevant sensor can be over a period of time while blood pressure could be measured by the users as required. Further, the computing power present in smartphones and sometimes in wearables allow for basic processing of the acquired data, such as, for example, de-noising, classification (of normal and abnormal signals), anomaly detection and efficient representation for storage purposes. Further, smartphones can also store some data for fast retrieval in an embedded database (DB).
2. **Server/Cloud:** Data captured by the measurement devices can be stored on remote servers or a cloud-based service. An important function of these systems is fast

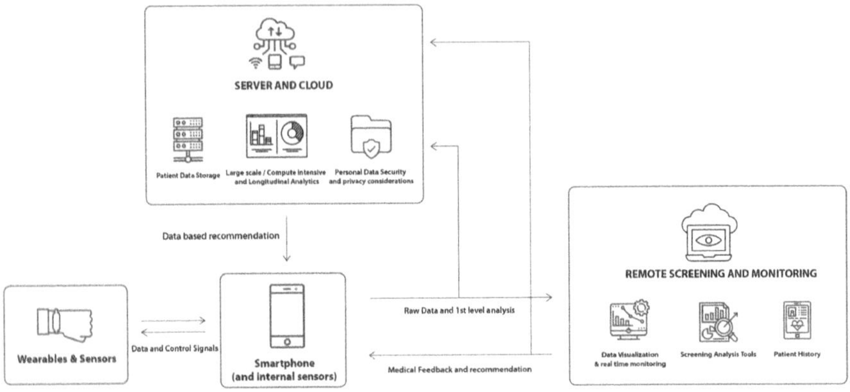

Fig. 7.2 A general system for wearable health monitoring

storage and retrieval of data as requested by the user. Further, these systems are concerned with ensuring that the data is securely stored and can only be accessed by legitimate agents. Further the high computing power enables more complex analysis such as longitudinal analysis of individual's data or analytics on data to understand trends and predict behaviours over a larger populace (after ensuring that all data used in the computations is legitimately accessed after obtaining the right permissions). Further data across users can be used to update and refine models dynamically or periodically.

3. **Remote Screening and Monitoring:** This module consists of tools that aid medical practitioners and caregivers in their monitoring, screening and diagnostic work. These include sub-modules to visualize the raw data and patient's history and demographic information. The data used could be straight from the measurement devices (raw and processed data) as in the case of real-time monitoring or could be combined with the data stored and analysis conducted on the server.

The evolution of smartphones and growing number of features accommodated in them have been an important mover for sensor evolution. The limited form factor of a wearable or a smartphone has resulted in smaller sensors. At the same time, sensors are evolving to become more sensitive to their stimuli. An example of such sensor is the ever-evolving camera sensor now used for sophisticated photography as well as teleopthalmology to capture retinal images in conjunction with an add-on lens. Also, the photoplethysmography (PPG) sensor is now used to capture data that is used to estimate health parameters such as heat rate variability, SpO_2 and blood pressure. The PPG sensor is now a part of smartphones as well as smartwatches because of its shrinking form factor. Further, several sensor-based applications now utilize the increasing computing power and storage available on smartphones or wearables such as smartwatches to effect on device computations. This has led to neural network models specifically tailored to their architecture. Advances in computation coupled with the evolution in architecture and fabrication processes have resulted in more and more sophisticated sensors embedded in or linked to smartphones and wearable. The relationship between sensors, wearables and smartphones is thus symbiotic, being propelled forward through the engine comprising human innovation, requirement and vision.

The next section presents examples of screening and monitoring applications that incorporate and illustrate some or all aspects of these modules.

7.3 Non-invasive Sensor-Based Health Screening and Monitoring

In contrast to conventional caregiver-based sensing and monitoring mechanisms, the evolving landscape of non-invasive sensors available on wearables and smart devices offers greater possibilities to measure and access an individual's health parameters. While diagnostic capabilities remain the Holy Grail for such systems, they have placed a certain amount of autonomy and agency in an individual's hands. Through such sensors, users can monitor their state of health and make active decisions related to their lifestyle. This will allow them to pre-empt and prevent onset of adverse health conditions. We thus describe here the application of sensors form different categories as presented in Sect. 7.1.2. The aim here is to demonstrate through examples, the role of each sensor type in measuring and monitoring vital health parameters.

7.3.1 Motion Sensors

In recent years, wearable devices and smartphones with integrated motion sensors like accelerometer, magnetometer and gyroscope have proven to be very efficient in tracking various fitness-related activities of an individual [8, 27]. Data from these motion sensors has helped in uncovering health and fitness-related information thus encouraging a user take up and follow a healthy lifestyle [7, 28]. In this section, we describe systems developed around motion sensors that are relevant to measuring the fitness and exercise-related activities of individuals. Specifically we focus on the following broad categories:

1. Methods to recognize and track typical activities undertaken by an individual (smartphone based).
2. Algorithms to automatically detect gym exercises and count repetitions (wearable based).

The next two subsections describe typical methods for above.

7.3.1.1 Smartphone-Based Activity Recognition

We aim to identify the type and intensity of activity based on the data captured by motion sensors in a smartphone. The data is first pre-processed and features are extracted that can be used in a detection framework such as a neural network or logistic regression. In what follows, a representative example [8] of developing an activity recognition system is described. The steps to develop such a system are as follows:

a. Define the activity types to be detected.
b. Develop an experiment to gather sufficient data for each of the activity types.
c. Train a machine learning system to automatically identify the activity type. This will include steps to pre-process the sensor data, feature extraction and engineering and developing appropriate machine learning models.
d. Test the developed system on new captured data.

We now discuss specific experiments for each of these steps.
Defining the types of activities: The activities considered in this example are running, walking, jogging, resting cycling and driving. These were chosen as the most frequent activities that are undertaken by an individual.
Data gathering: Data can be gathered through volunteers after obtaining relevant consents for their participation. The types of sensors present on the device play an important role in the design of a data gathering experiment. In the present example, data gathering experiment was first described to the volunteers. Further, they were instructed to record specific activities and label them. In the experiment performed, data was collected from 16 volunteers for 16,570 s through a smartphone app. Linear acceleration (l), acceleration due to gravity (g) and the angular velocity (w) were the prominent sensor-based inputs that were used by us in the activity recognition algorithm. Table 7.2 gives the amount of data collected (in seconds) for each of these activities.
Pre-processing and training: This step involves enhancing the data by suppressing noise and unwanted artefacts. In our work, this was accomplished through filtering and de-noising. Five discrete time signals were considered: horizontal and vertical linear acceleration and the three orthogonal components of angular velocity, and five more signals derived from low-pass versions of above were generated using Butterworth low-pass filter (cut-off frequency 2.5 Hz). All the signals were divided into windows for feature extraction. Several statistical features such as mean, median, skew, kurtosis, minimum, maximum, standard, range, etc. and pairwise correlation-based features were computed to characterize each window. Each feature was normalized with zero mean and unit standard deviation. In the training dataset, we had

Table 7.2 Amount of data (in seconds) captured for each activity

Activity	Rest	Walk	Jog	Run	Cycle	Drive
Time(s)	2800	3120	3000	2650	2900	2100

13,281 windows and features extracted from these windows were used to train a decision tree classifier, each node of which was an ANN (artificial neural network). Along with one input layer for features, each ANN had two hidden layers with 30 and 10 perceptrons, respectively. Corresponding to the six activities, the output layer consisted of six nodes. ANNs were trained using linear search algorithm with back-propagation as the cost function for 500 iterations. During the classification stage, the extracted features were fed to the corresponding ANN and the output vector was normalized to give a class probability vector.

7.3.1.2 Wearable Based-Gym Exercise Recognition and Repetition Counting

It is often observed that selecting individual gym activity and counting repetitions manually to keep track of calories burnt or to build stamina are cumbersome. The likelihood of false registration of an activity is very high if a user forgets to change the activity manually in a tracking and recording device. Therefore, automatic gym activity detection is very useful and it will allow the users to monitor their gym training more efficiently. The purpose of this section is to present methods to automatically detect gym exercises and count repetitions for better management of fitness.
Gym Exercise Recognition: Presented here is a framework and design to automatically detect gym exercises [29]. This involved eight different gym activities that encompass the major muscles of the body. These were further sub-divided into Type-I and Type-II categories. Type-I activities include push-ups, sit-ups, free squats and skipping which generally involves user's own body weight. Type-II activities include bicep curls, lateral pull-down, chest fly and bench press which are performed using additional weights.

250 min of data was collected from 20 unique volunteers using a data collection application on a smart watch. The application was designed to collect the accelerometer (ax, ay, az) and gyroscope (rx, ry, rz) sensor readings for each of the eight activities along the three axes. In addition to these signals, we derived four more signals, namely, acceleration magnitude and its component along the x-y-plane and angular velocity magnitude and its component along the y-z-plane. These planes were chosen because significant signal variations were observed in these planes for some of the exercise categories.

Each of the obtained ten signals were divided into overlapping windows of duration 5 s, and a 4 s overlap. The obtained windows of these ten signal axes were then low-pass filtered with a cut-off frequency 8 Hz. The features extracted from each window are presented in Fig. 7.3.

Using the features derived from the dataset, a two-level ANN was trained in which first level (two hidden layers with four and two perceptrons) classified the activities between active and passive states and second level (two hidden layers with seven and four perceptrons) classified activities between the eight gym exercises mentioned above. The training of ANNs was done using scaled conjugate gradient algorithm.

I. Signal magnitude features
1: Mean value of the window
2: Standard deviation value of the window
3: Skewness value of the window
4: Kurtosis value of the window
5: Maximum value in the window
6: Minimum value in the window
7: Root mean square value of the window
II. FFT based energy features
8: Parseval's Energy in the window
9: Parseval's Energy in the frequency range 0 – 0.4 Hz
10: Parseval's Energy in the frequency range 0.4 – 1 Hz
11: Ratio of energies of feature 9 and feature 10.
III. Peaks and Valleys features
12: Median value of the magnitude of peaks in the window
13: Median value of the magnitude of valleys in the window
14: Mean value of the magnitude of peaks in the window
15: Mean value of the magnitude of valleys in the window
16: Average distance between two consecutive peaks in the window
17: Average distance between two consecutive valleys in the window

Fig. 7.3 Features used in gym-based exercise recognition

The classifier detected activity with ~96% sensitivity and ~96% specificity on an average.

After detecting gym exercise, the next step is to count the number of repetitions for those exercises. This helps to estimate the number of calories expended in performing each exercise set [29]. This is described next.

Repetition Counter: For counting repetitions, we considered that axis of accelerometer and gyroscope data along which the variation in the signal was most prominent. These were: Push-ups—rz, Sit-ups—rz, Free Squats—az, Skipping—rz, Bicep curls—square root of sum of ry and rz squared, Lat Pull-down—rz, Chest fly—energy of gyroscope vector and Bench Press—square root of sum of ay and az squared.

Each signal was low-pass filtered using Butterworth filter of suitable order and cut-off frequency depending on exercises in order to remove any noise. The cleaned data still contained false peaks which occur due to vibrations while performing gym activities and are not removed by the low-pass filter. In addition, the data also exhibits changes from user to user. Therefore, in order to further streamline the repetition counting mechanism, we used a derivative method to calculate a threshold (derived empirically for each activity). We also removed false peaks by employing

Table 7.3 Repetition count accuracies for Type-I and Type-II exercises

Type-I		Type-II	
Exercise	Accuracy (%)	Exercise	Accuracy (%)
Push-ups	95.2	Bicep curls	96.6
Sit-ups	97.6	Lat Pull-down	96.4
Free squats	98.8	Chest fly	97.8
Skipping	98.2	Bench press	95.8

thresholding. Finally, there will be a limit to the speed with which each activity can be performed. Consequently, all filtered peaks were subjected to a time threshold which symbolizes the maximum time difference between true successive peaks. The remaining peaks after filtering were used to calculate the exact repetition count. The repetition count accuracies for all the eight activities are summarized in Table 7.3.

7.3.2 Physiological Sensors

Going by the philosophy of a healthy mind residing in a healthy body, sensors can play an invaluable role to assess both mental and physical parameters of a human body. Alertness and stress are examples of parameters indicative of mental well-being, while physical well-being can be indicated through parameters such as blood pressure, SpO_2 (blood oxygen content) and electrical activity of the heart [2]. Quantifying and tracking these parameters help to analyse the physical and mental state of a person. This, in turn, can be used by applications to help an individual manage their personal well-being more effectively.

Sensors embedded in smartphones and wearable devices are playing an important role in continuous monitoring of vital physiological parameters [30]. As the form factor of these sensors becomes smaller and their accuracy superior, measurements from personal devices are becoming increasingly reliable. An important challenge in this regard is to extract the right parameters from the sensor signal to estimate a desired physiological phenomenon.

In this section, we will discuss and explain a representative method to measure blood pressure (BP) using a physiological sensor called the photoplethysmograph (PPG) [9, 31]. The PPG sensor is essentially an optical transceiver that emits electromagnetic signal of known magnitude and wavelength and receives the reflected signal. A typical PPG sensor system is shown in Fig. 7.4. This radiation is typically in the infrared or visible part of the spectrum. When the sensor is placed over a part of an individual's body (such as a fingertip or wrist), the sensor can detect blood flow-related changes in the blood vessels. These changes are capable of capturing many different haemodynamics parameters such as Blood Pressure (BP) and Heart

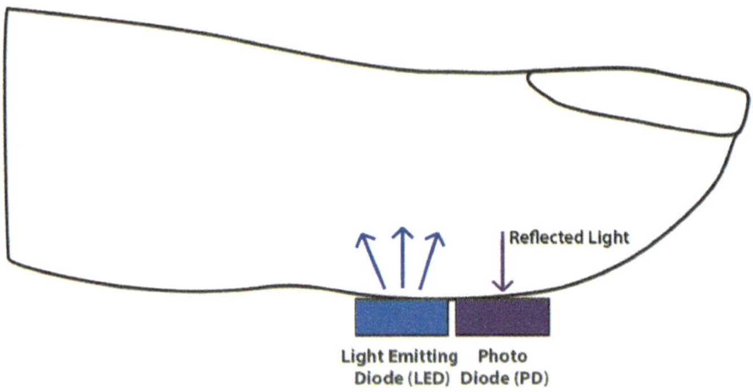

Fig. 7.4 A typical PPG sensor system

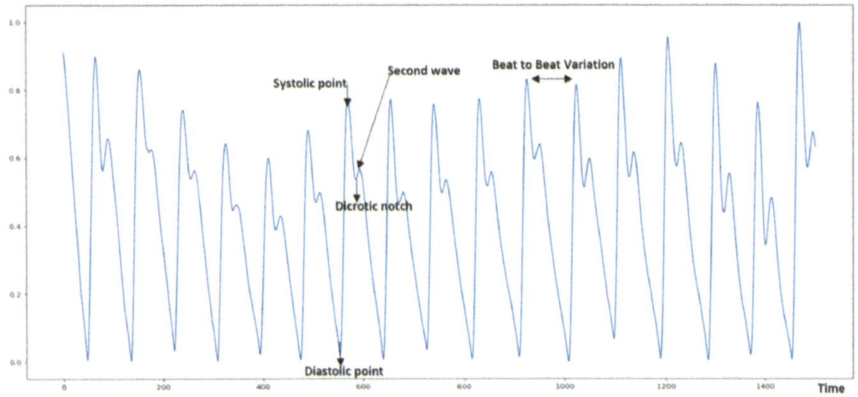

Fig. 7.5 Example of the PPG signal

Rate Variability (HRV). Figure 7.5 shows typical PPG waveform captures on a finger tip using red light.

7.3.2.1 A Brief Overview of Blood Pressure Physiology

Blood Pressure (BP) is a periodic signal with a frequency of heart rate and is typically defined as the pressure exerted on the blood vessel walls when the blood flows through it. The upper limit of BP is called Systolic Blood Pressure (SBP) and its lower limit is called Diastolic Blood Pressure (DBP). For all practical purposes, normal BP range in adult population is 120 mmHg SBP and 80 mmHg DBP. Worldwide prevalence of hypertension (raised BP) in adults is 26.4% (men) and 23.1% (women), respectively. Often hypertension can lead to serious implications and hence it is important to periodically monitor BP.

In the next section, we discuss the processing and modelling of acquired PPG signal towards estimation of Systolic (SBP) and Diastolic (DBP) BP.

7.3.2.2 Method

In recent times, extracting HRV-based features from PPG signal have been a topic of active research. Further, studies have shown that there is a strong correlation between HRV, stress and BP. This section thus presents a methodology to estimate BP from a PPG signal [9, 31]. While existing methods have used PPG from multiple locations [32–34] and have also used other signals in conjunction such as the electrocardiogram (ECG) [35] to estimate BP, and the subsequent text describes a method to do so from a single PPG signal. In what follows, we describe the processes in each block.
Data acquisition
Acquiring the data correctly for building the model is one of the most crucial steps. Several smartphone and wearable have a PPG sensor. PPG signals were acquired using this sensor. All the processing was done using the IR component of the PPG sensor. The red light component merely acts as the indicator for the user to identify the sensor. There are multiple frequencies at which the signal can be measured. In this example, we have set the sampling frequency 100 Hz. The data was collected from 235 volunteers of diverse physiological and demographic profiles of which 185 were used for training and 50 for testing. The testing was done on a set mutually exclusive with training set to enhance model reliability. For all experiments, prior consent was taken from subjects, the data was kept anonymized and was used only for the intended research purpose. Additionally, principles outlined in the Helsinki declaration were adhered to while experimentation. The ground truth for estimating systolic and diastolic BP was measured by using a mercury-based cuff device.
Pre-processing
Each data sample size consisted of 15 s signal window. Subsequently, peaks were detected using wavelet smoothing followed by trend removal and dynamic peak search. From the signal window, pulses were chosen as the signal between two successive minima containing a single maximum. All other pulses were discarded. Additionally, template matching was also performed on each of the pulses over a range of PPG templates chosen from those pulses which had no noise interference. These templates were chosen across subjects with diverse physiological and demographic profiles. In order to maintain uniformity of pulses, the signals were interpolated to a fixed length using cubic spline interpolation and normalized using min-max normalization before determining the correlation coefficient. Two criteria were used to determine the validity of a PPG signal window as given below:

1. Pulses that had Pearson correlation coefficient less than 0.8 with all the templates were rejected.
2. After this pre-processing, the window was retained if the number of pulses n extracted from a window of N seconds satisfied $n >= N/2$. This ensured that the window had significant PPG information.

Sl. No.	Feature
1	PPG pulse length
Frequency Domain Features	
2-36	$Sig[1:7]$, Sig = 256 point FFT of PPG, VPG, APG, PPG3, PPG4
37-76	Mean of $Sig[i:i+2]$ for $i=8:3:31$, Sig = 256 point FFT of PPG, VPG, APG, PPG3, PPG4
77-91	Mean of $Sig[2^i:2^i+1]$ for $i=5:7$, Sig = 256 point FFT of PPG, VPG, APG, PPG3, PPG4
Time Domain Features	
92-95	Maxima locations of PPG, VPG, APG & PPG4
96-98	Minima locations of APG, PPG3 & PPG4
99-103	Mean of $Sig[m-10:m+10]$ where Sig = PPG, VPG, APG, PPG3 & PPG4, and m = maxima location
104-158	Relative frequencies in Magnitude histogram with 11 equal bins of $Sig[1:m]$, Sig = PPG, VPG, APG, PPG3 & PPG4, and m = maxima location
159-213	Relative frequencies in Magnitude histogram with 11 equal bins of $Sig[m:l]$, Sig = PPG, VPG, APG, PPG3 & PPG4, m = maxima location, l = PPG length
214-223	Mean of $Sig[i:i+2]\forall i=1:3:31$, Sig = USDC
224-233	Mean of $Sig[i:i+5]\forall i=1:5:51$, Sig = DSDC

Fig. 7.6 List of features used in the described method to estimate BP from PPG pulse

Feature Extraction From a single PPG pulse, a total of 233 features were extracted in time and frequency domains. The first four derivatives of the PPG pulse (VPG—velocity plethysmogram, APG—acceleration plethysmogram, PPG3—third derivative and PPG4—fourth derivative) were also considered for extraction of relevant features in time and frequency domains. To mitigate issues related to varying pulse lengths, interpolated and normalized pulses were used along with the pulse length as a feature. The features from individual pulses were averaged over the window to increase robustness. Apart from the PPG features, the demographic and physiological profile values were also used as independent features. A comprehensive list of features is listed in Fig. 7.6 and explained below.

Blood Pressure Estimation: Approximately, 20% of the pulses were removed during the pre-processing step because they did not comply with the quality requirements to estimate BP. After this, a total of 53000 samples were available for training the BP estimation model, while the testing data comprised of 9000 samples.

In order to estimate BP from the extracted features, ridge regression modelling was used. Here, the loss function used was linear least square and regularization was given by L_2-norm. The objective function of ridge regression used was

Table 7.4 Partitioning matrix based on age, gender and Body Mass Index (BMI)

Age	Gender	BMI
<40 Years	Male	<24 kg/m^2
≥40 Years	Male	≥24 kg/m^2

$$E(\mathbf{w}) = \frac{1}{2}\sum_{n=1}^{N}\{t_n - \mathbf{w}^T\phi(x_n)\}^2 + \frac{\lambda}{2}\sum_{j=1}^{M}|w_j|^2, \quad (7.1)$$

where E is the loss function, N is the number of samples, t_n corresponds to the target values, ($\mathbf{w}$) is the weight vector, ϕ is the basis function in input space of x_n, M is the number of features and λ is the regularization parameter.

The regularization parameter λ controls the strength of the penalty term, and it is crucial for the model performance to choose the best λ. For the proposed model, the regularization parameter ranged from 0.00001 to 60 in 500 logarithmic steps, and the best performing models were retained.

After training a BP estimation model using PPG and demographic features, we aimed to leverage the effect of the latter to enhance our models. Towards this, the entire dataset was partitioned independently on the basis of age, gender and BMI (see Table 7.4). Individual Lasso models were trained on each of the partitioned data using same regularization parameters. Finally, only best performing models were retained.

All the trained diastolic models were sorted according to their mean absolute error of the predicted DBP and the top 50 models with least mean absolute error were identified. The output of these best performing diastolic models was fed into each systolic model along with the features. After BP prediction from the multiple models for all the partitions (based on age and gender), the SBP and DBP outputs from all the models were scattered in a 2D plane, and the Mahalanobis distance of each point (a predicted SBP-DBP pair) from the mean of all the SBP and DBP pairs. All points within Mahalanobis distances of two standard deviations (corresponding to a 95% confidence interval) were retained while others were rejected as outliers.

7.3.2.3 Results

We incorporate a set of distinguishing PPG features into our models, which results in test accuracies of 4.8 mmHg Mean Absolute Error (MAE) for Diastolic BP (DBP), and 5.7 mmHg MAE for Systolic BP (SBP). This is a marked improvement over generalized models (12.2% for DBP and 13.9% for SBP) [9, 36], demonstrating that single PPG sensor-based methods can predict BP with a high degree of accuracy.

7.3.3 Ambient Sensors

Ambient sensors measure the state of the atmosphere in terms of parameters such as temperature, humidity and different types of electromagnetic radiations. A good example of the latter is UV radiation. The World Health Organization (WHO) reports that over exposure of UV radiation can lead to health conditions like sunburn, premature skin ageing, cataract, and in long term can also lead to skin cancer, while under exposure can potentially cause a vitamin D deficiency [37]. Traditionally, UV sensors equipped with photodiodes are used to estimate UV exposure and wearables equipped with UV sensor are available to measure personalized UV exposure level [38]. However, the accuracy of these solutions is often dependent on the orientation of the sensor with respect to the sun [39].

Ambient sensors in smartphones are often used to measure parameters that are directly measurable such as surrounding temperature, light intensity and humidity. In addition, there are other parameters which directly affect human health. An example is the dose of ultraviolet (UV) radiation which indirectly regulates biochemical reactions in human body. Dedicated sensors to measure UV exposure tend to be costly and require a level of compliance from the user (the rays of the sum have to be directed at a particular angle with respect to the UV sensor). It is therefore desirable if the UV exposure can be estimated through the data gathered by other ambient sensors already present in the smartphone or wearable. In addition, it is also desirable if the estimation can be made direction invariant and therefore more adoptable by users. In this section, we describe a method to estimate the UV index using the measurements of ambient light intensity. In this way, the data captured by light sensors that are typically present in a smartphone can be utilized.

UV index estimation finds application in estimating the optimal time a user may continue outdoors without risking sunburn. After estimating user's skin type [19, 20], a live cumulative UV dosage for the day and a live estimate of remaining time duration for which user can continue to be in the outdoor state without getting sun burnt can be provided [21]. Furthermore, this system can be used to generate personalized recommendations such as use of protective gears like cap, sun glasses, shirt and sunscreen lotion's SPF level based on the user's current UV exposure.

A system that enables us to estimate UV exposure dose through measured ambient light by sensors in a mobile/wearable device is shown in Fig. 7.7 [19]. It consists of three major parts: phone state detection, orientation-invariant lux estimation and lux versus UV index correlation.

First, the state of the phone is classified into one of the following categories: indoor state, outdoor sunny state and outdoor cloudy state. This classification is performed in two steps. The first level classification is to identify if the measurements are being taken indoors or outdoors. This is achieved by applying an empirical threshold on lux value obtained from the phone's ambient light sensor. In the latter case, an SVM classifier is used to further differentiate between outdoor sunny and outdoor cloudy states. This SVM classifier takes the lux value, the angle of position of the sun (the solar elevation angle) and the angle between the sun and the sensor (the angle of

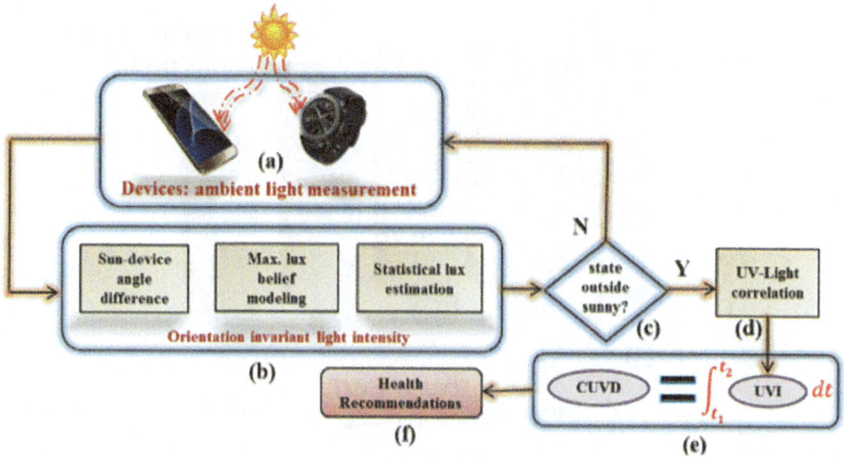

Fig. 7.7 Proposed framework. **a** Ambient light measurement. **b** Sun-device angular difference calculation. **c** Light intensity-based device state detection. **d** UV light correlation for UV index estimation. **e** Cumulative UV dose (CUVD) estimation. **f** UV index and CUVD-based recommendations

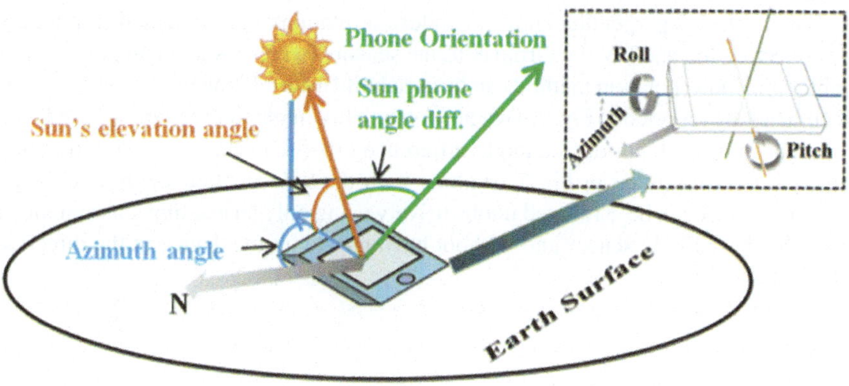

Fig. 7.8 Orientation of sun (azimuth and elevation angles) and device (azimuth, roll and pitch angles) on earth's coordinate frame

deviation) as the input. For training and to estimate the optimal threshold, the lux and angle values data can be gathered from representative smartphone/wearable devices for different states. For orientation-invariant lux estimation, data can be collected using a light sensor attached to a rotating motor which collects lux values at various angles ranging from 0 to 90° between the sensor and the sun position. This data can be used to generate a belief value model for different states which will depend on the sun's elevation angle and the angle of deviation as shown in Fig. 7.8.

To calculate orientation-invariant lux, a bootstrap sampling method can be used to estimate lux corresponding to 0° angle by a weighted combination of the belief value

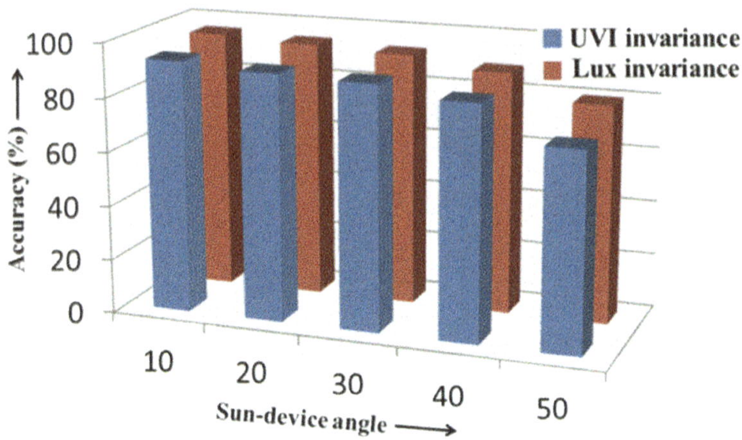

Fig. 7.9 Average accuracies for different sun-device angles

and the observed lux value. With this method, a good estimate for lux perpendicular (i.e. lux value when device is placed in the direction facing sun) can be obtained.

Finally, the lux perpendicular to UV index correlation can be obtained empirically by collecting lux and UV data from external sensors (with sensors facing the sun). A cubic spline interpolation method can be used build a correlation model. The average accuracies are reported in Fig. 7.9 when bootstrap lux-angle samples are selected with different minimum sun-device angle difference (10–50°). The lux-UVI correlation function is also plotted in Fig. 7.10 along with the 95% confidence interval band. Thus UV index can be estimated unobtrusively using only an existing smartphone or wearable-based light sensor and without having to point the device in the direction of the sun.

7.3.4 *Display Sensors*

This section discusses how a mobile display (such as a smartphone) often in conjunction with other wearables and accessories can be used for human disease screening and monitoring [22, 40, 41]. While cameras in smartphones can be used to capture images of diseased parts of the body (e.g. skin, eyes) that can then be analysed to detect disease characteristics, the display screen can be a natural rendering mechanism for diagnosing and monitoring conditions of the visual system. For example, Habtamu et al. [42] have developed smartphone-based screening Apps (Peek) that enable caregivers to make referral decisions about patients with eye problems particularly in developing countries and low-income areas. In doing so, they attempt to address the issue of eye health and wellness in such regions. Applications such as these are a result of the ever-improving quality and resolution of a mobile display

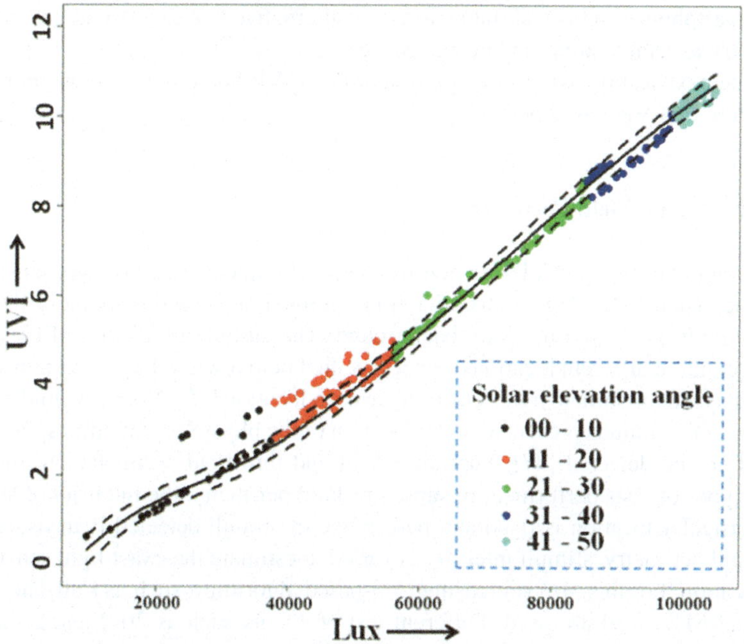

Fig. 7.10 Lux UVI correlation plots

screen. We present here, a virtual reality-based application of how visual display can be used to test for vision defects, specifically loss of visual field [23, 43].

According to the World Health Organization, diseases such as glaucoma, macular degeneration and diabetic retinopathy account for 30% of moderate to severe vision impairment cases [44]. These diseases lead to visual defects such as vision loss and can cause irreversible damage to the eyes. Detection of visual field defects due to these diseases is therefore vital for effective and timely management.

Visual Field Perimetry (VFP), conducted on the Humphrey Visual Field Analyser (HFA), is a standard diagnostic test that enables early detection of visual field defects. The test analyses patient responses to light stimuli to estimate their sensitivity over the visual field [45]. However, its large size and cost limits its reach especially in remote areas where bulky equipment cannot be easily transported. Also privately run clinics may not have adequate space for a full HFA. As a result, numerous other head-mounted perimeters such as imo [24], VirtualEye [25] and Kasha [46] have been proposed. These solutions are highly customized devices meant only for perimetry. This leads to high costs though somewhat less than a standard the HFA. Costs can be further reduced if a perimetry system can be implemented on a system constructed from commercially available apparatus.

We observe that virtual reality systems provide a natural mechanism to test the visual field and we discuss in this section a smartphone-based head-mounted perimeter system [22]. It is based on commercially available Samsung GearVR and Samsung smartphones and is a portable, low-cost alternative that can extend the reach of perimetry to remote areas and home settings.

In the next section, we give a brief overview of VFP before describing an implementation of a VR-based perimeter.

7.3.4.1 Visual Field Perimetry

Visual Field Perimetry (VFP) is a non-invasive technique to measure light sensitivity over the visual field. Patients fixate at a pre-defined target and respond to flashing light stimuli on a fixed intensity background. The stimuli are displayed for a few milliseconds at different locations over a specified field of view. Patient responses are then analysed to estimate light sensitivity at each location [45]. Note that, under ideal fixation, any stimulus presented at the location of the physiological blind spot (PBS) should not be detected [47]. Suprathreshold and threshold perimetry are the two major types of tests performed. In suprathreshold perimetry, the intensity of stimuli at a particular location is fixed and percentage of stimuli detected is measured. In threshold perimetry, stimuli intensity is varied to estimate the exact light sensitivity at a location. For this, 4-2 staircasing [48]-based algorithms such as Full Threshold [45] and SITA [49] are used. Different test protocols such as 30-2 and 24-2 [45] (representing the field of view and number of locations tested) exist. Also, stimuli of fixed size such as Goldmann III and Goldmann I are used [22, 50].

7.3.4.2 Methodology

In this section, we discuss the consideration in designing a VR-based perimeter. Specifically, we discuss ideas related to how the intensity of the stimuli can be changed, the test methodology and estimation of fixation loss on a patient's part.

7.3.4.3 Interactions and Interface

First patients choose the test to be taken [22], patients can select suprathreshold or full threshold perimetry test along with 30-2 or 24-2 protocol [45]. Once the test starts, light stimuli are presented to the patient at random locations on the visual field for 200 ms with random intervals (1200–1800 ms) between consecutive stimuli. Patients record their responses by pressing a button on a standard controller each time they see a stimulus (Fig. 7.11) that comes with a typical VR apparatus. Once the test is over, an auto-generated report displays the heatmap of visual field measurements along with reliability parameters. These reports are stored on-device and can be shared with an ophthalmologist for in-depth analysis. Further, a mechanism can be added that enables the ophthalmologist to examine each individual location to view exactly which stimuli were missed by the patient (Fig. 7.12).

Fig. 7.11 Patient taking the perimetry test using a VR head-mounted apparatus is required to fixate at the black centre dot and respond to white flashing stimuli by pressing a button on the VR controller [22]. (Left-hand side image source: FAVPNG)

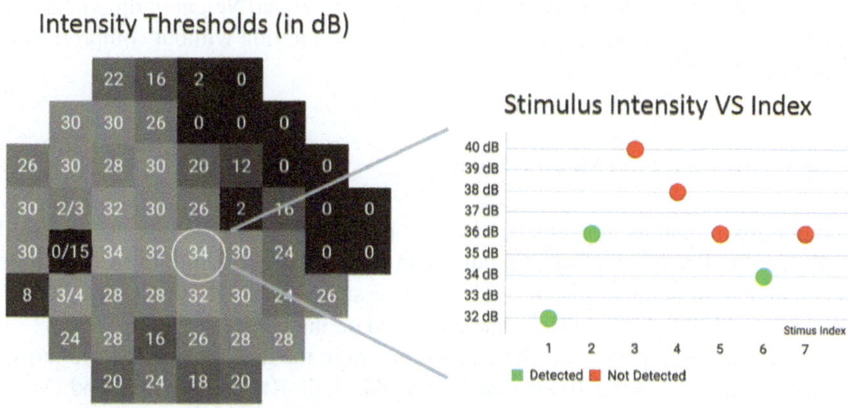

Fig. 7.12 Heatmap of thresholds estimated by a VR-based perimeter along with intensities of stimuli presented and patient responses at the selected location

7.3.4.4 Threshold Estimation

In perimetry, stimulus intensity (I_{dB}) is measured in decibels (dB) as follows:

$$I_{dB} = 10 \log_{10} \left(\frac{L_{max}}{L_T - L_B} \right), \tag{7.2}$$

where L_{max} is the maximum stimulus luminance, L_B is the background luminance and L_T is the stimulus luminance [50]. Note that an in increase in intensity in decibels (I_{dB}) indicates a decrease in luminance.

According to the ISO standards for perimetry, L_B is set to 10 cd/m^2 and L_T is varied during the test [50]. The value of L_{max} depends on the device display. Typically, L_{max} is 3183 cd/m^2 on the Humphrey VFA [50] and is much lower for a smartphone. For example, it is 440 cd/m^2 on the Samsung Galaxy S8 [51]. The resulting threshold heatmap will therefore look different. However, through a simple arithmetic operation, the threshold intensities displayed as output of a VR-based perimeter can be made HFA equivalent [43].

Threshold stimulus luminance refers to the luminance of a stimulus which has a 50% probability of being detected [50]. Full threshold 4-2 staircasing strategy with one reversal is a standard method for threshold estimation. In this, intensity is increased (luminance is decreased) in fixed increments until the stimulus is no longer seen, and then decreased at fixed increments until it is seen again. The intensity of the last stimulus detected is taken to be the threshold [48]. However, the further away the initial estimate is from the actual threshold, higher the number of stimuli presentations required for accurate threshold estimation and consequently longer tests. Such issues are overcome by methods such as SITA where an initial estimate based on a prior threshold distribution obtained from historical patient data [49]. When such a patient database is not available, methods such as Correlated Neighbourhood Thresholding (CNT) algorithm can be used to reduce test duration without compromising accuracy [43].

7.3.4.5 Fixation Loss Measurement

Fixation loss attempts to evaluate when the eyes of the test-taker are not focussed on the centre dot and this loss of fixation can skew results resulting in an inaccurate heatmap and subsequent visual field loss assessment. In the HFA fixation, loss is typically tracked using a camera that is focused on the patient's eye.

In the VR-based perimeter, there is no camera in the headset to track eye movements. Fixation losses can be estimated using the Heijl–Krakau (HK) method. Alternatively, higher end portable perimeters employ a camera to track eye movements, thereby detecting any fixation losses. However, this drives up the cost of the portable perimeter apparatus. In either case, when the fixation loss as characterized by eye movement or number of detections at the PBS exceeds a particular medically defined threshold, the test may need to be reset and restarted.

The next section gives an insight into this through a validation experiment conducted with an actual hospital.

7.3.4.6 Experiments

Data collection was carried out at a leading eye hospital in India to compare the performance of the VR-based perimeter and HFA [43]. Tests were conducted by technicians who were trained to operate both the devices. The visual fields of 10 patients (both eyes)—healthy and glaucomatous (7 Male, 3 Female, Ages 25–75 years)—were compared. Patients selected had prior experience with the HFA. They were given clear instructions and a short demonstration on how to use the VR-based perimeter system. Each patient was tested on the HFA and VR-based perimeter on the same day with the 24-2 full threshold and CNT protocol, respectively. The average test time per eye (including breaks) was 11.45 min. The coherence between visual field measurements was evaluated based on the average mean absolute difference in thresholds (MAD) and sensitivity/specificity in identifying defective regions, i.e. regions where the threshold estimated by the HFA is less than 20 dB. Note that VR-based perimeter can measure a minimum threshold of 8 dB. Hence, for the purpose of comparison, thresholds below 8 dB on the Humphrey were taken to be 8 dB.

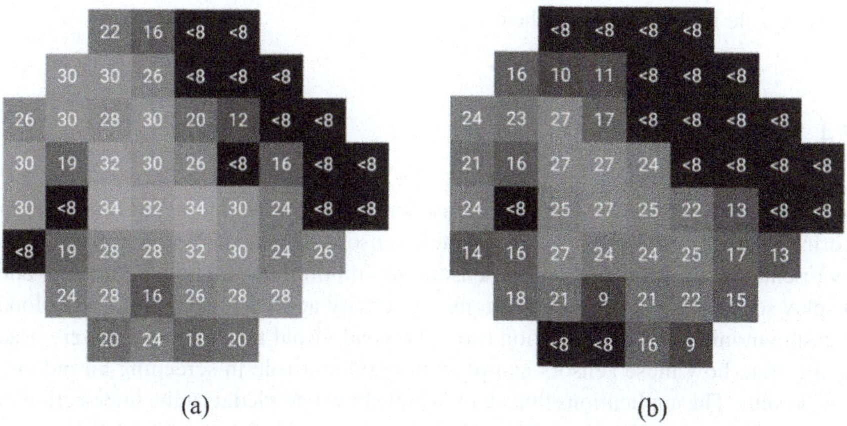

Fig. 7.13 Thresholds measured over the visual field of a left eye with advanced glaucoma. Darker shade indicates low sensitivity. **a** Threshold heatmap obtained on VR-based perimeter **b** Threshold heatmap obtained on the HFA

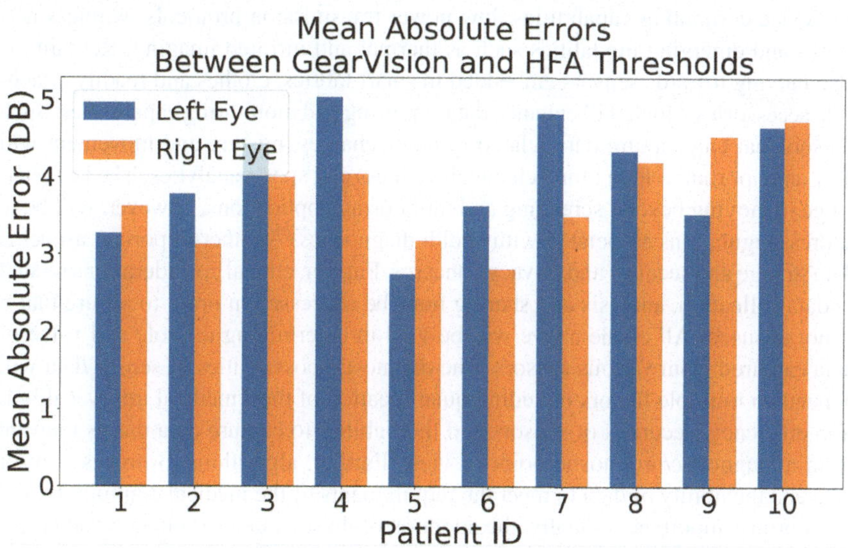

Fig. 7.14 Mean absolute error between VR-based perimeter and HFA thresholds (dB)

7.3.4.7 Results

Figure 7.13 shows the heatmaps of estimated thresholds obtained on VR-based perimeter and HFA for eyes affected by advanced glaucoma. Clear coherence between the visual field measurements can be observed. The average Mean Absolute Error (MAD) in threshold estimates was observed to be 3.77 ($\pm$1.64) dB across 35 eyes (18 patients) (Fig. 7.14).

The above points to the feasibility of using VR apparatus to estimate visual field. As the VR apparatus becomes increasingly sophisticated, it is expected to play a growing role in vision and eye health [40].

7.4 Conclusion

In this chapter, we discussed different aspects of health, wellness and fitness monitoring involving smartphone and wearable sensors. The discussion was based on a representative classification of the sensors into motion, physiological, ambient and display sensors. Examples applications for activity and exercise monitoring, blood pressure monitoring, UV radiation estimation and visual field estimation were used to illustrate how these sensors can play an important role in screening an individual's health. The applications thus demonstrate how knowledge at the intersection of health and well-being, computing, data processing and information technology can be effectively leveraged. Moving forward advancements are expected to augment the current paradigms in monitoring and screening health parameter such as the use of flexible sensors and non-contact sensors with decreasing form factors, increasing on-device computing capabilities, low power transmission protocols, wireless networks and emerging modalities such as thermal and infrared imaging. Research is also moving towards sensors embedded in smart fabrics, clothes and readily attachable accessories which will enhance data capturing and monitoring capabilities. Such sensors that can capture data related to biomechanics, posture and movement will play an important role in biomechanical wellness and sports analytics. The key challenge to moving beyond screening and monitoring applications, however, will be to address regulations associated with health diagnostics. Another important aspect is data storage and security and privacy concerns. Further, ethical considerations related to data collection, analysis and sharing must be addressed in order to ensure that it is not misused. All of the above will be vital in determining the role and value of data captured from various sensors. The diagnostic possibilities of sensor data will depend on multiple factors including quantification of their medical utility, calibration efficiency, accuracy of sensors and their ability to capture data that is relevant from a diagnostic/prognostic point of view. Further, algorithms to process sensor data and the ability of data to meet the requirements of the medical community will be of prime importance. Finally, the adoption of these systems (helped by advances in mobile communication devices) and acceptance in the general populace and medical community will play an important role in their proliferation. The success of any sensor-based system in graduating towards diagnostic applications will thus involve successfully addressing the above aspects.

Acknowledgements The authors wish to thank Khaidem Suman for his contributions towards preparing various figures in the chapter.

References

1. Flodgren G, Rachas A, Farmer AJ, Inzitari M, Shepperd S (2015) Interactive telemedicine: effects on professional practice and health care outcomes. Cochrane Database Syst Rev 2015(9)
2. Majumder S, Mondal T, Jamal Deen M (2017) Wearable sensors for remote health monitoring. Sensors 17(1)
3. Martin Cooper Wikipedia entry. https://en.wikipedia.org/wiki/martin_cooper_(inventor)
4. GSM Arena Website. https://www.gsmarena.com/
5. Sharma M, Anand A, Srivastava R, Prakash D, Kaligounder L, Tiwari VN, De A (2018) Method and system for comprehensive tennis serve analysis using wearables. Patent Application: India 10/2018
6. Narayanan R, Tiwari VN, Sahoo S, Nayak MM, Venkatesan SM, Aloknath DE, Jilla V, Lee CH, Ramakrishnan S, Narasimhamurthy R, Prasad A (2016) Method and apparatus for health care. Patent Application: US20160081620A1
7. Chan CB, Ryan DA, Tudor-Locke C (2004) Health benefits of a pedometer-based physical activity intervention in sedentary workers. Prev Med 39(6)
8. Bajpai A, Jilla V, Tiwari VN, Venkatesan SM, Narayanan R (2015) Quantifiable fitness tracking using wearable devices. In: 37th annual international conference of the IEEE engineering in medicine and biology society (EMBC)
9. Dey J, Gaurav A, Tiwari VN (2018) Instabp: cuff-less blood pressure monitoring on smartphone using single ppg sensor. In: 40th annual international conference of the IEEE engineering in medicine and biology society (EMBC)
10. Sahoo S, Tiwari VN, Narayanan R (2016) Endurance based personalized fitness planner. In: 38th annual international conference of the IEEE engineering in medicine and biology society (EMBC)
11. Mayya S, Jilla V, Tiwari VN, Nayak MM, Narayanan R (2015) Continuous monitoring of stress on smartphone using heart rate variability. In: IEEE 15th international conference on bioinformatics and bioengineering (BIBE)
12. Datta S et al (2017) Novel statistical post processing to improve blood pressure estimation from smartphone photoplethysmogram. In: Proceedings of the first international workshop on human-centered sensing, networking, and systems
13. Gavish B, Ben-Dov IZ, Bursztyn M (2008) Linear relationship between systolic and diastolic blood pressure monitored over 24 h: assessment and correlates. J Hypertens
14. Bhowmik T, Dey J, Tiwari VN (2017) A novel method for accurate estimation of hrv from smartwatch ppg signals. In: 2017 39th annual international conference of the IEEE engineering in medicine and biology society (EMBC)
15. Venkatesan SM, Tiwari VN (2019) Method for generating 3d biometric model of body part of user and electronic device thereof. Patent Application: WO2019017686A1
16. Bansal A, Kumar S, Bajpai A, Tiwari VN, Nayak M, Venkatesan S, Narayanan R (2015) Remote health monitoring system for detecting cardiac disorders. IET Syst Biol
17. Raghavendra BS, Bera D, Bopardikar AS, Narayanan R (2011) Cardiac arrhythmia detection using dynamic time warping of ecg beats in e-healthcare systems. In: IEEE international symposium on a world of wireless, mobile and multimedia networks. IEEE

18. Nayak MM, Narayanan R, Tiwari VN, Lee HC, Saluja P (2016) Method and system for displaying electrocardiogram (ecg) data on a display device. Patent Application: India 10/2016
19. Dey S, Sahoo S, Agrawal H, Mondal A, Bhowmik T, Tiwari VN (2017) Personalized cumulative uv tracking on mobiles and wearables. In: 39th annual international conference of the IEEE engineering in medicine and biology society (EMBC)
20. Tiwari VN, Venkatesan SM, Narayanan R (2018) Electronic device and method for providing information related to skin type of object. Patent Application: US20180199856A1
21. Tiwari VN, Dey S, Narayanan R, Sahoo S, De A (2018) Electronic device and method for providing information of uv dose thereof. Patent Application: WO2018052271A1, US20180080818A1
22. Sircar T, Mishra A, Bopardikar A, Tiwari VN (2018) Gearvision: smartphone based head mounted perimeter for detection of visual field defects. In: 2018 40th annual international conference of the IEEE engineering in medicine and biology society (EMBC). IEEE, pp 5402–5405
23. Tiwari VN, Bose J, Bopardikar A, Sircar T, Raykar DA, Mishra A, Bhaduri A, Nandan A (2018) Systems and methods for determining defects in visual field of a user. Patent Application: WO 2018169330A1
24. Matsumoto C, Yamao S, Nomoto H, Takada S, Okuyama S, Kimura S, Yamanaka K, Aihara M, Shimomura Y (2016) Visual field testing with head-mounted perimeter imo. PLoS ONE 11:08
25. Wroblewski D, Francis BA, Sadun A, Vakili G, Chopra V (2014) Testing of visual field with virtual reality goggles in manual and visual grasp modes. BioMed Res Int
26. Kumar S, Bansal A, Tiwari VN, Nayak MM, Narayanan RV (2014) Remote health monitoring system for detecting cardiac disorders. In: IEEE international conference on bioinformatics and biomedicine (BIBM)
27. Janssen I, LeBlanc AG (2007) Systematic review of the health benefits of physical activity and fitness in school aged children and youth. Int J Behav Nutr Phys Activ 7(40)
28. O'Donovan G et al (2010) The abc of physical activity for health: a consensus statement from the british association of sport and exercise sciences. J Sports Sci 28(6)
29. Maheedhar M, Gaurav A, Jilla V, Tiwari VN, Narayanan R (2016) Stayfit: a wearable application for gym based power training. In: 38th annual international conference of the IEEE engineering in medicine and biology society (EMBC)
30. Varughese GI, Lip GY (2005) Goodbye mercury? blood pressure measurement and its future. J R Soc Med
31. Gaurav A, Maheedhar M, Tiwari V, Narayanan R (2017) Method and electronic device for cuff-less blood pressure (bp) measurement. US20170238818A1
32. Dastjerdi AE, Kachuee M, Shabany M (2017) Non-invasive blood pressure estimation using phonocardiogram. In: IEEE international symposium on circuits and systems (ISCAS)
33. McCombie DB, Reisner AT, Asada HH (2006) Adaptive blood pressure estimation from wearable ppg sensors using peripheral artery pulse wave velocity measurements and multi-channel blind identification of local arterial dynamics. In: 28th annual international conference of the IEEE engineering in medicine and biology society (EMBC)
34. Tiwari VN, Venkatesan SM (2018) System and method for measuring vital parameters using dual camera. Patent Application: India 48/2018
35. Fung P, Dumont G, Ries C, Mott C, Ansermino M (2004) Continuous noninvasive blood pressure measurement by pulse transit time. In: 26th annual international conference of the IEEE engineering in medicine and biology society (EMBC)
36. Gaurav A, Maheedhar M, Tiwari VN, Narayanan R (2016) Cuff-less ppg based continuous blood pressure monitoring: a smartphone based approach. In: 38th annual international conference of the IEEE engineering in medicine and biology society (EMBC)
37. 2002 Global Solar UV index: a practical guide. World Health Organization. https://www.who.int/uv/publications/en/uviguide.pdf
38. Ales TM, Timmers R, Long AM, Sullivan SJ, Ranganathan S, Obermann TR, Johnson ED (2011) Uv detection devices and methods. Patent no US8044363B2

39. Cheuk M, Xu D, McLean R (2014) Delivery of personal ultraviolet radiation information to smartphones. In: NIWA UV workshop
40. Goyal A, Bopardikar A, Tiwari VN (2018) Estimation of spherical refractive errors using virtual reality headset. In: 2018 40th annual international conference of the IEEE engineering in medicine and biology society (EMBC). IEEE
41. Goyal A, Tiwari VN, Sircar T, De A (2019) Method for determining refractory power of eye using immersive system and electronic device thereof. Patent Application: WO2019156485A1
42. Habtamu E, Bastawrous A, Bolster NM, Tadesse Z, Callahan EK, Gashaw B, Macleod D, Burton MJ (2019) Development and validation of a smartphone-based contrast sensitivity test. Transl Vis Sci Technol 8(13)
43. Sircar T, Pradhan Z, Bopardikar AS, Rao HL, Agrawal H, Tiwari VN (2019) Development and clinical validation of gearvision- a smartphone based head mounted perimeter. In: IEEE India council international conference (INDICON)
44. WHO Blindness and Visual Impairment. http://www.who.int/news-room/fact-sheets/detail/blindness-and-visual-impairment
45. Standard Automated Perimetry. https://eyewiki.aao.org/standard_automated_perimetry
46. Hollander D, Volpe N, Moster M, Liu G, Balcer L, Judy K, Galetta S (2000) Use of a portable head mounted perimetry system to assess bedside visual fields. Br J Ophthalmol 84(10):1185–1190
47. Zeyen T (1997) Interpretation of automated perimetry. Bulletin de la Societe belge d'ophtalmologie 267:191–197
48. Johnson CA, Chauhan BC, Shapiro LR (1992) Properties of staircase procedures for estimating thresholds in automated perimetry. Invest Ophthalmol Vis Sci 33:2966–2974
49. Bengtsson B, Olsson J, Heijl A, Rootzen H (1997) A new generation of algorithms for computerized threshold perimetry, sita. Acta Ophthalmologica Scandinavica 75:368–375
50. ISO Standards (1999) Ophthalmic instruments - perimeters
51. Samsung S8 Luminance. https://m.gsmarena.com/samsung_galaxy_s8-review-1603p3.php

Chapter 8
Sleep Monitoring in Adults Using Wearables and Unobtrusive Technology

Margot Deviaene, Dorien Huysmans, Ivan D. Castro, Pascal Borzée, Dries Testelmans, Bertien Buyse, Sabine Van Huffel, and Carolina Varon

Abstract Sleep is a complex physiological process that plays a fundamental role in maintaining homeostasis and overall health. It has an internal structure characterized by sleep stages, which is often affected by either the high demands of the current 24-h society or by different sleep disorders such as sleep apnea. These disturbances to the regular sleep structure have been strongly associated with reductions in cognitive and behavioral performance, attention deficit, depression, nocturia, memory loss, snoring, and cardiovascular diseases. Therefore, it is crucial to identify sleep problems in an early stage before the overall health is compromised in an irreversible way. Currently, sleep disorders are diagnosed using polysomnography (PSG), which is the gold-standard sleep test usually recorded in a sleep laboratory. This test is

M. Deviaene–D. Huysmans—These authors have equal contributions to the work.

M. Deviaene · D. Huysmans · S. Van Huffel (✉) · C. Varon
KU Leuven, Department of Electrical Engineering (ESAT), STADIUS Centre for Dynamical Systems, Signal Processing and Data Analytics, Leuven, Belgium
e-mail: sabine.vanhuffel@esat.kuleuven.be

M. Deviaene
e-mail: margot.deviaene@esat.kuleuven.be

D. Huysmans
e-mail: dorien.huysmans@esat.kuleuven.be

C. Varon
e-mail: carolina.varon@esat.kuleuven.be

I. D. Castro
IMEC-Leuven, Circuits and Systems for Health, Leuven, Belgium
e-mail: ivand.castro@imec.be

P. Borzée · D. Testelmans · B. Buyse
UZ Leuven, Department of Pneumology, Leuven University Centre for Sleep and Wake Disorders, Leuven, Belgium
e-mail: pascal.borzee@uzleuven.be

D. Testelmans
e-mail: dries.testelmans@uzleuven.be

B. Buyse
e-mail: bertien.buyse@uzleuven.be

G. D. Gargiulo and G. R. Naik (eds.), *Wearable/Personal Monitoring Devices Present to Future*, https://doi.org/10.1007/978-981-16-5324-7_8

often associated with elevated costs and reduced comfort. With this in mind, many studies have focused on the development of wearables and unobtrusive technologies that can be used at home and that can monitor sleep during more than one single night. This chapter discusses unobtrusive state-of-the-art sensors and algorithms for sleep monitoring in adults, with a special focus on heart rate, respiration, and blood oxygenation monitoring.

8.1 Physiological Background of Human Sleep

Sleep is a natural, though complex process, which follows an internal architecture of alternating states. As such, sleep cannot be described by a single state of the body. Instead, it consists of different sleep stages. These sleep stages are associated with characteristic patterns at cerebral, cardiac, and respiratory levels. However, sleep disorders could alter these characteristics. Therefore, the detailed analysis of one's sleep architecture could serve the detection of these sleep disorders.

8.1.1 Sleep Stages

Sleep stages and their characteristics were first defined by Rechtschaffen and Kales (R&K) in 1968 [132]. Later in 2007, the American Academy of Sleep Medicine (AASM) updated these R&K rules and published a manual for sleep scoring and associated events [26]. These sleep scoring rules are based on patterns and wave characteristics found in the electroencephalogram (EEG),[1] the electrooculogram (EOG),[2] and the chin electromyogram (EMG).[3] To facilitate the analysis, input signals are scored in consecutive windows of 30s, which are referred to as epochs [132]. Every epoch is scored with one of the five sleep stages defined by the AASM. These stages are Wakefulness (W), Rapid Eye Movement sleep (REM sleep), and non-REM (NREM) sleep 1, 2, and 3 (respectively N1, N2, and N3). Usually, stages N1 and N2 are referred to as *light* sleep and N3 as *deep* sleep [138].

Apart from patterns in the EEG, EOG, and EMG signals, differences in sleep stages are reflected in the regulation of both branches of the autonomic nervous system (ANS), namely, the parasympathetic nervous system (PNS) and the sympathetic nervous system (SNS) [104]. As such, distinct characteristics can be observed as well at cardiac, respiratory, and cardiorespiratory levels during NREM and REM sleep. These characteristics have been exploited for the development of ambulatory systems

[1] The EEG captures the electrical activity from the brain, commonly obtained from the scalp using surface electrodes.

[2] The EOG records the electrical signal caused due to the opposite polarity between the front and back of the eye, which acts as a dipole.

[3] The EMG records the electrical activity of the muscles.

for sleep monitoring. The reason for this is that cardiac and respiratory information can be extracted from the Electrocardiogram (ECG),[4] which can be easily recorded using wearable and unobtrusive technology. This is exactly the focus of this chapter; therefore, an overview of these characteristics or physiological changes during sleep stages is presented next.

8.1.1.1 NREM Sleep

It is very well-known that during NREM sleep, the PNS activity dominates over the SNS [104]. Compared to wake, the breathing frequency increases, though the variability is reduced. This is accompanied by a reduction in tidal volume,[5] resulting in a breathing which is more shallow and rapid [57]. Between different NREM stages, there is no significant difference in tidal volume and breathing frequency; however, respiration becomes more regular during deep sleep compared to light sleep [57].

Due to the high activity of the PNS during NREM, bradycardia emerges and the heart rate reaches a minimum during N3 [104, 153]. Additionally, heart rate variability (HRV) is lower during N3 compared to REM and wake, and regular oscillations can be observed during N3. These oscillations are associated with respiratory sinus arrhythmia (RSA), which is the modulation of the heart rate (HR) with respiration. Spectral analysis of the tachogram during NREM reveals a decrease in the low-frequency (LF) band (0.04-0.15 Hz) and an increase in the high-frequency (HF) band (0.15–0.4 Hz) of the HRV [165].

8.1.1.2 REM Sleep

REM sleep is characterized by muscle atonia, increased physiological activity, and the act of dreaming. During this sleep stage, the PNS is more active compared to the wake. Nevertheless, phasic fluctuations in SNS and PNS activity occur during REM sleep. As such, important distinctions exist between tonic and phasic REM sleep epochs [114]. During tonic REM, the SNS activity drops even below NREM levels, while during phasic REM, the SNS becomes very active and variable [104]. Characteristic rapid eye movements are also only present during phasic periods.

On a respiratory level, both tonic and phasic REM exhibit a decrease in ventilation due to a reduction in ventilatory drive. In general, the respiratory system becomes unstable as the depth of breathing becomes highly variable [57].

A general increase in cardiovascular instability is an important feature of REM sleep. During tonic REM sleep, there is marked bradycardia and hypotension, resulting in a decrease in HR and blood pressure (BP) even below levels of NREM sleep. On the other hand, phasic REM sleep epochs are characterized by great transient increases in HR and BP, produced by a phasic increase in the SNS [20].

---

[4] The ECG records the electrical activity of the heart.

[5] Tidal volume corresponds to the volume of air inspired/expired with each breathing cycle.

These phasic fluctuations during REM sleep result in the instability of cardiovascular and respiratory parameters. Therefore, REM sleep epochs can exhibit a great variety in cardiac and respiratory characteristics among each other. As a consequence, the detection of REM sleep epochs presents a more challenging task for automated sleep scoring algorithms compared to deep sleep.

8.1.2 Sleep Architecture

For healthy persons, a normal night of sleep consists of five to six cycles, where REM and NREM phases are alternated with occasional awakenings. A hypnogram is the visualization of a person's sleep architecture over time. The hypnogram of a healthy adult is depicted in Fig. 8.1a. One cycle typically lasts for 90–110 min. An initial wake period is followed by light sleep, where a person transients from N1 to N2. Thereafter, deep N3 sleep is reached. The cycle is terminated with a phase of REM sleep. In this fashion, NREM and REM alternate throughout the night,; however, their relative distribution changes. The duration of NREM sleep decreases and is compensated by an increase in REM sleep. Within NREM sleep, the portion of N3 will drop, though replaced by lighter N2 sleep. During final cycles, N3 may not even occur. The average distribution of different sleep stages throughout the night is given in Table 8.1. A longer sleep time will lead to an increase in REM sleep. The effect of aging will lead to a decrease in total sleep time as awakenings will occur more often. In patients with obstructive sleep apnea (OSA), the amount of REM and N3 sleep will be heavily affected due to arousals, which result in fragmented sleep. A hypnogram of an adult with severe OSA is depicted in Fig. 8.1b.

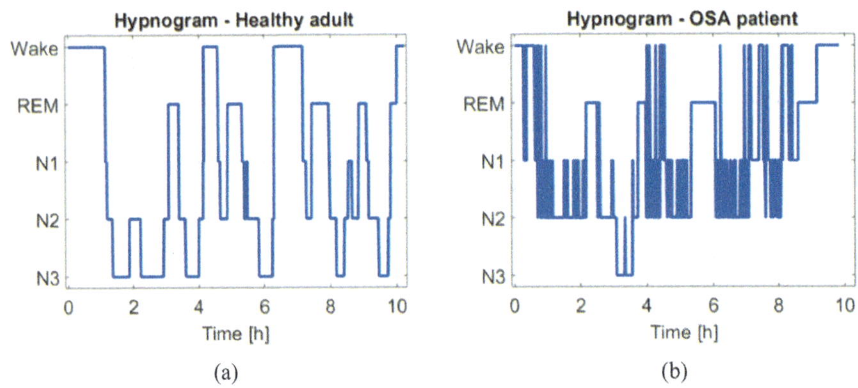

Fig. 8.1 **a** Hypnogram of a healthy adult. **b** Hypnogram of an adult with severe OSA

Table 8.1 Average sleep stage distribution for a healthy young adult

Sleep stage	Wake	REM	NREM	N1	N2	N3
Time in bed	5%	20–25%	75–80%	2–5%	45–55%	13–23%

8.2 Polysomnography at a Sleep Laboratory

The gold-standard test in sleep medicine is polysomnography (PSG). This test is typically carried out for one night at a sleep laboratory during which several physiological signals are measured while the subject is sleeping with the aim of evaluating one or more aspects of their sleep. Within the most common purposes of a PSG study, one can find the evaluation of sleep quality, the identification of sleep stages, sleep-wake activity, and the diagnosis of sleep disorders such as OSA.

It is still common clinical practice to evaluate the PSG by manual scoring of sleep stages and events (e.g., apneas) according to the AASM rules. By using multiple physiological and contextual signals recorded in a PSG, characteristic patterns can be identified, allowing to score these events. The most common signals recorded during a PSG test are listed below. In some cases, these signals can vary depending on the specific sleep laboratory.

EEG: This set of signals is mainly useful in the identification of sleep stages, and hence helps to determine arousals and the wakefulness of the patient. This is done by evaluating characteristic waves in different frequency domains: Delta (3 Hz or lower), Theta (3.5–7.5 Hz), Alpha (7.5–13 Hz), and Beta (14 Hz or greater). For example, Delta activity is observed during N3 sleep, Theta activity is characteristic of REM sleep, Alpha activity is used as a marker of relaxed wakefulness, and Beta activity is often observed during wakefulness and drowsiness.

EMG: In a PSG test, the EMG can be obtained from different locations such as the chin and the limbs. The measurement of chin EMG provides an indication of muscle tone, which is reduced with sleep onset and is very low during REM sleep [15], whereas limb EMG can help to identify periodic limb movements, which can be an additional cause of sleep disturbance [15].

EOG: The potential obtained in this signal provides an objective measurement of eye movement and is useful in the identification of REM sleep and the sleep onset. Sleep onset is characterized by slow and rolling eye movements with a symmetric onset and offset, whereas in REM sleep, sharp rapid eye movements can be observed which are more asymmetric with a fast onset and a slow offset [15].

ECG: At least 1 ECG lead (commonly lead II) is recorded during PSG. ECG signals are useful for the identification of cardiac comorbidities but can also provide information on disturbances of the heart rhythm caused by sleep-related disorders.

Audio: Recording sounds by using a microphone placed in the neck near the trachea can be used to detect snoring and other tracheal sounds [15]. Recording sound in the PSG room (e.g., the audio registration of the video), on the other hand, can be useful as additional information to verify the nature of arousals and other events.

Respiration: A direct measurement of airflow via a pneumotachograph is the gold-standard method to measure respiration and define apneas [62]. For this method, the use of a mask is necessary; therefore, it is only used in combination with therapeutic positive airway pressure (PAP) devices. In clinical practice, the combination of a nasal pressure sensor and an oronasal thermistor is the standard used for diagnostic PSGs. These measurements are more comfortable for the patient and provide an approximation of the airflow exchange. The end-tidal CO_2 is also an interesting parameter, this is the concentration of CO_2 at the end of an exhaled breath. Reliably measuring the end-tidal CO_2 is, however, difficult. Therefore, transcutaneous CO_2 measurements are often used.

Respiratory effort: This is commonly measured by belts around the abdomen and thorax using respiratory inductance plethysmography (RIP). These signals provide information related to the respiratory effort, and hence also play an important role when diagnosing breathing-related sleep disorders.

Oximetry: The standard measurement of the blood oxygen saturation (SpO_2) during a PSG is done by means of an optical measurement in transmission mode. This is typically done with a sensor at the fingertip but can also be measured in alternative locations (e.g., earlobe). The SpO_2 measurement has been defined as a mandatory signal for quantifying the apnea-hypopnea index (AHI) used to measure OSA severity [44] (together with respiration and EEG). This measurement is accompanied by a pulse photoplethysmography (PPG) signal, which also provides cardiac-related information such as HR.

Body position and behavioral observation: This provides additional diagnostic information, as some sleep disorders are influenced by the orientation during sleep [15]. Position can be monitored with sensors, but also by video recording, the latter being the most common in standard PSG.

The signals listed above are an important source of information that allows to obtain a complete picture of the person's physiology during sleep. This facilitates the task of diagnosing and following up sleep-related conditions. Nevertheless, it has several disadvantages related to patient comfort, the test cost, and the availability of sleep laboratory beds. The rather obtrusive sensors that need to be attached to the patient, combined with the unusual sleep setting and possible sensor verification from the nursing staff, significantly reduce patient comfort during the night. This leads to a suboptimal setting when aiming to evaluate a typical night of sleep. In addition, the PSG procedure implies a high cost, not only because of the equipment required for the recording of the signals and the sleep laboratory installations, but also because of the need for the test to be supervised and annotated by trained personnel. This high cost causes a limited number of beds available for PSG studies when compared to the high prevalence of sleep-related disorders. These limitations have motivated the development of sleep monitoring systems for home monitoring, which will not only tackle these challenges but will also allow multiple night monitoring. Devices for home monitoring have the potential to enable a broader patient screening, early detection of sleep-related conditions, and a longer follow-up.

8.3 Wearables and Unobtrusive Technologies for Sleep Monitoring at Home

8.3.1 Out of Center (Home) Poly(somno)Graphy Devices

Extended home sleep monitoring can be done using PSG-like devices adapted in size to enable its use at home. These measure a reduced set of the most important physiological signals from a standard in-lab PSG and can be denominated "Out of Center" (OOC) PSG portable devices [44]. When no EEG is included for the correct measurement of sleep, the devices are classified as polygraphy (PG) rather than PSG.

Multiple of these are currently available in the market and can be classified according to the number of signals that are measured. An initial classification scheme based on this principle was proposed by the AASM in 1994 [40]. An updated AASM classification scheme was proposed by Collop et al. [44] in 2011, with the aim to better classify new emerging portable sleep monitoring devices. This classification is done based on measurements of Sleep, Cardiovascular, Oximetry, Position, Effort, and Respiratory (SCOPER) parameters. For a review on some of the OOC devices, the reader is referred to [44, 62].

An example of a PG OOC device is shown in Fig. 8.2. These devices allow to monitor some of the main PSG signals at the cost of a certain degree of discomfort for the patient, as most require strapping devices to the chest or head and even using a nasal cannula. This remaining discomfort can limit the monitoring time and affect the measurement.

Fig. 8.2 Example of portable PG OOC device for home sleep monitoring. Courtesy of Philips Respironics, all rights reserved

The need for an even higher comfort than the one provided by PG OOC devices, lower cost, and longer monitoring time has led to the recent development of wearables and sensors placed around the patient. These sensors not only measure less signals than an in-lab PSG, but also measure these either in a less obtrusive way or by applying an indirect measurement of some of the physiological parameters. This has been accompanied by the development of application-specific algorithms based on robust signal processing and machine learning techniques, to monitor some of the main sleep disorders such as OSA.

8.3.2 *Home Sleep Monitoring with Unobtrusive Sensors*

With the advance and miniaturization of electronics over the last decades, more compact and less obtrusive devices have been developed. These have further enabled the monitoring of physiological signals in non-clinical settings in the form of wearables and sensors around the patient. Application-specific integrated circuits have played a crucial role in the electronics being used for ambulatory healthcare monitoring.

The main areas of wearables and unobtrusive technologies that are available for sleep-related monitoring in non-clinical settings (i.e., at home) with an increased comfort include the following: Actigraphy, portable EEG devices, portable ECG devices and patches, ballistocardiography (BCG) devices, radar-based monitoring, and capacitively coupled biopotentials. It is worth noting that these technologies vary in the accuracy of the target measurements and their relevance to specific sleep-related disorders. These bring different levels of increased comfort, while in some cases compromising the sensitivity to artefacts caused by motion, sensor positioning, and other uncontrolled factors typical in a home setting.

Although important work has been done in the evaluation of the usefulness of some of these technologies related to specific sleep monitoring purposes, additional validation is still required in order to assess the performance of these technologies on their own, or combined with others, and of application-specific algorithms that have been developed for these. An overview of the different wearable sensor technologies listed above is provided next.

8.3.2.1 Actigraphy

Actigraphy devices aim to measure the movement from the limbs and/or torso of the person during sleep. The measurement of movement has been identified as important in sleep-related conditions as it can provide information regarding the state of the person, which can be related to physiological changes [105]. In this way, different types of movements can be identified as normal or abnormal and can give information which, if correctly analyzed, can aid in specific diagnosis and sleep characterizations. Parameters that can be estimated from actigraphy measurements include sleep quality, latency, duration, efficiency, fragmentation, circadian rhythm, sleep-wake periods, and activity levels [105].

The AASM indicates in their guidelines that actigraphy is reliable in measuring sleep for healthy adults [120]. Nevertheless, these devices only allow to gather general sleep information, as detailed data of sleep-wake physiology are not included [27]. In addition, there are several limitations in different areas including sensor placement, the number of axes and sensors, sensitivity for wake periods [27, 105], data quantification modes, validation, and scoring algorithms. A more detailed description of these limitations is provided in [105], as well as an overview of the main contributors to actigraphy scoring algorithms.

The most common measurement of actigraphy is the use of a multi-axial accelerometer worn around the wrist (i.e., wrist actigraphy). Other locations that have been studied include the diaphragm, the chest, the leg/ankle, and the trunk [21, 79]. In addition, some smartphone applications use the information from the accelerometer included in the phone when the latter is placed on top of the mattress, but these are expected to be less accurate [27].

Examples of commercially available wrist-based actigraphy devices include the Apple Watch, Biostrap, Empatica, Fitbit, Garmin Vivosmart, Whoop, Xiaomi, Lark, and Sleep Tracker. Currently, there is a high number of actigraphy wrist-worn devices in the market [105, 140], most of which measure at least one additional physiological parameter, with PPG-based heart rate being the most common one. The main differentiators between actigraphy devices are as follows: (a) the availability of raw data for the development of new algorithms, (b) the type of scoring algorithm tackling a specific sleep-related condition, and (c) proper clinical validation against the PSG gold standard.

8.3.2.2 Portable EEG Devices

In contrast with actigraphy measurements, EEG can provide more physiological information. The challenge in portable EEG monitoring is that the unobtrusiveness of these devices is rather limited or implies a compromise in the type of EEG signals that can be acquired, depending on electrode positioning and electrode-tissue interface. Standard EEG monitoring at the PSG laboratory is performed with glued-on electrodes. Frontal, central, and occipital derivations are recorded. A full EEG with a complete "10–20" electrode set is barely performed in sleep labs. Nonetheless, the reduced electrode set that is used still implies discomfort to the patient and requires installation by trained personnel.

Because of the lower comfort of using a hat-like device at home during multiple nights, the difficulty for correct electrode placement, and considering the added value of monitoring EEG for sleep-related disorders, more compact devices have been designed. These devices aim to monitor a limited set of EEG data with a reduced discomfort for the patient. Some of these are available as individually glued electrodes [45] and collect one or two EEG channels together with other PSG signals, but still require training for a correct installation; these could be categorized as OOC PSG-like devices as described in Sect. 8.3.1.

Other more portable options include the use of headbands [2, 4–6, 9] which also record EEG signals from a subset of dry electrodes distributed in the band. Although these collect EEG, not all of them provide the raw data, as some focus on their functionality as a sleep tracker or "sleep coach", with the purpose of providing feedback on the activities previous to sleep, guide respiration exercises, and emit tones in different frequencies, among other feedback mechanisms that aim to increase the quality of sleep.

Hardware implementations with even higher comfort have been studied by acquiring the EEG signals from around-ear [28, 50] and in-ear [69, 82, 113] electrodes. Some commercial products include these implementations with purposes of productivity increase [172] or as an application-independent platform [107]. Ear EEG has even been tested for sleep monitoring purposes. Results indicate that the automatic sleep scoring using these sensors can reach an accuracy close to that achieved by manual scoring of scalp EEG [113].

8.3.2.3 Portable ECG Devices and Patches

Portable ECG monitoring has seen a big advancement in the last decade. Holter monitors that allow to record up to a 12-lead ECG have become smaller, and hence these could be used as part of a solution to perform sleep monitoring at home. Similarly, small form-factor PG devices currently offer the capability of monitoring multiple ECG leads. The main disadvantage in comfort that a Holter or PG device implies relies on the use of wires connecting individual contact electrodes with the recorder unit. To overcome this, miniature ECG recorders have been developed in the form of a chest patch. These allow to conveniently monitor one lead (or few leads) without the discomfort of using cables by using a relatively small patch as the one shown in Fig. 8.3. This type of patch can be worn for up to 1 or 2 weeks and enable a more comfortable way of ECG monitoring, which could be of added value for home sleep monitoring. Within the available ECG patches in the market, one can find the MCOT [75] from Biotelemetry inc., Zio XT [157] from iRythm, and an investigational device from VivaLNK [152], among others.

As an alternative to ECG patches, there are also solutions based on tight chest bands and t-shirts that use dry electrodes, which could also be a source of ECG during a night of sleep monitoring.

These signals also have the potential to be a source of respiratory activity by computing the ECG-derived respiration (EDR), when aiming to perform home sleep monitoring with a reduced subset of signals.

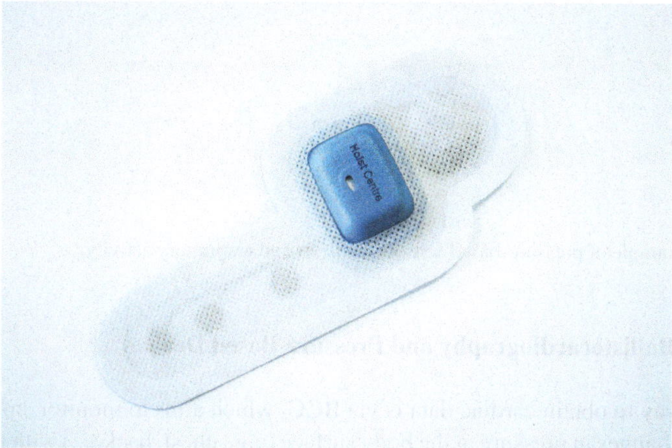

Fig. 8.3 Example of ECG chest patch

8.3.2.4 Optical Pulse Monitoring: PPG and SpO_2

Another way of measuring cardiac activity is by means of an optical pulse readout using PPG in either transmission or reflection mode. This allows to obtain the pulse information, hence providing a source of heart rate monitoring. In addition, given an adequate location of the sensor, the use of the correct wavelengths, and dedicated algorithms, SpO_2 can also be monitored.

Besides the standard finger-based PPG/SpO_2 monitors, wearables monitoring PPG generally have the form factor of a smartwatch or smart bracelet with an optical readout in the back. There are currently multiple commercial offerings of smartwatches that monitor PPG, including brands such as Fitbit, Apple, Huawei, Samsung, Garmin, and Polar among others. Some of these have lately added SpO_2 functionality to their offering and are even seeking FDA approval. This addition could play an important role in home monitoring of sleep-related conditions such as OSA. A review of developments and challenges of wearable PPG is provided in [32]. Other investigational devices available in the market that measure PPG signals include small wearable units such as the ones offered by Byteflies [3], which also offer units that can be used for ECG monitoring in the form of a patch.

Compared to ECG signals, PPG signals have a less sharp characteristic and may be more challenging to process when trying to obtain accurate beat-to-beat HR and HRV metrics, but these have the advantage of potentially enabling SpO_2 measurements, which is of added value in sleep-related conditions.

Fig. 8.4 Example of pressure-based sensors for heart and respiratory activity

8.3.2.5 Ballistocardiography and Pressure-Based Devices

Another way to obtain cardiac data is via BCG, which aims to monitor small movements or changes in pressure at the body surface (e.g., chest, back, ...) with the aim of deriving HR and even trying to obtain beat-to-beat HR and HRV. The same principle can be used to monitor respiratory activity, as this causes a change in pressure of the torso in the bed, for the case of sleep monitoring.

The advantage of being able to obtain respiratory and cardiac-related signals from one type of sensor has led to multiple companies offering pressure-based sleep monitoring solutions. The company Beddit (now bought by Apple) offers a piezoelectric-based sensor [1] that monitors average HR, average breathing rate, and snoring sounds (recorded via the microphone of the connected smartphone) and provides metrics of sleep time, bedtime, time to fall asleep, time away from the bed, wake-up time, and sleep efficiency. Similarly, Withings offers a sleep tracking mat [170] to be placed under the mattress, which detects snoring, performs heart rate and respiration rate tracking, and provides a sleep quality assessment within a coaching program. The company mentions it can also help identify signs of OSA. Another mat-based solution is offered by Emfit, with a sensor also placed below the mattress that aims to measure HRV, respiratory activity, and respiratory effort [59]. This sensor is shown in Fig. 8.4.

BCG setups for heart activity and pressure-based respiratory activity have thus the advantage of not requiring direct contact with the body. Nevertheless, the extracted heart activity tends to be less accurate than that of an ECG signal (and even of PPG signal) due to the motion-based measurement, as opposed to the electrical or optical-based measurements of the ECG and PPG. In addition, motion during sleeping can distort the measurements, and patients with high BMI are likely to cause sensor saturation [72]. The question of whether these unobtrusive signals can be used for monitoring specific sleep-related conditions needs to be answered with real-life validation studies.

8.3.2.6 Radar-Based Monitoring

A different technology that also aims to monitor the chest movement to extract respiratory and cardiac activity is the use of radar signals. This technology sends radiofrequency waves that are reflected in the person's skin. The phase of the signal is modulated by the physiological movement, generating a phase difference between the emitted and received signals, which is then used to calculate the distance changes. The waves are mainly pointed to the chest or the back of the torso. Electromagnetic signals of different frequencies have been used for this purpose, with higher frequency and power resulting in higher sensitivity to small displacements [123]. Carrier frequencies between hundreds of MHz up to more than 200 GHz have been used [92]. More details on the use of radar signals for vital signs monitoring can be found in the work of Kranjec et al. [86], Li et al. [92], and Mercuri et al. [111].

Radar technology has been demonstrated to be able to monitor respiratory and cardiac activity [91, 123, 128]. It has the advantage of monitoring through non-metallic obstacles [86] placed relatively far from the subject. Nevertheless, it is commonly affected by motion artefacts [92], and the acquisition of beat-to-beat HR is more challenging than when using an ECG signal.

Some radar-based solutions are available commercially, including the S+ device by ResMed [135], which monitors breathing and movement, and provides a sleep report based on these measurements. The device has not been tested for the monitoring of specific sleep-related conditions at the moment of writing this chapter, but it is mentioned that it monitors the sleep stages to create a personalized sleep chart. Similar radar-based devices have been tested with promising results regarding sleep stage classification [41].

8.3.2.7 Capacitively Coupled Biopotentials: ECG and Bioimpedance-Based Respiratory Activity

Considering the discomfort of using contact electrodes to monitor biopotentials (e.g., ECG), the capacitively coupled acquisition of these signals is a technology that has gained interest. The main physiological signals that could be useful in a sleep monitoring setting and can be acquired in a capacitively coupled manner include the capacitively coupled ECG (ccECG) and capacitively coupled bioimpedance (ccBIOZ) for respiration monitoring.

These measurements are done by replacing the skin-electrode galvanic contact with capacitive coupling, hence enabling monitoring through clothing and bedsheets. In this coupling, the skin forms one "plate" of the capacitor, a conductive surface forms the second "plate" of the capacitor, and any non-conductive materials between these conductive surfaces form the dielectric. This completes the standard structure of a capacitor, and hence the connection from the acquisition circuit to the skin is replaced from a galvanic connection to capacitive coupling. An illustration of a capacitively coupled electrode interface is shown in Fig. 8.5.

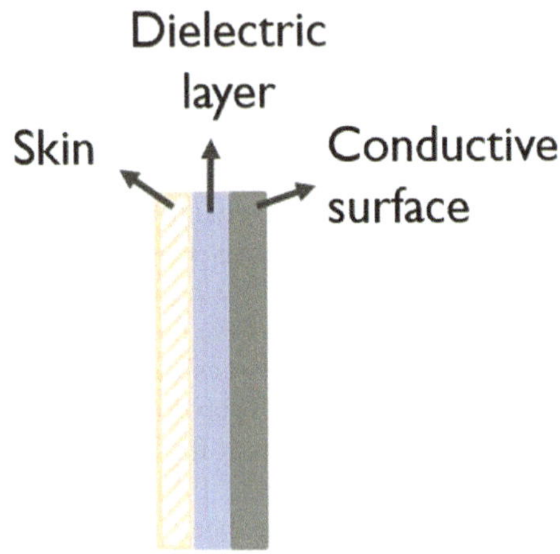

Fig. 8.5 Illustration of the structure formed by a capacitively coupled electrode interface for biopotential acquisition

ccECG acquisition has been explored since 1967 [137], but it is only in the last few decades that it has been more widely explored for a broad number of applications. These applications include sensors placed in the bathroom seat [19, 83, 97], in a wheelchair [129], in a car seat [38, 90, 106, 149], in an airplane sea [146, 147], in an office chair [13, 18, 94, 108], and in a bed [76, 95, 166, 171]. In the field of sleep monitoring, bed implementations have been tested during multiple hours [89, 95, 162], including comparison against polysomnography signals with reported HR coverages of up to 98% [89]. An initial evaluation toward the extraction of features that could be used in the identification of sleep apnea epochs was also done [34].

In the case of ccECG, the signal can be acquired using at least one pair of electrodes, typically accompanied by a third electrode for active noise canceling, denominated *driven right leg* (DRL). On the other hand, the acquisition of ccBIOZ requires 4 electrodes to perform a "4-point measurement", in which 2 electrodes are used to inject a known amount of high-frequency current through the body, while the remaining 2 electrodes perform a voltage readout. The 4-point measurement has the advantage that the injected current does not flow through the same electrodes used for voltage sensing, hence the impedances at the electrode-tissue interfaces are not included in the measurement.

The acquisition of ccBIOZ has been less studied than the acquisition of ccECG signals. Within the reported research in this field, Abad [11] explored in 2009 the use of contactless BIOZ for bioimpedance spectroscopy (BIS) purposes. In this work, he demonstrated that commercial BIOZ devices are unsuitable for contactless measurements and proposed a multi-frequency current source to be used in ccBIOZ BIS measurements. With the purpose of measuring both ventilation and HR values, Macias et al. [103] reported a ccBIOZ system integrated into a car seat. Here, a 4-point measurement was implemented using textile electrodes on the back of the seat and in the steering wheel (measurements were in galvanic contact at the steering wheel point). Although both cardiac and respiratory activities were acquired under

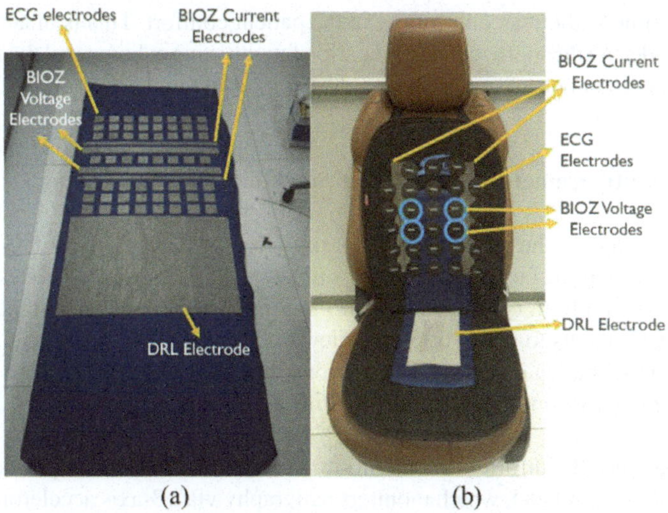

Fig. 8.6 Example of prototypes acquiring ccECG and ccBIOZ measurements simultaneously. **a** For sleep monitoring (covered by a normal bedsheet when in use). **b** For driver monitoring. Replicated from [37] with permission from the authors

controlled conditions, it was concluded that the system did not achieve acceptable performance due to the capacitive behavior of the electrode-tissue interface. A more recent system integrating both ccECG and ccBIOZ [37] was demonstrated to successfully acquire these signals in prototypes in the form factor of a car seat and a bed mattress. In addition, the system was shown to be able to provide a flexible interconnection that enables the real-time selection of up to 8 simultaneous ccECG electrodes (i.e., 4 ccECG channels) from an array of up to 64 electrodes, as a solution for the varying quality of the ccECG depending on user position. The prototypes presented in [37] are shown in Fig. 8.6 as an example of the possible implementations of ccECG and ccBIOZ measurements. It is worth noting that in the specific case of a mattress with the sensors, the mattress can be covered by normal bed linen and the patient can wear standard pyjamas, which significantly increases the comfort when compared to contact-based methods.

Such a multi-electrode approach, together with quality-based signal processing algorithms [33, 34] and optimizations in the electronic design [35, 36, 96, 155], aims to overcome the main challenges of capacitively coupled signals: the sensitivity to motion related to the varying electrode coupling for different positions or body shapes [17, 46, 167, 175] and the variability of signal quality depending on the electrostatic charges in the surroundings of the patient [46].

Although ccECG and ccBIOZ signals are likely to provide less coverage (in terms of time with high signal quality) than their contact-based counterparts, signal processing and system optimization approaches such as the ones mentioned above are expected to enable the use of the technology for home monitoring during extended periods of time without a compromise in the patient comfort. This is of added value when considering the more realistic scenario of monitoring at home and the increased analysis that an unobtrusive monitoring during multiple nights can allow.

8.3.2.8 Multiparametric Devices Dedicated for Sleep Monitoring

Taking advantage of the patch form factor used in some of the latest ECG monitors as well as other form factors enabled by miniaturized electronics, devices are now available which aim to combine some of the sensors/techniques mentioned above. This subsection aims to give a brief overview of the less obtrusive devices currently available in the market or being developed, which are specifically tailored for sleep monitoring at home and are significantly different than the PSG-like and PG OOC devices.

The company Beddr offers a small form-factor device called SleepTuner [23] to be placed in the forehead, which monitors actigraphy via a 3-axis accelerometer and uses optical sensors to monitor PPG (including PPG-derived HR) and SpO_2. Based on these sensors, the company's software performs sleep-related analysis which includes sleep duration, position, and stopped breathing events. Another patch-based solution is offered by the company Tatch [10], which aims to monitor respiratory effort, flow, oxygen level, heart rate, body position/movement, and snoring sounds. The company Onera [7] is currently developing a patch-based solution together with data analytics to enable the "first at-home medical grade sleep diagnostic patch system".

The WatchPAT devices offered by Cardio Sleep Solutions [151] allow to monitor actigraphy, SpO_2, chest motion, HR, body position, snoring, and peripheral arterial tonometry (PAT), which have been demonstrated to be adequate for the detection of sleep apnea [44]. Other solutions focus solely on the use of applications from smartphones [8, 136]; these mainly aim to quantify the quality of sleep or provide an early assessment of risk for sleep-related disorders such as OSA without directly monitoring physiological signals.

8.4 Machine Learning Algorithms for Sleep Staging at Home

Sleep scoring standards are developed mainly based on EEG signals. The current EEG sensor technologies, however, pose a certain level of obtrusiveness. This has motivated the search for alternative sensors and signals, which allow reliable and comfortable monitoring of sleep physiology. As a consequence, the development of novel algorithms for automated sleep staging based on these unobtrusive signals has been an active topic of research. As described in Sect. 8.3, cardiac and respiratory signals can indeed be more comfortably acquired by emerging unobtrusive sensor technologies compared to EEG-based monitoring. Therefore, sleep staging based on cardiac and respiratory signals presents the first leap toward home-based sleep monitoring. These sleep staging approaches are discussed in the following sections. First, the focus is on state-of-the-art algorithms, based mainly on ECG and/or RIP signals extracted from the PSG. Next, an overview of sleep staging algorithms based on data from wearable or unobtrusive sensor technologies is given with special attention to stand-alone actigraphy, BCG, and PPG. Finally, an outlook on the signal processing challenges commonly encountered when working with wearable data and future research in sleep staging is presented.

Algorithms are compared based on the performance of a 3-class classification task of Wake versus NREM versus REM (WNR). Nevertheless, many studies report a 2-class sleep staging performance, generally being sleep versus Wake. However, different ways exist of combining sleep stages, as Wake and REM share some characteristics and one could define "active sleep" (Wake, REM, N1) versus "quiet sleep" (N2 and N3) [98]). As REM is, therefore, difficult to classify, a 3-class WNR is preferred for better comparison. Studies in which a 4-class classification task reached superior performance to 3-class classification are also discussed. Typically, this classification task is defined as Wake versus REM versus Light sleep versus Deep sleep (WRLD).

Furthermore, the discussed studies and algorithms have been trained subject independent unless mentioned otherwise. This implies that the training data set does not contain data from subjects which have been included in testing. The studies also report performances by at least an average accuracy and Cohen's kappa κ score [43]. As such, algorithmic performances can be compared by equal measures. The average accuracy is the percentage of epochs correctly classified compared to the gold-standard annotations. The κ score is a measure of agreement that corrects for the level of agreement achieved by chance.

8.4.1 State-of-the-art Algorithms Based on Cardiac and Respiratory Signals

As PSG provides the gold standard for sleep staging, cardiac and respiratory-based sleep staging algorithms have been developed based on these PSG-derived signals. It allows direct comparison of all PSG signals without synchronization issues. Sleep stage annotations are directly applicable and high-quality data is assured, thus enabling state-of-the-art performances. In general, sleep staging algorithms are built from a feature extraction phase followed by a classification phase. A multitude of cardiac and respiratory features have been developed in the literature, typically modeling the ANS variation in the temporal and spectral space.

For a long time, algorithms reaching state-of-the-art performance used a combination of signals as input. For instance, Harper et al. showed in 1987 that classification performance can improve when combining modalities [71]. Some earlier studies explored sleep staging based on single modalities, such as ECG [173] or RIP [101], with the advantage of requiring less sensors. However, until 2018, performances of these studies were indeed inferior compared to co-occurring studies with multimodal input.

Therefore, these studies published before 2018 and based on a single modality are not mentioned in this chapter. Although, a complete review on automated sleep stage scoring was made by Faust et al. [61]. An overview of discussed papers which reached state-of-the-art performance is found in Table 8.2.

In 2006, Redmond and Heneghan tackled the challenging task of sleep staging based on cardiac and respiratory signals [134]. This was achieved by extraction of temporal and spectral features of the ECG RR-interval (time elapsed between two successive R-waves) and of the respiratory effort signal in standard 30 s epochs.

Furthermore, EDR features and spectral features from the cross-spectrum of the RR and EDR were calculated. As such, the feature set consisted of the power in the LF and HF band of the RR-interval, the EDR, the RR-EDR cross-spectrum, and the respiratory effort signals. Other features were the LF/HF power ratio of RR, mean RR, standard deviation of RR, difference between the longest and shortest RR-interval in the epoch, breath-by-breath correlation, and breath length variation. These features served as inspiration for subsequent studies. The accuracy and κ score of the subject-independent algorithm were, respectively, 67% and 0.32 for a 3-class WNR sleep staging task on healthy subjects. The authors improved the algorithm's performance by a linear discriminant classifier model using a time-dependent a priori probability. The accuracy and κ score then reached 76% and 0.46, respectively [133].

Willemen et al. improved sleep staging performance in 2014 by the combination of ECG, RIP, and actigraphy [168]. The study was performed on 36 healthy subjects and a total of 85 nights. A set of 13 feature groups was defined for an ECG, respiratory 1 Hz movement signal, extracted per epoch of 60 s. By transformations, a total of 750 features was obtained and subsequently reduced to 40 task-specific features by forward feature selection. These one-minute epochs were applied for classification, which is different from other studies that commonly classify sleep per 30 s epochs. This one-minute window accounts for the slow dynamics of the breathing rate and heart rate variability, as the HRV Task Force recommends interval lengths of at least 10 times the wavelength of the lowest frequency bound [30]. However, 60 s would only be a reliable choice for the HF band and too short to be fully reliable in the LF band. To validate the classified epochs, the 30 s epochs of the PSGs hypnogram were transformed to 60 s interval values by a set of decision rules. One RBF-kernel support vector machine (SVM) was optimized for different binary classification tasks. Three-class sleep staging of WNR achieved a mean accuracy and kappa of 81% and 0.62, respectively. It is noted that the study population's average age was relatively low: 22.1 ± 3.2 years. Similar results were obtained by Domingues et al., who performed a similar study [56].

Willemen et al. also developed a sleep staging algorithm for OSA patients [169], in which RR interbeat interval (IBI) series, the breathing signal, inter breath interval series, and the inspiration-to-expiration ratio interval series were extracted from the ECG and respiratory belt signals for 25 subjects. Sixteen feature groups were extracted from these signals in 60 s epochs. This window length was found to achieve

Table 8.2 Overview of state-of-the-art sleep stage classification algorithms based on cardiac, respiratory, and actigraphy signals. QDA: Quadratic discriminant analysis, LDA: Linear discriminant analysis, SVM: Support vector machines, HMM: Hidden Markov models, CRF: Conditional random fields, Deep NN: Deep neural network, LSTM: Long short-term memory networks

Author	Year	Data set	Signals	Classif.	# Rec.	Subjects		Task	Results	
						Healthy	OSA	# Classes	Acc. [%]	κ
Redmond et al.	2006	Private	ECG, RIP	QDA	37	23	14	3 WNR	67	0.32
Redmond et al.	2007	Private	ECG, RIP	LDA, QDA	31	31	0	2 WS	89	0.60
					31	31	0	3 WNR	76	0.46
Willemen et al.	2014	Private	ECG, RIP, ACT	SVM	85	36	0	2 WS	92	0.69
					85	36	0	3 WNR	81	0.62
					85	36	0	4 WRLD	69	0.56
Domingues et al.	2014	Private	ECG, RIP, ACT	HMM	20	20	0	3 WNR	78	0.58
Long et al.	2014	SIESTA	RIP	LDA	48	48	0	3 WNR	76.2	0.45
					48	48	0	4 WRLD	63.8	0.38
					25		25	2 WS	74.2	0.37
Willemen et al.	2015	UCD	ECG, RIP	LDA	25		25	3 WNR	67.0	0.41
					102	102	0	3 WNR	81.8	0.59
					102	0	102	3 WNR	77.0	0.50
					102	102	0	4 WRLD	71.0	0.51
Fonseca et al.	2018	Private, SIESTA	ECG, RIP	LDA, HMM, CRF	102	0	102	4 WRLD	70.6	0.46
Wei et al.	2018	SLPDB	ECG	Deep NN	18	16	0	3 WNR	77	0.56
					5793	5793	0	2 NREM vs REM+W	85	0.68
					5793	5793	0	3 WNR	82.0	0.63
Li et al.	2018	SLPDB, CinC2018, SHHS	ECG	SVM	5793	5793	0	4 WRLD	65.9	0.47
					195	195	0	4 WRLD	76.5	0.63
Radha et al.	2019	SIESTA	ECG	LSTM	51	0	51	4 WRLD	78.5	0.60

the best results to distinguish apneic from healthy breathing in a study by de Chazal et al. [39]. By detrending the first input time series over different intervals and afterward transforming extracted features, a total of 510 features was defined. A triple layer validation scheme was constructed to train the classifier parameters, perform feature selection ,and define a test set. The study obtained an accuracy and κ score of, respectively, 70% and 0.41 for WNR classification of OSA patients.

Fonseca et al. compared three of their developed sleep staging methods based on conditional random fields (CRF), hidden Markov models (HMMs), and Bayesian linear discriminants (LDs) [64]. Features were extracted using windows centered on non-overlapping epochs of 30 s, where the window length depended on the feature type. Additionally, they explored the incorporation of time information in their classifiers and applied their methods on 102 healthy subjects as well as 102 patients with OSA. In general, the best performing classifier was a CRF boosted with time information (CRFt). Although, CRFt performed not significantly better than standard CRF in the case of OSA patients. This can be subscribed to the fact that OSA leads to the decrease of REM and N3 presence and an increase in sleep fragmentation due to arousals associated with respiratory events (i.e., apneas) [139]. Therefore, the presence and progression of their sleep stages might depend more on the occurrence of disordered breathing events than on a healthy sleep architecture. Three-class sleep staging of WNR achieved a mean accuracy and κ of, respectively, 81.8% and 0.59 for healthy subjects compared to 77% and 0.50 for OSA patients. On a healthy data set, the developed CRFt algorithm performs comparable to [168], both for WNR and WLRD classification tasks. With respect to OSA patients, Fonseca et al. reached a substantial improvement in performance compared to [168], in which an accuracy and κ of 70% and 0.41, respectively, was reported. Moreover, the study of Fonseca et al. included only two out of three modalities, which can be seen as an advantage for long-term home monitoring.

In 2018, Li et al. developed a sleep staging algorithm based on a single lead ECG signal from extensive public data sets [93]. They were able to surpass state-of-the-art algorithms using a single modality, however, by extraction of respiratory information from the ECG. First, the authors derived spectrograms of the cardiorespiratory coupling in 5 min windows centered on each 30 s epoch. They applied convolutional neural networks (CNN) on the spectrograms for subsequent feature extraction (i.e., representation learning). Then, the extracted features were combined with handcrafted ECG features into an SVM model. With an accuracy and κ of 81.6% and 0.63 for WNR classification, this model is competitive to [64, 168], albeit with the application of a single modality.

The current state-of-the-art sleep staging model is described by Radha et al. [131], applying exclusively a single lead ECG signal as an input. A set of 132 handcrafted HRV features was fed into an Long Short-Term Memory (LSTM) network. This feature set consisted of time- and frequency-domain features, entropy and regularity features, and miscellaneous features. To extract the feature vector of a 30 s epoch, a window of 4.5 minutes of IBI data centered around this epoch was considered. The LSTM network type is chosen for its ability to capture long-term temporal dependencies. To determine the optimal number of LSTM layers and cells per layer,

18 combinations were trained and compared. The final model consisted of $2.6 \cdot 10^5$ parameters. This potentially involves a substantial time complexity; however, the study does not mention training time and memory consumption. The model was validated on 195 healthy subjects and 51 OSA patients. The algorithm reached an accuracy and κ of 76.5% and 0.63 on the former and 78.5% and 0.60 on the latter for a 4-class WRLD classification task. Interestingly, authors observed a negative correlation between performance and age, presumably by changes in autonomic function [177] and alteration in sleep architecture [148].

8.4.2 Sleep Staging Approaches Based on Wearable and Unobtrusive Sensor Technologies

The sleep staging algorithms described in the previous section report state-of-the-art performances, yet these were developed on PSG data. In order to monitor patients at home, different wearable or unobtrusive sensor technologies were implemented as described in Sect. 8.3. Among these, actigraphy, BCG, and PPG have played a more important role in sleep staging research and specific algorithms have been developed.

As actigraphy is purely motion based, it is not suitable for refined sleep monitoring. Nevertheless, it presents an established method for sleep/wake classification. These studies are discussed in Sect. 8.4.2.1 and summarized in Table 8.3. On the contrary, BCG enables the recording of multiple physiological signals: cardiac, respiratory, and movement information. This modality has been explored for sleep staging by several studies, which are described in Sect. 8.4.2.2. Furthermore, PPG has gained interest in sleep research as the classic finger-based recording shifted to a smartwatch configuration. Studies applying PPG in sleep staging are discussed in Sect. 8.4.2.3. Table 8.4 gives an overview of discussed papers on sleep staging approaches based on BCG and PPG. Other suitable modalities for wearable or unobtrusive sleep staging such as arterial blood pressure, peripheral arterial tonometry, oximetry, audio, video, and temperature are discussed in [138]. Furthermore, radar technology for sleep staging in OSA has been explored by [47].

8.4.2.1 Actigraphy

Actigraphy or activity-based sleep tracking is a reliable and valid methodology for monitoring sleep-wake and circadian rhythm patterns in healthy adults [99]. The sleep staging capacity of actigraphy is limited as it is known to overestimate sleep time. This is because it cannot differentiate motionless periods of wakefulness from sleep. On the other hand, it presents a potential tool for the unobtrusive screening of certain sleeping disorders. However, the technology is not able to diagnose sleep disorders that involve altered motility during sleep such as OSA [126, 142], and performance will be impacted by disorders altering the ANS [66]. Therefore, actigraphy

Table 8.3 Overview of state-of-the-art sleep/wake classification algorithms based on actigraphy signals. ACT*: Surrogate actigraphy; LDA: Linear discriminant analysis, QDA: Quadratic discriminant analysis

Author	Year	Data set	Signals	Classif.	# Subjects			Results	
						Healthy	Insomnia	Acc. [%]	κ
Devot et al.	2010	Private	ECG, RIP, ACT		9	9	0	96.1	0.70
			ECG, RIP, ACT		27	0	27	84.5	0.61
			ACT		9	9	0	93.8	0.51
			ACT	LDA, QDA	27	0	27	78.0	0.39
Long et al.	2013	Private	RIP, ACT	LDA	15	15	0	95.7	0.66
Fonseca et al.	2016	Private	ECG, RIP, ACT*		15	15	0	93	0.66
			ECG, RIP, ACT*		40	15	25	87	0.56
			RIP, ACT*		15	15	0	93	0.64
			RIP, ACT*	LDA	40	15	25	85	0.5

is usually combined with cardiac and respiratory signals [55, 56, 168]. Devot et al. compared sleep staging using cardiac, respiratory, and actigraphy signals to sleep staging based solely on actigraphy. As expected, they obtained superior results with the former approach [55]. In order to minimize obtrusive sensors while preserving classification performance, Long et al. retained the respiratory signal in combination with actigraphy [100] and achieved comparable results as Devot et al. In [63], a surrogate actigraphy signal was estimated from body motion artefacts derived from the ECG and respiratory effort signals. The surrogate signal was combined with RIP or ECG+RIP in a sleep-wake classifier. This approach achieved similar results as classification in combination with the reference actigraphy signal, both in a healthy as mixed population including insomniacs. The authors concluded that in setups where RIP is the only modality, as it is one of the most applied modalities in home sleep monitoring, actigraphy posed a significant added value. In case both RIP and ECG are acquired, the application of actigraphy is redundant.

Table 8.4 Overview of state-of-the-art sleep stage classification algorithms based on wearables, except for the last two studies including PPG from the PSG. (*) Test and training set contain nights from the same individuals

Author	Year	Data set	Signals		# Subjects		Task	Results	
					Healthy	OSA	# Classes	Acc. [%]	κ
Kortelainen et al.	2010	Private	BCG	18	9	0	3 WNR	79*	0.44*
Migliorini et al.	2010	Private	BCG	17	11	0	3 WNR	76.8*	0.55*
Kurihara and Watanabe	2012	Private	BCG	20	10	0	3 WNR	78	0.48
Hwang et al.	2016	Private	Bed sensors	25	12	13	4 WRLD	70.9	0.48
Beattie et al.	2017	Private	ACT, PPG (wrist)	60	60	0	4 WRLD	69.0	0.52
				215	152	0	2 SW	91.5	0.55
				215	152	0	3 WNR	72.9	0.46
Fonseca et al.	2017	Private, SIESTA	PPG (wrist)	215	152	0	4 WRLD	59.3	0.42
Uçar et al.	2018	Private	PPG (PSG)	10	0	10	2 WS	73.4	0.59
Casal et al.	2019	SHHS	PPG (PSG)	5804	5804		2 WS	85.2	–

8.4.2.2 Ballistocardiography

The following studies have applied a BCG-based bed sensor for sleep staging. Kortelainen et al. used commercial Emfit material to configure a BCG system [85]. From the acquired signal, they extracted the IBI and movement activity. IBI features trained a hidden Markov Model for a WNR task, while the motion signal served as an additional input for wake stage detection. Nine healthy subjects were included in the study, of which in total 18 sleep recordings were acquired. The three-class WNR classification task resulted in an accuracy and κ of 79% and 0.44, respectively. It is noted, however, that the training and test set contain recordings from the same subjects as a leave-one-out cross-validation (LOOCV) was performed on these 18 recordings. This approach could lead to an overestimation of the subject-independent classification performance.

In the same year, 2010, a similar study by Migliorini et al. was published [112]. Seventeen recordings from 11 healthy subjects were acquired, using the commercial Emfit sensor material as well. Similarly, a LOOCV was performed for parameter optimization. The achieved accuracy score was comparable to [85]; however, a higher κ of 0.55 was reached. As opposed to Kortelainen et al., the authors included features from the respiratory component, which is inherently present in the BCG signal.

Kurihara and Watanabe implemented a pneumatic system based on an air tube and pressure sensor to acquire the BCG. They obtained similar sleep stage performances as Kortelainen et al. using a comparable data set [87].

In 2016, Hwang et al. used a polyvinylidene fluoride sensor for sleep staging in 12 healthy and 13 OSA patients [73]. The motion signal was applied for wake detection, while information extracted from the respiratory signal was investigated for REM and deep sleep (N3). The 4-class classification had an average accuracy of 70.9% and κ of 0.48, where no significant difference was found between the control and OSA populations. Comparing to a later study of Fonseca et al. [64] in 2018, who also applied a WRLD classification on OSA patients, the current method reached similar performance, though by the application of an unobtrusive device.

8.4.2.3 Pulse Photoplethysmography

Studies on wearable PPG sleep staging can be traced back to 2017. Beattie et al. [22] from Fitbit research performed sleep staging in 60 healthy subjects based on a wrist-worn device, measuring three-dimensional accelerometry and PPG. The interval between peaks of the PPG wave was taken as a surrogate for an ECG-derived IBI. As such, motion, breathing variability, and HRV could be extracted from this modality. However, the PPG signal is more sensitive to movement artefacts compared to ECG, especially when worn as wearable at the wrist. Similar to BCG research by [85], no IBI information could be extracted in periods of heavy motion. In the case of unlikely sleep architecture patterns, the authors applied a post-processing step by smoothing, e.g., an isolated wake epoch during a long period of deep sleep is converted to the annotation of its surrounding epochs. This assumption is only reasonable when monitoring healthy subjects. After LOOCV, the overall accuracy was 69% with a κ of 0.52 for a 4-class WRLD task.

A similar study was published in 2017 by Fonseca et al. [66]. Although, the study included a larger data set of 152 healthy subjects, from which a validation set was held out for testing. Features and machine learning techniques were similar to their earlier study [65] (later discussed in [64] and described in Sect. 8.4.1), with the exclusion of respiratory signals. The performance of the current study for both WNR and WRLD tasks was lower and could partially be subscribed to the reduced number of input signals. However, as the authors have pointed out themselves, a respiratory rate could technically be deduced from the PPG signal [88]. This is potentially beneficial, as it could provide features capturing changes in sympathetic tone, important for detection of REM [66].

Previous studies have performed sleep staging based on wearable PPG in a healthy population. However, sleep staging in a pathological population is essential as total sleep time is an important outcome for severity assessment, e.g., OSA. Uçar et al. [160] and Casal et al. [31] applied sleep-wake classification in an OSA population. As data processing becomes more challenging in a pathological population, non-wearable PPG signals were used in these studies. However, it offers the potential of integrating these algorithms with signals from wearable devices.

8.4.3 Signal Processing Challenges Presented by Wearable Systems

The signal processing of wearable data is associated with specific challenges. This includes sensitivity to motion and synchronization between the wearable device and PSG system.

First, the presence of motion is ambiguous. On the one hand, movement induces excessive noise in the overall BCG signal, which impedes the IBI extraction and HRV analysis. On the other hand, it provides valuable information on the patient's sleep architecture, similar to actigraphy. Heavy motion can mainly be subscribed to wake stages and thereby it compensates for the loss of information regarding HRV. Moreover, the separation between wake and REM is improved by the inclusion of motility information, as the cardiac activity presents similar characteristics during both stages [85].

Furthermore, synchronization between wearable and standard devices is beneficial as it enables quality control of the wearable signal. Comparison of the HR or respiration rate extracted from both devices can act as a quality indicator. Synchronized signals allow direct comparison of predicted and ground truth hypnograms as well.

The synchronization procedure is usually achieved by alignment of tachograms, derived from the heartbeats detected in the wearable and from the R-peaks detected in the ECG signal from the standard device [22, 66]. However, in case of heavy

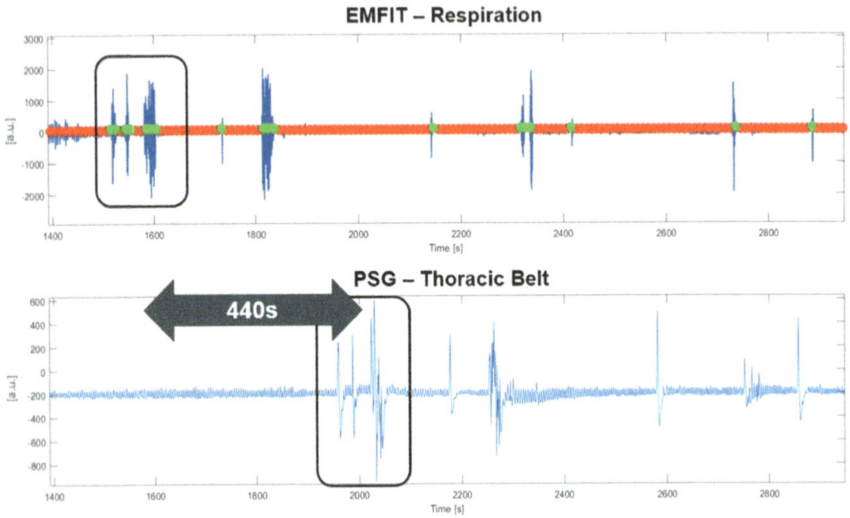

Fig. 8.7 Synchronization by artefact patterns of an Emfit BCG signal with a thoracic belt signal from the PSG. As wearables are sensitive to motion, tachogram-based synchronization might not always be possible as heartbeat detection might be impeded by large artefacts

motion due to restlessness or sleep disorders (such as OSA), the tachogram is hard to derive and thereby troubling synchronization. In [72], a method for synchronization of BCG recordings of OSA patients was proposed, based on artefacts rather than IBI. First, artefacts were detected in the BCG recordings without training the algorithm on artefact annotations. Then, a segment including several subsequent artefacts defined the artefact template. Next, a corresponding data pattern was sought in the thoracic belt of the PSG to align the segments. Figure 8.7 displays a notable artefact pattern found in the BCG signal, which can be linked to a corresponding pattern in the thoracic belt.

Furthermore, it is noted that the equivalence of tachograms derived from the ECG and the PPG or BCG device is only valid in the absence of cardiovascular health problems. For instance, ectopic beats during arrhythmia can be traced by the ECG. In some cases, these beats do not affect the pumping mechanism of the heart. Therefore, these ectopics are sometimes not acquired by the BCG or PPG [66, 85].

8.4.4 Future Research in Sleep Staging at Home

On a future prospect, cardiac and respiratory-based sleep staging will be explored by deep learning networks, as intended by [131]. Currently, deep learning-based algorithms are more actively developed within the field of EEG-based sleep staging [122, 156, 159]. This entails an algorithmic pipeline where both feature extraction and classification are optimized by extensive neural networks. The advantage is the automated training procedure of the complete pipeline and often superior classification performances. However, one of the requirements to properly develop these algorithms is the availability of large amounts of data. In the field of EEG-based signal processing, data is often provided by the publicly available Physionet Sleep-EDF database [68, 81]. Additionally, it contains EOG, EMG, oronasal respiration signals, and body temperature. Equivalently, the availability of a large data set containing ECG, respiration, and sleep stage scoring could benefit the development of these cardiac and respiratory-based algorithms. This would, moreover, allow benchmarking of algorithms, which is currently difficult due to the variety in data sets.

Another potential issue is the suitability of these cardiac and respiratory-based sleep staging algorithms for real-time applications. At present, most studies focus on increasing sleep staging performance, which has a theoretical maximum defined by the inter-rater agreement of 82.6% for 5 sleep stages [141]. The developed algorithms often require several minutes or more of sleep data for pre- and post-processing or to include time information. Fewer studies investigate the suitability of algorithms or design them specifically for real-time applications. Real-time processing is, however, not necessary for the diagnosis of many sleep disorders, nor do these sleep disorders pose a threat that requires real-time monitoring. Considering the fact that offline monitoring is sufficient for these sleep disorders relaxes the constraints of newly developed wearable hardware as internal algorithmic processing is not a priority. In contrast, online monitoring is critical in the field of neonatal care. First, it serves to optimize the timing of nurse intervention so as to minimize the sleep disturbance of the neonate. This was achieved by sleep staging based on EEG monitoring, though presenting a non-wearable approach [16]. Second, online monitoring is required to generate alarms as motor responses often precede changes in vital signs, such as seizures and apneas. Movements could be real-time detected based on a BCG approach [77]. As unobtrusiveness and real-time sleep staging are crucial in this field, research in neonatal care could complement the field of sleep disorders and its advances in unobtrusive techniques and wearability.

The last challenge concerns the classic 30 s epoch length for annotation, analysis, and validation. This epoch length was optimized for EEG-based sleep analysis on paper [102]. At a paper speed of 10mm/s, one page meant a 30 s recording, which served well to visualize spindles and delta waves [51]. The 30 s epoch length was further recommended by the R&K manual [132]. The sleep stage annotation of such an EEG epoch is defined as the sleep stage which comprises the largest portion of this 30 s. This procedure is efficient for hand scoring and a reasonable approach for a healthy population, in which sleep stages have a certain stability and persist over several epochs [138, 145]. Nevertheless, half-minute epochs are less suited for a population with fragmented sleep (e.g., OSA), which is associated with short-term awakenings, arousals, and critical respiratory events. As such, classic sleep staging is less reliable in this population and they might benefit from a smaller time scale in sleep scoring and sleep analysis [145]. Additionally, the probability distribution of a short-term epoch over different sleep stages conveys more information than classical hypnograms, as proposed in [154]. Therefore, current sleep staging algorithms might not only require further validation and adaptation in a variety of age groups and disorders, but also a shift in valuing the gold standard.

8.5 Detection, Screening, and Phenotyping of Sleep Apnea in an Ambulatory Setting

Obstructive sleep apnea is the most common sleep-related breathing disorder, and it is estimated that worldwide almost 1 billion people are affected by this disorder [25]. However, most of these subjects remain undiagnosed, and consequently untreated. OSA patients experience repetitive complete or partial cessations of breathing during the night which are caused by a narrowing of the upper airway. In many countries, diagnosis of sleep apnea is currently based on manually scoring these events from an overnight in-hospital polysomnography. The AASM has defined a set of scoring rules which are considered as the gold standard for scoring OSA [26]. According to the AASM2012 rules, events are scored if they last longer than 10 s. An apnea is scored when an airflow amplitude decrease of more than 90% occurs, a hypopnea, on the other hand, only requires a decrease in airflow amplitude of at least 30%, but accompanied by either an oxygen desaturation of more than 3% or an arousal. The apnea-hypopnea index is computed as the number of apneas and hypopneas per hour of sleep. Subjects are diagnosed with OSA if they either have an AHI larger than 5 accompanied with symptoms, or if their AHI is larger than 15, independent of the presence of symptoms [143].

Many researchers have been developing methods for automated in-home screening and diagnosis of sleep apnea. In this section, an overview will be given of methods using signals that can be easily acquired in a home environment. The methods will be ordered according to the SCOPER system, which was introduced in Sect. 8.3.1. A short overview of the use of these five categories of sensors in the diagnosis of sleep apnea is given below:

1. **Sleep:** A measurement of sleep, for example, using actigraphy, will enable the calculation of the hours of sleep; using this measure instead of the recording time leads to a better estimation of the AHI [120]. These methods were the topic of Sect. 8.4.
2. **Cardiovascular:** This category includes ECG, PPG, and PAT as well as all other measures of the heart rate. Bradycardia can be observed during apneas, followed by tachycardia when breathing is retaken [70], as can be observed in Fig. 8.8. Moreover, these signals can be used to derive an estimate of respiration and detect autonomic arousals.
3. **Oximetry:** The SpO_2 signal is very useful for OSA screening, since apneas often result in an oxygen desaturation as can be seen from the example in Fig. 8.8.
4. **Position:** Visual and non-visual measures (e.g., using an accelerometer combined with a gyroscope [80]) can determine the body position of a subject. Studies have

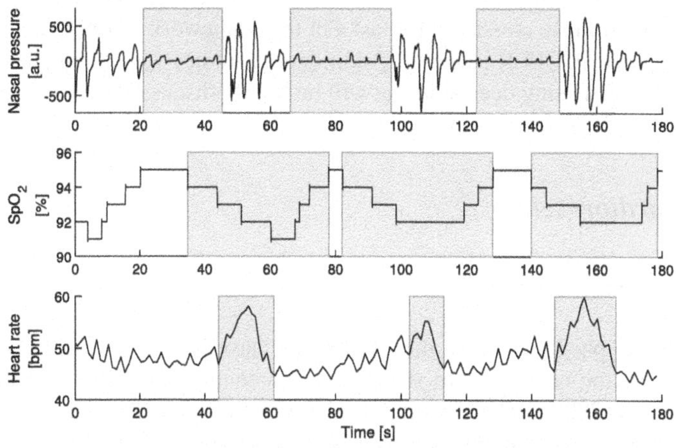

Fig. 8.8 Example of a segment with apneic events. From top to bottom, the nasal pressure, SpO_2, and heart rate signals are plotted, with, respectively, the annotated apneic events, oxygen desaturations, and sympathetic activation depicted by the shaded areas

shown that OSA severity can be position dependent. In fact, a higher AHI and more severe apneas have been observed during supine sleep [124]. Differentiating between lateral and supine sleep could thus help to improve automated OSA detection algorithms. No studies, however, propose to use position measurement on its own for the detection of OSA. Therefore, the analysis of position sensors will not be discussed further.

5. **Effort and Respiration:** The reduction in airflow is the primary effect of apneas. Sensors measuring respiratory effort provide extra information which helps to differentiate between central and obstructive apneas.

An overview of current research and commercial devices for OSA detection categorized according to the SCOPER system can be found in [109]. In this review, sound recording devices were defined as the sixth category, in addition to sleep, cardiovascular, oximetry, position, effort, and respiratory parameters. In this section, the focus will be on the detection algorithms, rather than the wearable sensors, which were already discussed in Sect. 8.3.

Most OSA detection algorithms consist of four main steps: starting with the signal preprocessing phase, next feature extraction, feature selection, and finally classification. Two classification problems are studied: event-based or subject-based classification. In the case of event-based classification, the goal is to detect all apneic events within the recording, and this is often done by splitting the recording into 1 min windows. Based on the number of windows classified as apneic, the AHI can be estimated. Subject-based classification, on the other hand, extracts features over the whole recording which will then be correlated to the AHI and used to predict the OSA severity category of a patient. In this section, features commonly used for either one of those classification tasks in the framework of OSA detection will be discussed per SCOPER category. Additionally, studies describing automatically generated features using deep learning will be briefly discussed in Sect. 8.5.5.

8.5.1 Cardiovascular

8.5.1.1 ECG

The effects of apneic events on the ECG signal have been known since the 1980s [70, 116], but the research into sleep apnea screening using ECG signals really took a boost in 2000 with the Computers in Cardiology Challenge [115] and the subsequent release of the Physionet Apnea-ECG database [127]. This data set is still one of the most used data sets for ECG-based OSA screening. An extensive overview of ECG-based methods for OSA screening can be found in [60, 110]. The ECG-based screening approaches proposed in [110] obtain subject-based classification accuracies of 72–100% and an area under the curve (AUC) of 89–100%.

The used features are generally based either on changes in HR [70] or changes in respiration, captured using the EDR [116]. The respiration can be extracted from the ECG since ECG waveform characteristics are altered by respiratory-induced movements of the chest electrodes and changes in the electrical impedance of the thoracic cavity. The simplest measure of EDR is the amplitude of the R-waves, but many more elaborate methods have been proposed. A comparative study of 10 EDR algorithms is discussed in [164], where it was shown that the EDR signals computed from changes in the QRS slopes are most robust in the presence of artefacts, changing respiratory rates, and different ECG-recording systems.

OSA detection algorithms typically use statistical, frequency-domain, and nonlinear features extracted from both the HR and the EDR. Recently, measures of cardiovascular interactions have also been identified as relevant for OSA detection.

Statistical features: Different time-domain features of the RR-interval time series[6] and EDR have been proposed by De Chazal et al. [48] and are often used in the literature. An overview of these features can be found in Table 8.5.

Frequency-domain features: Multiple methods have been applied to extract frequency information from the heart rate, but the power spectral density and wavelet decomposition [29] are the most popular. Often the power in the standard HRV frequency bands is considered: very low-frequency range (0.003–0.04 Hz), low-frequency range (0.04–0.15 Hz), and the high-frequency range (0.15–0.4 Hz).

Nonlinear features: An overview of nonlinear parameters used on cardiorespiratory signals for sleep apnea detection is given in [165]. Analysis of the sample entropy of the HRV signal has shown a reduced complexity for apneic subjects compared to controls [12]. This can be linked to the cyclic pattern of bradycardia and tachycardia during apneas. Detrended fluctuation analysis has also been applied on

[6] Time differences between consecutive R-peaks in the ECG.

Table 8.5 Overview of commonly used ECG time-domain features

RR	
Mean	μ_{rr}
Standard deviation	σ_{rr}
Serial correlation coefficients (of order k)	$r_k = \frac{\sum_{i=1}^{m}(rr_i - \mu_{rr})(rr_{i+k} - \mu_{rr})}{\sum_{i=1}^{m}(rr_i - \mu_{rr})^2}$
NN50	# adjacent RR-intervals that differ more than 50 ms
pNN50	pNN50=$\frac{NN50}{m}$
Standard deviation of interbeat intervals	SDSD = σ_{rd}, with $rd_i = rr_{i+1}$
Root mean square of interbeat differentials	RMSSD = $\sqrt{E(rd_i^2)}$
Allan factor	A(T)=$\frac{E[(N_{i+1}(T)-N_i(T))^2]}{2E[N_{i+1}(T)]}$
	with $N_i(T)$= # beats in the ith window of T seconds
EDR	
Mean	μ_{edr}
Standard deviation	σ_{edr}

the HRV of apneic subjects, but only slight changes in long-term control mechanisms of the heart rate could be observed [165].

Cardiorespiratory interactions: The HRV is modulated by respiration through the mechanisms called respiratory sinus arrhythmia. It has been observed that the information shared between respiration and HRV decreases during apneas and these reductions have been used as features to detect apneas [163]. An overview of different methods to estimate the RSA from the HRV and a respiratory measurement (e.g., the EDR or a respiratory belt) can be found in [118].

8.5.1.2 PPG

From the PPG signal, the pulse rate (PR) can be extracted as an alternative for the HR. All features proposed for the HR could, therefore, also be estimated from the PR. Moreover, decreases in the amplitude fluctuation of the PPG (DAP) have been linked to vasoconstriction due to sympathetic arousals at the end of apneas. The use of these DAP events has been studied for the detection of OSA in children [67]. Additionally, characteristics of the PPG waveform such as pulse amplitude, width, and slope transit time have been linked to respiration and their behavior can be useful for sleep apnea detection [53].

8.5.1.3 PAT

The peripheral arterial tonometry [144] sensor was developed for the detection of sleep apnea. It is based on a pulsatile plethysmography signal measured on the finger which can capture vasoconstriction and tachycardia related to apneas. Decreases in PAT amplitude of 33% in combination with a tachycardia of 15% were used in [144] for the detection of apneas.

8.5.1.4 Contactless Cardiovascular Measurements

Several contactless alternatives for ECG recording have been tested for sleep apnea detection. In [34], a multi-channel contactless capacitively coupled electrocardiography embedded in a mattress was proposed. The ECG-based apnea features defined in [163] were extracted from ccECG signals of healthy volunteers. The HR features achieved high similarity to the features extracted from the reference ECG signal. Signal morphology features, on the other hand, showed lower similarity with the reference resulting in problems with the EDR extraction. The similarity in HR features shows the potential of ccECG sensors for the detection of sleep apnea.

Ballistocardiography has also been studied in sleep apnea subjects. In [178], for instance, Zink et al. optimized the algorithm for BCG beat-to-beat detection in a sleep apnea population and obtained a correlation coefficient R^2 of 0.95 between the beat-to-beat cycle lengths extracted from the BCG and ECG. Whereas Hwang et al. investigated the correlation between the oxygen desaturation index (ODI, see Sect. 8.5.2) and HR features extracted from three different BCG sensors [74].

Researchers have also focused on extracting heart rate from radar or infrared (IR) video recordings next to the breathing movements. Zhu et al. have validated IR-based heart rate monitoring in sleep apnea subjects and discussed the open challenges in this field [176].

8.5.2 Oximetry

The SpO_2 signal has been widely studied for sleep apnea screening since it can easily be acquired in a home environment. In clinical practice, the SpO_2 signal is often characterized by simple statistics such as the ODI, which counts the number of times a desaturation larger than 4% occurs per hour of sleep, or time spent below an oxygenation level of 90% (T90). It has been shown that these simple parameters can reliably confirm moderate to severe OSA, but that a negative test result does not rule out mild OSA [158]. Therefore, researchers have been developing a range of more elaborate SpO_2 parameters in order to detect subtle changes due to apneas. In [158], four categories of commonly used SpO_2 parameters were defined: desaturation characteristics, time series variables, frequency spectrum variables, and nonlinear variables.

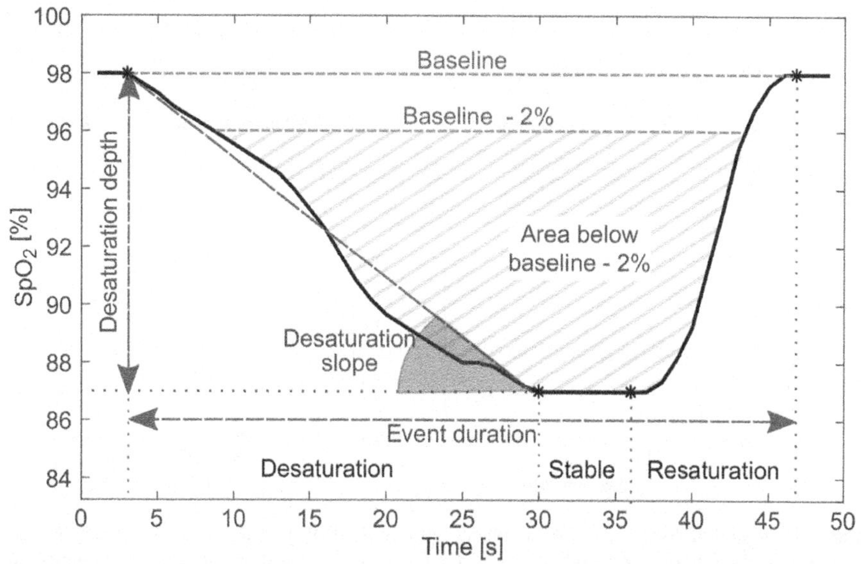

Fig. 8.9 Overview of desaturation characteristics which can be extracted from oxygen desaturations

Desaturation characteristics: The desaturation characteristics include the desaturation depth, duration, and area. These can be computed compared to a baseline level or another reference point. Additionally, these parameters can be calculated on subparts of the desaturation event, e.g., on the downward and upward part separately. Some studies also considered desaturation slopes or higher order derivatives of the signals [54]. In Fig. 8.9, an overview of these features is given for an example oxygen desaturation.

Time series analysis variables: These parameters include statistical measures that can be computed over the full SpO_2 signal or window based. These statistics include the following: the mean, minimum, quantile values, standard deviation, higher order statistical moments, and cumulative time spent below certain SpO_2 values (e.g., T90) [158].

Frequency spectrum variables: Studies have shown that OSA subjects have an unstable or overly sensitive respiratory control system [58]. Therefore, instead of going back to a stable breathing pattern after an apnea, often another apnea is triggered. As such, apneas tend to be grouped together, as can be seen from Fig. 8.10. When the apneas follow each other closely, a periodic pattern of desaturations can occur as is shown in the lower right plot of Fig. 8.10. Different studies have used the power in the 0.01–0.033 Hz frequency band to capture this periodicity [158]. Other measures of the PSD have been proposed, such as wavelets [117], autocorrelation analysis, or the use of phase rectified signal averaging (PRSA) [54].

Nonlinear analysis variables: The recurrent desaturations during the night also cause the SpO_2 signals to have a higher irregularity, variability, and complexity. These

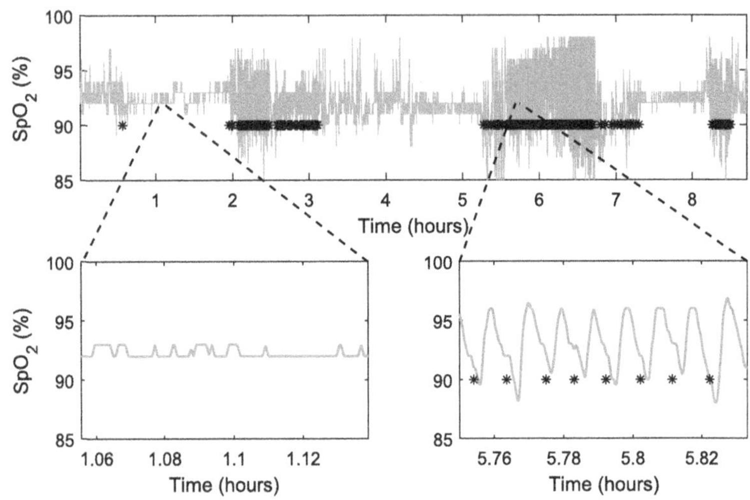

Fig. 8.10 Example SpO_2 signal of a subject with an AHI of 39.4 in which the apneas (*) are grouped together. In the lower right plot, a 5 min apneic segment is shown in which periodicity can be observed in the SpO_2 signal, whereas the left plot shows no periodicity during normal sleep

can, respectively, be measured using the sample entropy, central tendency measure, and Lempel-Ziv complexity [14].

Terrill et al. concluded in their review [158] that the studies which obtained the best OSA detection results used features from multiple of these four categories. These categories thus contain complementary information. However, including more than three to four features in the classifier did not lead to a significant increase in performance. SpO_2-based OSA screening methods typically obtain accuracies of 80–95% with an AUC of 90–95%, depending on the used features, data set, and AHI threshold [158].

8.5.3 *Effort and Respiration*

8.5.3.1 Airflow

The AASM rules for visuals scoring of apneic events rely mainly on the airflow measured using an oronasal thermal and nasal pressure sensor [26]. Therefore, several researchers have developed methods for the automatic detection of sleep apnea on these signals. All drops in peak signal amplitude of more than 90% are scored as apneas, whereas drops of more than 30% could represent a hypopnea. These drops are relative to pre-event baseline breathing which is defined as "the mean amplitude of stable breathing and oxygenation in the 2 min preceding onset of the event or the mean amplitude of the 3 largest breaths in the 2 min preceding onset of the event"

[26]. This baseline can be difficult to define for subjects with high AHIs and its definition is ambiguous to be translated into algorithms. Moreover, breathing and movement artefacts will further complicate this baseline extraction for automated algorithms [42]. Therefore, Ciołek et al. proposed a robust airflow envelope tracking using nonlinear filters. Using this reliable baseline, rule-based methods can then be applied to detect apneas and hypopneas when a decrease in airflow amplitude larger than a predefined threshold is observed.

These rule-based methods, however, have as disadvantage that they cannot differentiate if hypopneas are associated with arousals or desaturations. Therefore, more elaborate respiratory features have been proposed in order to capture changes that might have a correlation with arousals or desaturations. Koley et al. proposed statistical features of the respiratory amplitudes and intervals over windows of 8 s [84]. Additionally, they extracted the area and length of the respiratory curve and frequency-domain features using the PSD. In this study, the deviation of each of the feature values from the subject's mean values was also considered.

8.5.3.2 Respiratory Effort

Similar processing methods can be applied to respiratory effort signals obtained by RIP abdominal and thoracic belts. Despite the absence of airflow during obstructive apneas, the respiratory effort is still present, which challenges the detection of apneas when using this signal modality. However, when the effort is measured in combination with other modalities like airflow or SpO_2, it can be used to differentiate between central and obstructive events [119]. When both the abdominal and thoracic belts are monitored, the thoracoabdominal paradox[7] [26] can also be used as marker for apneic events.

A more extensive overview of methods using airflow and respiratory effort signals to detect OSA can be found in [110, 161]. The methods based on respiration reported in [110] obtained subject-based classification accuracies of 82–96% and an AUC of 88–90%.

8.5.3.3 Contactless Measures of Respiration

A lot of research has been conducted toward contactless sensors to measure respiratory effort. Respiratory movements can be extracted by pressure sensors in the mattress. An example of such a sensor is the Emfit, discussed in Sect. 8.3.2.5. In [72], the use of this sensor for the detection of sleep apnea was investigated. Rather than focusing on the breathing cessations, this study used unsupervised learning (i.e.,

[7] During obstructive apneas, the respiratory effort signals from the thoracic and abdominal belts are often out of phase, and this phenomenon is called the thoracoabdominal paradox.

clustering) to detect breathing distortions in the respiratory signal which are linked to arousals and large chest movements when breathing is regained. The amount of breathing distortions could be correlated to the AHI.

Kagawa et al. [78] proposed a system in which two Doppler radars are installed underneath the mattress in order to measure thoracic and abdominal effort. The system was tested for the detection of sleep apnea, a rule-based algorithm including the detection of paradoxical breathing showed promising results.

In [176], it was shown that the breathing rate can also be extracted accurately from IR video recordings in OSA subjects.

Castro et al. [37] proposed to measure the respiration using a capacitively coupled bioimpedance sensor embedded in a mattress. This system, however, still needs to be tested on OSA subjects.

8.5.4 Sound

Snoring is one of the most common symptoms of OSA, and differences in acoustic characteristics of snoring have been observed between OSA patients and simple snorers [150]. Moreover, due to apneas, subjects often make grasping and choking sounds. Researchers have developed acoustic features that can be used for the automatic detection of sleep apnea using sound recordings. Solà-Soler et al. have proposed snore intensity, pitch, frequency, and spectral envelope parameters [150]. Their study showed that the snore-to-snore variability of these features was increased in OSA subjects. Moreover, multiscale entropy [24] of audio recordings has also been shown to be useful to detect apneas.

8.5.5 Automatically Generated Features Using Deep Learning

All features mentioned before are handcrafted, they are extracted based on previous knowledge of the effect apneas have on the signals. The signals might, however, contain more information than these features. Therefore, more and more researchers have moved toward automatic feature extraction using deep learning, avoiding the challenging task of feature engineering. In [121], 21 studies detecting sleep apnea using deep learning were compared. Most studies were based on the ECG signal, but also SpO_2 and respiration were considered. The convolutional neural network (CNN) was the most popular network. It is, however, not clear if this is the best choice since some studies suggested that recurrent neural networks (RNNs) outperformed CNN [121]. The included ECG-based methods report global screening accuracies of 79.5–100%.

8.5.6 Multimodal OSA Detection

Since the scoring rules for sleep apnea are based both on a reduction in airflow and the presence of oxygen desaturation, the combination of respiration and oximetry has frequently been studied [119, 161]. Moreover, cardiac signals are often added to capture the presence of arousals. Xie et al. [174] investigated the combination of SpO_2 and ECG-based features. They concluded that the SpO_2-based classifier outperformed the ECG-based one and that only a slight increase in performance could be obtained when both feature sets were combined. Similar results were obtained in our own study when PPG features were added to a SpO_2-based classifier [53]. Behar et al., on the other hand, tested the combination of oximetry, sound, and actigraphy. When comparing the different modalities, oximetry outperformed the other two with a classification accuracy of 85.1%, but an increase in performance of 3.3% was obtained when all three modalities were combined.

8.5.7 Comparison of Methods and Modalities

When comparing the OSA detection performances which are reported by studies using different modalities, automatic classification using respiratory signals obtained lower performances than ECG and SpO_2-based methods. Although, manual scoring of OSA is mainly based on respiration. This might be due to a higher noise level in the respiratory signals [110].

ECG-based methods seem to outperform other modalities in single sensor applications. But this might be due to the fact that many ECG-based algorithms use public data sets, such as the Apnea-ECG database, which are less distorted by noise and might be easier to classify [110]. When different modalities are compared on the same data set, the SpO_2 signal performs best most of the times [53, 174]. In order to validate the potential added value of deep learning compared to feature-based algorithms, comparative studies on more challenging data sets than the Apnea-ECG data set should be performed.

Another observation is the fact that combining different sensor modalities often does not result in a relevant improvement of performance, probably because one of the sensors performs quite well and dominates the analysis [110]. When SpO_2-based methods are combined with other modalities, they tend to dominate the performance [53, 174]. Not all apneic events can, however, be detected using the SpO_2 signal since the AASM rules allow scoring apneas when no desaturation is present and hypopneas with only an arousal [26]. Therefore, sometimes the SpO_2 remains stable during apneas. For the databases used in [54], 11.5% of annotated apneas could not be linked to an oxygen desaturation larger than 1%. Therefore, further research should be conducted to better merge different modalities in order to detect apneas that were missed by individual modalities. Additionally, combining different modalities improves the classification of apneas into event subtypes (central vs. obstructive) [119].

Although the Apnea-ECG database has stimulated the research into apnea detection methods greatly, most of the recordings only contain ECG signals. The database can, therefore, not be used to develop multimodal algorithms nor can it be used to compare the performances of different modalities. Moreover, the control group was composed of healthy subjects, instead of subjects with other sleep-related complaints which are seen in the sleep lab. Papini et al. have shown in their study [125] a drop in accuracy of 15% and a drop in sensitivity of 30% for algorithms trained on the apnea-ECG data set which were tested on a database including a wider spectrum of sleep disorders. These results show that the performance of apnea detection algorithms is strongly dependent on the data set used to train the algorithm. Therefore, one should consider the target population for which the application is being developed and train on a similar data set with the same pretest probability to have OSA.

Next to the Apnea-ECG data set, the National Sleep Research Resource (NSRR) [49] also provides PSG data sets, of which the Sleep Heart Health Study (SHHS) [130] has been used the most for the automatic detection of sleep apnea. This data set, however, was recorded in the general population, whereas sleep experts advise against OSA screening in the general population, but rather advise to screen groups with a high pretest probability in order to ensure a high post-test probability [44].

The availability of large public data sets, ideally including wearable data recorded simultaneously with the gold-standard PSG, is even more important for algorithms based on deep learning. These algorithms are completely data-driven and their performance tends to improve if more data is available.

Another important factor when comparing different algorithms is the used testing method. In order to obtain a fair generalization performance, the test set should be patient independent. However, studies often do not mention specifically if subject independence of the test set was imposed [121]. Especially for cardiovascular signals, the classification performance can increase significantly if part of the data from a test set patient was included in the training set since cardiovascular features can be highly patient dependent.

8.5.8 Future of Sleep Apnea Screening: Beyond the AHI?

Most of the studies developing automated screening algorithms for OSA validate their methods based on the AHI, which is used to define OSA severity in clinical practice. The AHI, however, only takes into account the number of apneas, but not their severity. This might explain why the AHI does not correlate well with the sleepiness of patients nor with cardiovascular outcome [52]. Moreover, when an AHI cutoff of 15 events per hour of sleep is considered, prevalences of up to 50% can be observed in middle-aged male populations [25]. Therefore, researchers are searching for alternatives for the AHI in order to better phenotype OSA subjects and prioritize them for treatment.

In [58], Eckert reviews phenotypic approaches for OSA, with a focus on the PALM[8] score. This score describes four traits causing OSA: an impaired upper airway anatomy, a low respiratory arousal threshold, an unstable ventilatory control system, and ineffective upper airway dilator muscles. According to the most dominant trait, other treatments could be proposed. Measures of each of the causes were defined based on gold-standard PSG and PAP drops, but should still be tested on wearable measurements.

In [52, 158], on the other hand, SpO_2 severity parameters are proposed which correlate better than the AHI with the cardiovascular risk of OSA subjects.

This area of research is promising to enhance the phenotyping of OSA subjects, and in the future, these developed parameters could also be extracted from wearables in a home environment.

8.6 Conclusion

Sleep monitoring used to be a clearly defined domain based on EEG analysis. Bringing sleep monitoring to a home environment resulted in the emergence of a multitude of sensors and sleep staging algorithms. Within this chapter, unobtrusive state-of-the-art sensors and algorithms for sleep monitoring in adults were discussed as alternatives to the costly and uncomfortable in-hospital PSG. These novel technologies have the potential to enable a broader patient screening, early detection of sleep-related conditions, and a long-term follow-up at home.

Recently developed wearables and sensors placed around the patient were introduced, which measure PSG signals in a less obtrusive way or apply indirect measurements of the physiological parameters. The focus was on devices monitoring heart rate, respiration, and blood oxygenation, but actigraphy and portable EEG devices were also briefly discussed. An important aspect of these recently developed devices is that, although these provide increased comfort and a higher amount of data, the acquired data is of variable quality. This is caused by both the use of these technologies in an uncontrolled environment, as well as by the inherent characteristics of the technologies. In order to be able to exploit these solutions optimally, it is of high importance that research around this topic also includes the development and evaluation of algorithms for an automatic assessment of the quality of the data, as well as compensation methods for data of intermediate quality. Moreover, these novel devices require the development of application-specific algorithms for sleep monitoring.

The evaluation and comparison of sleep staging algorithms depend on many parameters, apart from studied population and algorithmic details. It includes sensor types, applied input signals, and mutual combinations. In general, good performances have been achieved based on solely the cardiac signal, being ECG or PPG,

[8] PALM is an acronym for the measured parameters for each of the four studied traits: **P**crit, **a**rousal threshold, **l**oop gain, and **m**uscle responsiveness.

as it conveys respiratory information as well. The latest state-of-the-art sleep staging algorithms are furthermore based on deep learning techniques, which are expected to become even more important for future implementations. However, a validation of wearable or unobtrusive signals is often lacking. Therefore, the field of sleep staging at home would benefit from the development of algorithms that consider the limitations of wearable/unobtrusive data from the start.

Obstructive sleep apnea is the most common sleep-related breathing disorder, and it, however, often remains undiagnosed. Therefore, researchers have been developing OSA detection algorithms using signals that can be easily acquired in a home environment. ECG and SpO_2-based algorithms have obtained the most promising results. However, most of the studies did not include experiments at home. These algorithms should thus be further tested on signals acquired unobtrusively from a home environment. Moreover, not much performance gain has been shown from multimodal algorithms, and smarter ways to combine different signal modalities should thus be investigated.

Acknowledgements This research received funding from Agentschap Innoveren en Ondernemen (VLAIO) 150466 OSA+, imec sbo funds Wearablehealth, and the Flemish Government (AI Research Program). SVH, CV, DH, and MD are affiliated to Leuven.AI - KU Leuven institute for AI, B-3000, Leuven, Belgium. KU Leuven Stadius acknowledges the financial support of imec.

References

1. Beddit. https://www.beddit.com/. [Accessed 3 Feb 2020]
2. Bitbrain dry eeg. https://www.bitbrain.com/neurotechnology-products/dry-eeg. [Accessed 15 Jan 2020]
3. Byteflies. https://www.byteflies.com/. [Accessed 3 Feb 2020]
4. Dreem. https://dreem.com/en. [Accessed 15 Jan 2020]
5. Muse. https://choosemuse.com/. [Accessed 15 Jan 2020]
6. Neurovigil. https://neurovigil.com/. [Accessed 15 Jan 2020]
7. Onera health. https://www.onerahealth.com/. [Accessed 21 Jan 2020]
8. Sleep score labs. https://www.sleepscore.com/. [Accessed 21 Jan 2020]
9. Sleep sheperd. https://sleepshepherd.com/. [Accessed 15 Jan 2020]
10. Tatch health. https://www.tatchhealth.com/. [Accessed 21 Jan 2020]
11. Abad DG (2009) Development of a capacitive bioimpedance measurement system. MS thesis
12. Al-Angari HM, Sahakian AV (2007) Use of sample entropy approach to study heart rate variability in obstructive sleep apnea syndrome. IEEE Trans Biomed Eng 54(10):1900–1904
13. Aleksandrowicz A, Leonhardt S (2007) Wireless and non-contact ecg measurement system -the aachen smartchair-. Acta Polytech 47(4–5)
14. Alvarez D, Hornero R, Marcos JV, Wessel N, Penzel T, Glos M, Del Campo F (2013) Assessment of feature selection and classification approaches to enhance information from overnight oximetry in the context of apnea diagnosis. Int J Neural Syst 23(05):1350020
15. American Association of Sleep Technologists: Sleep technology technical guideline. standard polysomnography (2012)
16. Ansari AH, De Wel O, Pillay K, Dereymaeker A, Jansen K, Van Huffel S, Naulaers G, De Vos M (2019) A convolutional neural network outperforming state-of-the-art sleep staging algorithms for both preterm and term infants. J Neural Eng

17. Atallah L, Serteyn A, Meftah M, Schellekens M, Vullings R, Bergmans J, Osagiator A, Oetomo SB (2014) Unobtrusive ecg monitoring in the nicu using a capacitive sensing array. Physiol Meas 35(5):895
18. Baek HJ, Chung GS, Kim KK, Park KS (2011) A smart health monitoring chair for nonintrusive measurement of biological signals. IEEE Trans Inf Technol Biomed 16(1):150–158
19. Baek HJ, Kim JS, Kim KK, Park KS (2008) System for unconstrained ecg measurement on a toilet seat using capacitive coupled electrodes: the efficacy and practicality. In: 2008 30th annual international conference of the IEEE engineering in medicine and biology society. IEEE, pp 2326–2328
20. Bassetti C, Dogas Z, Peigneux P (2014) Sleep medicine textbook. European sleep research society (2014)
21. Bates A, Ling MJ, Mann J, Arvind DK (2010) Respiratory rate and flow waveform estimation from tri-axial accelerometer data. In: 2010 international conference on body sensor networks. IEEE, pp 144–150
22. Beattie Z, Oyang Y, Statan A, Ghoreyshi A, Pantelopoulos A, Russell A, Heneghan C (2017) Estimation of sleep stages in a healthy adult population from optical plethysmography and accelerometer signals. Physiol Meas 38(11):1968
23. Beddr: Sleep tuner. https://www.beddrsleep.com/how-it-works. [Accessed 21 Jan 2020]
24. Behar J, Roebuck A, Shahid M, Daly J, Hallack A, Palmius N, Stradling J, Clifford GD (2014) Sleepap: an automated obstructive sleep apnoea screening application for smartphones. IEEE J Biomed Health Inform 19(1):325–331
25. Benjafield AV, Ayas NT, Eastwood PR, Heinzer R, Ip MS, Morrell MJ, Nunez CM, Patel SR, Penzel T, Pépin JL et al (2019) Estimation of the global prevalence and burden of obstructive sleep apnoea: a literature-based analysis. Lancet Respir Med 7(8):687–698
26. Berry RB, Budhiraja R, Gottlieb DJ, Gozal D, Iber C, Kapur VK, Marcus CL, Mehra R, Parthasarathy S, Quan SF et al (2012) Rules for scoring respiratory events in sleep: update of the 2007 aasm manual for the scoring of sleep and associated events. J Clin Sleep Med 8(05):597–619
27. Bianchi MT (2018) Sleep devices: wearables and nearables, informational and interventional, consumer and clinical. Metabolism 84:99–108
28. Bleichner MG, Debener S (2017) Concealed, unobtrusive ear-centered eeg acquisition: ceegrids for transparent eeg. Front Hum Neurosci 11:163
29. Bsoul M, Minn H, Tamil L (2010) Apnea medassist: real-time sleep apnea monitor using single-lead ecg. IEEE Trans Inf Technol Biomed 15(3):416–427
30. Camm AJ, Malik M, Bigger JT, Breithardt G, Cerutti S, Cohen RJ, Coumel P, Fallen EL, Kennedy HL, Kleiger R et al (1996) Heart rate variability: standards of measurement, physiological interpretation and clinical use. Circulation 93:1043–1065
31. Casal R, Di Persia LE, Schlotthauer G (2019) Sleep-wake stages classification using heart rate signals from pulse oximetry. Heliyon 5(10):e02529
32. Castaneda D, Esparza A, Ghamari M, Soltanpur C, Nazeran H (2018) A review on wearable photoplethysmography sensors and their potential future applications in health care. Int J Biosens Bioelectron 4(4):195
33. Castro I, Varon C, Moeyersons J, Gomez AV, Morales J, Deviaene M, Torfs T, Van Huffel S, Puers R, Van Hoof C (2019) Data quality assessment of capacitively-coupled ecg signals. In: Proceedings of the 2019 computing in cardiology conference (CinC), Singapore, pp 8–11
34. Castro I, Varon C, Torfs T, Van Huffel S, Puers R, Van Hoof C (2018) Evaluation of a multichannel non-contact ecg system and signal quality algorithms for sleep apnea detection and monitoring. Sensors 18(2):577
35. Castro ID, Mercuri M, Torfs T, Lorato I, Puers R, Van Hoof C (2018) Sensor fusion of capacitively coupled ecg and continuous-wave doppler radar for improved unobtrusive heart rate measurements. IEEE J Emerg Select Top Circuits Syst 8(2):316–328
36. Castro ID, Morariu R, Torfs T, Van Hoof C, Puers R (2016) Robust wireless capacitive ecg system with adaptive signal quality and motion artifact reduction. In: 2016 IEEE international symposium on medical measurements and applications (MeMeA). IEEE, pp 1–6

37. Castro ID, Patel A, Torfs T, Puers R, Van Hoof C (2019) Capacitive multi-electrode array with real-time electrode selection for unobtrusive ecg and bioz monitoring. In: 2019 41st annual international conference of the IEEE engineering in medicine and biology society (EMBC). IEEE, pp 5621–5624
38. Chamadiya B, Heuer S, Hofmann U, Wagner M (2009) Towards a capacitively coupled electrocardiography system for car seat integration. In: 4th European conference of the international federation for medical and biological engineering. Springer, pp 1217–1221
39. de Chazal P, Penzel T, Heneghan C (2004) Automated detection of obstructive sleep apnoea at different time scales using the electrocardiogram. Physiol Meas 25(4):967–983
40. Chesson AL Jr, Berry RB, Pack A (2003) Practice parameters for the use of portable monitoring devices in the investigation of suspected obstructive sleep apnea in adults. Sleep 26(7):907–913
41. Chung KY, Song K, Shin K, Sohn J, Cho SH, Chang JH (2017) Noncontact sleep study by multi-modal sensor fusion. Sensors 17(7):1685
42. Ciołek M, Niedźwiecki M, Sieklicki S, Drozdowski J, Siebert J (2014) Automated detection of sleep apnea and hypopnea events based on robust airflow envelope tracking in the presence of breathing artifacts. IEEE J Biomed Health Inform 19(2):418–429
43. Cohen J (1960) A coefficient of agreement for nominal scales. Educ Psychol Measur 20(1):37–46
44. Collop NA, Tracy SL, Kapur V, Mehra R, Kuhlmann D, Fleishman SA, Ojile JM (2011) Obstructive sleep apnea devices for out-of-center (ooc) testing: technology evaluation. J Clin Sleep Med 7(05):531–548
45. Compumedics: Sompte sleep monitor. https://www.compumedics.com.au/products/somte/. [Accessed 15 Jan 2020]
46. Czaplik M, Eilebrecht B, Walocha R, Walter M, Schauerte P, Leonhardt S, Rossaint R (2012) The reliability and accuracy of a noncontact electrocardiograph system for screening purposes. Anesthesia Analgesia 114(2):322–327
47. De Chazal P, Fox N, O'hare E, Heneghan C, Zaffaroni A, Boyle P, Smith S, O'connell C, McNicholas WT (2011) Sleep/wake measurement using a non-contact biomotion sensor. J Sleep Research 20(2):356–366
48. De Chazal P, Heneghan C, Sheridan E, Reilly R, Nolan P, O'Malley M (2003) Automated processing of the single-lead electrocardiogram for the detection of obstructive sleep apnoea. IEEE Trans Biomed Eng 50(6):686–696
49. Dean DA, Goldberger AL, Mueller R, Kim M, Rueschman M, Mobley D, Sahoo SS, Jayapandian CP, Cui L, Morrical MG et al (2016) Scaling up scientific discovery in sleep medicine: the national sleep research resource. Sleep 39(5):1151–1164
50. Debener S, Emkes R, De Vos M, Bleichner M (2015) Unobtrusive ambulatory eeg using a smartphone and flexible printed electrodes around the ear. Sci Rep 5:16743
51. Dement W, Kleitman N (1957) Cyclic variations in eeg during sleep and their relation to eye movements, body motility, and dreaming. Electroencephalogr Clin Neurophysiol 9(4):673–690
52. Deviaene M, Borzée P, van Gilst M, van Dijk J, Overeem S, Buyse B, Testelmans D, Van Huffel S, Varon C (2020) Multilevel interval coded scoring to assess the cardiovascular status of sleep apnea patients using oxygen saturation markers. Accepted for publication in IEEE Transactions on Biomedical Engineering, available in early access
53. Deviaene M, Lázaro J, Huysmans D, Testelmans D, Buyse B, Van Huffel S, Varon C (2018) Sleep apnea detection using pulse photoplethysmography. In: 2018 computing in cardiology conference (CinC), vol 45. IEEE, pp 1–4
54. Deviaene M, Testelmans D, Buyse B, Borzée P, Van Huffel S, Varon C (2018) Automatic screening of sleep apnea patients based on the spo 2 signal. IEEE J Biomed Health Inform 23(2):607–617
55. Devot S, Dratwa R, Naujokat E (2010) Sleep/wake detection based on cardiorespiratory signals and actigraphy. In: 2010 annual international conference of the IEEE engineering in medicine and biology. IEEE, pp 5089–5092

56. Domingues A, Paiva T, Sanches JM (2014) Hypnogram and sleep parameter computation from activity and cardiovascular data. IEEE Trans Biomed Eng 61(6):1711–1719
57. Douglas N, White D, Pickett CK, Weil J, Zwillich C (1982) Respiration during sleep in normal man. Thorax 37(11):840–844
58. Eckert DJ (2018) Phenotypic approaches to obstructive sleep apnoea-new pathways for targeted therapy. Sleep Med Rev 37:45–59
59. emfit: Emfit sleep apnea monitor. https://www.emfit.com/emfit-qs-for-diagnosing-sleep-apnea. [Accessed 3 Feb 2020]
60. Faust O, Acharya UR, Ng E, Fujita H (2016) A review of ecg-based diagnosis support systems for obstructive sleep apnea. J Mech Med Biol 16(01):1640004
61. Faust O, Razaghi H, Barika R, Ciaccio EJ, Acharya UR (2019) A review of automated sleep stage scoring based on physiological signals for the new millennia. Computer methods and programs in biomedicine
62. Flemons WW, Littner MR, Rowley JA, Gay P, Anderson WM, Hudgel DW, McEvoy RD, Loube DI (2003) Home diagnosis of sleep apnea: a systematic review of the literature: an evidence review cosponsored by the american academy of sleep medicine, the american college of chest physicians, and the american thoracic society. Chest 124(4):1543–1579
63. Fonseca P, Aarts RM, Long X, Rolink J, Leonhardt S (2016) Estimating actigraphy from motion artifacts in ecg and respiratory effort signals. Physiol Meas 37(1):67–82
64. Fonseca P, Den Teuling N, Long X, Aarts RM (2018) A comparison of probabilistic classifiers for sleep stage classification. Physiol Meas 39(5):aabbc2. https://doi.org/10.1088/1361-6579/aabbc2
65. Fonseca P, Long X, Radha M, Haakma R, Aarts RM, Rolink J (2015) Sleep stage classification with ecg and respiratory effort. Physiol Meas 36(10):2027
66. Fonseca P, Weysen T, Goelema MS, Møst EI, Radha M, Lunsingh Scheurleer C, van den Heuvel L, Aarts RM (2017) Validation of photoplethysmography-based sleep staging compared with polysomnography in healthy middle-aged adults. Sleep 40(7):zsx097
67. Gil E, Vergara JM, Laguna P (2008) Detection of decreases in the amplitude fluctuation of pulse photoplethysmography signal as indication of obstructive sleep apnea syndrome in children. Biomed Signal Process Control 3(3):267–277
68. Goldberger AL, Amaral LA, Glass L, Hausdorff JM, Ivanov PC, Mark RG, Mietus JE, Moody GB, Peng CK, Stanley HE (2000) Physiobank, physiotoolkit, and physionet: components of a new research resource for complex physiologic signals. Circulation 101(23):e215–e220
69. Goverdovsky V, Looney D, Kidmose P, Mandic DP (2015) In-ear eeg from viscoelastic generic earpieces: Robust and unobtrusive 24/7 monitoring. IEEE Sens J 16(1):271–277
70. Guilleminault C, Winkle R, Connolly S, Melvin K, Tilkian A (1984) Cyclical variation of the heart rate in sleep apnoea syndrome: Mechanisms, and usefulness of 24 h electrocardiography as a screening technique. The Lancet 323(8369):126–131
71. Harper RM, Schechtman VL, Kluge KA (1987) Machine classification of infant sleep state using cardiorespiratory measures. Electroencephal Clin Neurophysiol 67(4):379–387. https://doi.org/10.1016/0013-4694(87)90126-X. http://www.sciencedirect.com/science/article/pii/001346948790126X
72. Huysmans D, Borzée P, Testelmans D, Buyse B, Willemen T, Huffel SV, Varon C (2019) Evaluation of a commercial ballistocardiography sensor for sleep apnea screening and sleep monitoring. Sensors 19(9):2133
73. Hwang SH, Lee YJ, Jeong DU, Park KS (2016) Unconstrained sleep stage estimation based on respiratory dynamics and body movement. Methods Inf Med 55(06):545–555
74. Hwang SH, Lee YJ, Jeong DU, Park KS et al (2016) Oxygen desaturation index estimation through unconstrained cardiac sympathetic activity assessment using three ballistocardiographic systems. Respiration 92(2):90–97
75. Inc B (2020) Mcot patch. https://www.myheartmonitor.com/device/mcot-patch/. [Accessed 3 Feb 2020]
76. Ishida S, Shiozawa N, Fujiwara Y, Makikawa M (2007) Electrocardiogram measurement during sleep with wearing clothes using capacitively-coupled electrodes. In: 2007 29th annual

international conference of the IEEE engineering in medicine and biology society. IEEE, pp 2647–2650
77. Joshi R, Bierling BL, Long X, Weijers J, Feijs L, Van Pul C, Andriessen P (2018) A ballistographic approach for continuous and non-obtrusive monitoring of movement in neonates. IEEE J Trans Eng Health Med 6:1–10
78. Kagawa M, Tojima H, Matsui T (2016) Non-contact diagnostic system for sleep apnea-hypopnea syndrome based on amplitude and phase analysis of thoracic and abdominal doppler radars. Med Biol Eng Comput 54(5):789–798
79. Kelly JM, Strecker RE, Bianchi MT (2012) Recent developments in home sleep-monitoring devices. ISRN Neurol 2012
80. Kemp B, Janssen AJ, Van Der Kamp B (1998) Body position can be monitored in 3d using miniature accelerometers and earth-magnetic field sensors. Electroencephal Clin Neurophysiol/Electromyography Motor Control 109(6):484–488
81. Kemp B, Zwinderman AH, Tuk B, Kamphuisen HA, Oberye JJ (2000) Analysis of a sleep-dependent neuronal feedback loop: the slow-wave microcontinuity of the eeg. IEEE Trans Biomed Eng 47(9):1185–1194
82. Kidmose P, Looney D, Jochumsen L, Mandic DP (2013) Ear-eeg from generic earpieces: A feasibility study. In: 2013 35th annual international conference of the IEEE engineering in medicine and biology society (EMBC). IEEE, pp 543–546
83. Kim KK, Lim YK, Park KS (2004) The electrically noncontacting ecg measurement on the toilet seat using the capacitively-coupled insulated electrodes. In: The 26th annual international conference of the IEEE engineering in medicine and biology society, vol 1. IEEE, pp 2375–2378
84. Koley BL, Dey D (2013) Automatic detection of sleep apnea and hypopnea events from single channel measurement of respiration signal employing ensemble binary svm classifiers. Measurement 46(7):2082–2092
85. Kortelainen JM, Mendez MO, Bianchi AM, Matteucci M, Cerutti S (2010) Sleep staging based on signals acquired through bed sensor. IEEE Trans Inf Technol Biomed 14(3):776–785
86. Kranjec J, Beguš S, Drnovšek J, Geršak G (2013) Novel methods for noncontact heart rate measurement: A feasibility study. IEEE Trans Instrum Meas 63(4):838–847
87. Kurihara Y, Watanabe K (2012) Sleep-stage decision algorithm by using heartbeat and body-movement signals. IEEE Trans Syst, Man, Cybern-Part A: Syst Humans 42(6):1450–1459
88. Lázaro J, Gil E, Bailón R, Mincholé A, Laguna P (2013) Deriving respiration from photoplethysmographic pulse width. Med Biol Eng Comput 51(1–2):233–242
89. Lee HJ, Hwang SH, Yoon HN, Lee WK, Park KS (2015) Heart rate variability monitoring during sleep based on capacitively coupled textile electrodes on a bed. Sensors 15(5):11295–11311
90. Leonhardt S, Aleksandrowicz A (2008) Non-contact ecg monitoring for automotive application. In: 2008 5th international summer school and symposium on medical devices and biosensors. IEEE, pp 183–185
91. Li C, Cummings J, Lam J, Graves E, Wu W (2009) Radar remote monitoring of vital signs. IEEE Microwave Mag 10(1):47–56
92. Li C, Lubecke VM, Boric-Lubecke O, Lin J (2013) A review on recent advances in doppler radar sensors for noncontact healthcare monitoring. IEEE Trans Microw Theory Tech 61(5):2046–2060
93. Li Q, Li Q, Liu C, Shashikumar SP, Nemati S, Clifford GD (2018) Deep learning in the cross-time frequency domain for sleep staging from a single-lead electrocardiogram. Physiol Meas 39(12):124005
94. Lim YG, Hong KH, Kim KK, Shin JH, Lee SM, Chung GS, Baek HJ, Jeong DU, Park KS (2011) Monitoring physiological signals using nonintrusive sensors installed in daily life equipment. Biomed Eng Lett 1(1):11–20
95. Lim YG, Kim KK, Park KS (2007) Ecg recording on a bed during sleep without direct skin-contact. IEEE Trans Biomed Eng 54(4):718–725

96. Lim YG, Lee JS, Lee SM, Lee HJ, Park KS (2014) Capacitive measurement of ecg for ubiquitous healthcare. Ann Biomed Eng 42(11):2218–2227
97. Lim YK, Kim KK, Park KS (2004) The ecg measurement in the bathtub using the insulated electrodes. In: The 26th annual international conference of the IEEE engineering in medicine and biology society, vol 1. IEEE, pp 2383–2385
98. Lisenby M, Richardson P, Welch A (1976) Detection of cyclic sleep phenomena using instantaneous heart rate. Electroencephalogr Clin Neurophysiol 40(2):169–177
99. Littner M, Kushida CA, Anderson WM, Bailey D, Berry RB, Davila DG, Hirshkowitz M, Kapen S, Kramer M, Loube D, Wise M, Johnson SF (2003) Practice parameters for the role of actigraphy in the study of sleep and circadian rhythms: an update for 2002. Sleep 26(3):337–341. https://doi.org/10.1093/sleep/26.3.337
100. Long X, Fonseca P, Foussier J, Haakma R, Aarts RM (2013) Sleep and wake classification with actigraphy and respiratory effort using dynamic warping. IEEE J Biomed Health Inform 18(4):1272–1284
101. Long X, Yang J, Weysen T, Haakma R, Foussier J, Fonseca P, Aarts RM (2014) Measuring dissimilarity between respiratory effort signals based on uniform scaling for sleep staging. Physiol Meas 35(12):2529
102. Loomis AL, Harvey EN, Hobart GA III (1938) Distribution of disturbance-patterns in the human electroencephalogram, with special reference to sleep. J Neurophysiol 1(5):413–430
103. Macías R, García M, Ramos J, Bragos R, Fernández M (2013) Ventilation and heart rate monitoring in drivers using a contactless electrical bioimpedance system. J Phys: Conf Ser, 434, 012047. IOP Publishing
104. Mancia G (1993) Autonomic modulation of the cardiovascular system during sleep
105. Matar G, Lina JM, Carrier J, Kaddoum G (2018) Unobtrusive sleep monitoring using cardiac, breathing and movements activities: an exhaustive review. IEEE Access 6:45129–45152
106. Matsuda T, Makikawa M (2008) Ecg monitoring of a car driver using capacitively-coupled electrodes. In: 2008 30th annual international conference of the IEEE engineering in medicine and biology society. IEEE, pp 1315–1318
107. MBrainTrain: ceegrids. https://mbraintrain.com/concealed-eeg/. [Accessed 3 Feb 2020]
108. McDonald NJ, Anumula HA, Duff E, Soussou W (2012) Noncontact ecg system for unobtrusive long-term monitoring. In: 2012 annual international conference of the IEEE engineering in medicine and biology society. IEEE, pp 1614–1618
109. Mendonça F, Mostafa SS, Ravelo-García AG, Morgado-Dias F, Penzel T (2018) Devices for home detection of obstructive sleep apnea: A review. Sleep Med Rev 41:149–160
110. Mendonça F, Mostafa SS, Ravelo-García AG, Morgado-Dias F, Penzel T (2018) A review of obstructive sleep apnea detection approaches. IEEE J Biomed Health Inform 23(2):825–837
111. Mercuri M, Lorato IR, Liu YH, Wieringa F, Van Hoof C, Torfs T (2019) Vital-sign monitoring and spatial tracking of multiple people using a contactless radar-based sensor. Nat Electron 2(6):252–262
112. Migliorini M, Bianchi AM, Nisticò D, Kortelainen J, Arce-Santana E, Cerutti S, Mendez MO (2010) Automatic sleep staging based on ballistocardiographic signals recorded through bed sensors. In: 2010 annual international conference of the IEEE engineering in medicine and biology, pp 3273–3276. https://doi.org/10.1109/IEMBS.2010.5627217
113. Mikkelsen KB, Tabar YR, Kappel SL, Christensen CB, Toft HO, Hemmsen MC, Rank ML, Otto M, Kidmose P (2019) Accurate whole-night sleep monitoring with dry-contact ear-eeg. Sci Rep 9(1):1–12
114. Molinari S, Foulkes D (1969) Tonic and phasic events during sleep: psychological correlates and implications. Percept Mot Skills 29(2):343–368. https://doi.org/10.2466/pms.1969.29.2.343 PMID: 4310916
115. Moody G, Mark R, Goldberger A, Penzel T (2000) Stimulating rapid research advances via focused competition: The computers in cardiology challenge 2000. In: Computers in cardiology 2000, vol 27 (Cat. 00CH37163). IEEE, pp 207–210
116. Moody GB, Mark RG, Zoccola A, Mantero S (1985) Derivation of respiratory signals from multi-lead ecgs. Comput Cardiol 12(1985):113–116

117. Morales JF, Varon C, Deviaene M, Borzée P, Testelmans D, Buyse B, Van Huffel S (2017) Sleep apnea hypopnea syndrome classification in spo 2 signals using wavelet decomposition and phase space reconstruction. In: 2017 IEEE 14th international conference on wearable and implantable body sensor networks (BSN). IEEE, pp 43–46
118. Morales Tellez JF, Deviaene M, Milagro J, Testelmans D, Buyse B, Willems R, Orini M, Van Huffel S, Bailón R, Varon C (2019) Evaluation of methods to characterize the change of the respiratory sinus arrhythmia with age in sleep apnea patients. In: Proceedings of EMBC. IEEE (2019)
119. Moret-Bonillo V, Alvarez-Estévez D, Fernández-Leal A, Hernández-Pereira E (2014) Intelligent approach for analysis of respiratory signals and oxygen saturation in the sleep apnea/hypopnea syndrome. Open Med Inform J 8:1
120. Morgenthaler T, Alessi C, Friedman L, Owens J, Kapur V, Boehlecke B, Brown T, Chesson A Jr, Coleman J, Lee-Chiong T et al (2007) Practice parameters for the use of actigraphy in the assessment of sleep and sleep disorders: an update for 2007. Sleep 30(4):519–529
121. Mostafa SS, Mendonça F, Ravelo-García AG, Morgado-Dias F (2019) A systematic review of detecting sleep apnea using deep learning. Sensors 19(22):4934
122. Mousavi S, Afghah F, Acharya UR (2019) Sleepeegnet: automated sleep stage scoring with sequence to sequence deep learning approach. PLoS ONE 14(5):e0216456
123. Obeid D, Sadek S, Zaharia G, El Zein G (2010) Multitunable microwave system for touchless heartbeat detection and heart rate variability extraction. Microw Opt Technol Lett 52(1):192–198
124. Oksenberg A, Khamaysi I, Silverberg DS, Tarasiuk A (2000) Association of body position with severity of apneic events in patients with severe nonpositional obstructive sleep apnea. Chest 118(4):1018–1024
125. Papini GB, Fonseca P, Margarito J, van Gilst MM, Overeem S, Bergmans JW, Vullings R (2018) On the generalizability of ecg-based obstructive sleep apnea monitoring: merits and limitations of the apnea-ecg database. In: 2018 40th annual international conference of the IEEE engineering in medicine and biology society (EMBC). IEEE, pp 6022–6025
126. Paquet J, Kawinska A, Carrier J (2007) Wake detection capacity of actigraphy during sleep. Sleep 30(10):1362–1369
127. Penzel T, Moody GB, Mark RG, Goldberger AL, Peter JH (2000) The apnea-ecg database. In: Computers in cardiology 2000, vol 27 (Cat. 00CH37163). IEEE, pp 255–258
128. Petkie DT, Benton C, Bryan E (2009) Millimeter wave radar for remote measurement of vital signs. In: 2009 IEEE radar conference. IEEE, pp 1–3
129. Pinheiro E, Postolache O, Girão P (2012) A practical approach concerning the capacitive acquisition of the electrocardiogram in a moving wheelchair. In: Proceedings of the XX IMEKO world congress
130. Quan SF, Howard BV, Iber C, Kiley JP, Nieto FJ, O'Connor GT, Rapoport DM, Redline S, Robbins J, Samet JM et al (1997) The sleep heart health study: design, rationale, and methods. Sleep 20(12):1077–1085
131. Radha M, Fonseca P, Moreau A, Ross M, Cerny A, Anderer P, Long X, Aarts RM (2019) Sleep stage classification from heart-rate variability using long short-term memory neural networks. Sci Rep 9(1):1–11
132. Rechtschaffen A, Kales A (1968) A manual of standardized terminology, techniques, and scoring system for sleep stages for human subjects. Natl Inst Health 204
133. Redmond SJ, de Chazal P, O'Brien C, Ryan S, McNicholas WT, Heneghan C (2007) Sleep staging using cardiorespiratory signals. Somnologie-Schlafforschung und Schlafmedizin 11(4):245–256
134. Redmond SJ, Heneghan C (2006) Cardiorespiratory-based sleep staging in subjects with obstructive sleep apnea. IEEE Trans Biomed Eng 53(3):485–496
135. ResMed: S+ sleep monitor. https://splus.resmed.com/. [Accessed 3 Feb 2020]
136. Resonea: Drowzle pro. https://www.resonea.com/drowzle-pro/. [Accessed 21 Jan 2020]
137. Richardson P (1967) The insulated electrode: a pasteless electrocardiographic technique. In: 20th Annual conference on engineering in medicine and biology, vol 9, pp 15–17

138. Roebuck A, Monasterio V, Gederi E, Osipov M, Behar J, Malhotra A, Penzel T, Clifford G (2013) A review of signals used in sleep analysis. Physiol Meas 35(1):R1
139. Roehrs T, Zorick F, Wittig R, Conway W, Roth T (1989) Predictors of objective level of daytime sleepiness in patients with sleep-related breathing disorders. Chest 95(6):1202–1206
140. Roomkham S, Lovell D, Cheung J, Perrin D (2018) Promises and challenges in the use of consumer-grade devices for sleep monitoring. IEEE Rev Biomed Eng 11:53–67
141. Rosenberg RS, Van Hout S (2013) The american academy of sleep medicine inter-scorer reliability program: sleep stage scoring. J Clin Sleep Med 9(01):81–87
142. Sadeh A, Acebo C (2002) The role of actigraphy in sleep medicine. Sleep Med Rev 6(2):113–124
143. Sateia MJ (2014) International classification of sleep disorders. Chest 146(5):1387–1394
144. Schnall RP, Shlitner A, Sheffy J, Kedar R, Lavie P (1999) Periodic, profound peripheral vasoconstriction as a new marker of obstructive sleep apnea. Sleep 22(7):939–946
145. Schulz H (2008) Rethinking sleep analysis comment on the aasm manual for the scoring of sleep and associated events. J Clin Sleep Med 4(02):99–103
146. Schumm J, Axmann S, Arnrich B, Tröster G (2010) Automatic signal appraisal for unobtrusive ecg measurements. Int. J. Bioelectromagn 12(4):158–164
147. Schumm J, Setz C, Bächlin M, Bächler M, Arnrich B, Tröster G (2010) Unobtrusive physiological monitoring in an airplane seat. Pers Ubiquit Comput 14(6):541–550
148. Scullin MK, Bliwise DL (2015) Sleep, cognition, and normal aging: integrating a half century of multidisciplinary research. Perspect Psychol Sci 10(1):97–137
149. Singh RK, Sarkar A, Anoop C (2016) A health monitoring system using multiple non-contact ecg sensors for automotive drivers. In: 2016 IEEE international instrumentation and measurement technology conference proceedings. IEEE, pp 1–6
150. Solà-Soler J, Fiz JA, Morera J, Jané R (2012) Multiclass classification of subjects with sleep apnoea-hypopnoea syndrome through snoring analysis. Med Eng Phys 34(9):1213–1220
151. Solutions CS (2020) Watch pat. https://cardiosleepsolutions.com/product/watchpat-one/. [Accessed 21 Jan 2020]
152. Solutions VCH (2020) Vivalnk ecg patch. http://www.vivalnk.com/ecg-monitor. [Accessed 3 Feb 2020]
153. Somers VK, Dyken ME, Mark AL, Abboud FM (1993) Sympathetic-nerve activity during sleep in normal subjects. N Engl J Med 328(5):303–307
154. Stephansen JB, Olesen AN, Olsen M, Ambati A, Leary EB, Moore HE, Carrillo O, Lin L, Han F, Yan H et al (2018) Neural network analysis of sleep stages enables efficient diagnosis of narcolepsy. Nat Commun 9(1):1–15
155. Sun Y, Yu XB (2016) Capacitive biopotential measurement for electrophysiological signal acquisition: A review. IEEE Sens J 16(9):2832–2853
156. Supratak A, Dong H, Wu C, Guo Y (2017) Deepsleepnet: A model for automatic sleep stage scoring based on raw single-channel eeg. IEEE Trans Neural Syst Rehabil Eng 25(11):1998–2008
157. iRhythm Technologies: Zio xt. https://www.irhythmtech.com/products-services/zio-xt. [Accessed 3 Feb 2020]
158. Terrill PI (2019) A review of approaches for analysing obstructive sleep apnoea-related patterns in pulse oximetry data. Respirology
159. Tsinalis O, Matthews PM, Guo Y, Zafeiriou S (2016) Automatic sleep stage scoring with single-channel eeg using convolutional neural networks. arXiv preprint arXiv:1610.01683
160. Uçar MK, Bozkurt MR, Bilgin C, Polat K (2018) Automatic sleep staging in obstructive sleep apnea patients using photoplethysmography, heart rate variability signal and machine learning techniques. Neural Comput Appl 29(8):1–16
161. Uddin M, Chow C, Su S (20148) Classification methods to detect sleep apnea in adults based on respiratory and oximetry signals: a systematic review. Physiol Meas 39(3):03TR01
162. Ueno A, Akabane Y, Kato T, Hoshino H, Kataoka S, Ishiyama Y (2007) Capacitive sensing of electrocardiographic potential through cloth from the dorsal surface of the body in a supine position: A preliminary study. IEEE Trans Biomed Eng 54(4):759–766

163. Varon C, Caicedo A, Testelmans D, Buyse B, Van Huffel S (2015) A novel algorithm for the automatic detection of sleep apnea from single-lead ecg. IEEE Trans Biomed Eng 62(9):2269–2278
164. Varon C, Morales J, Lázaro J, Orini M, Deviaene M, Kontaxis S, Testelmans D, Buyse B, Borzée P, Sörnmo L, Laguna P, Gil E, Bailón R (2020) A comparative study of ecg-derived respiration in ambulatory monitoring using the single-lead ecg. Submitted to scientific reports
165. Varon C, Van Huffel S (2017) Complexity and nonlinearities in cardiorespiratory signals in sleep and sleep apnea. In: Complexity and nonlinearity in cardiovascular signals. Springer, pp 503–537
166. Vehkaoja A, Salo A, Peltokangas M, Verho J, Salpavaara T, Lekkala J (2014) Unconstrained night-time heart rate monitoring with capacitive electrodes. In: XIII mediterranean conference on medical and biological engineering and computing 2013. Springer, pp 1511–1514
167. Weil MB, Oehler M, Schilling M, Maier LS (2012) First clinical evaluation of a novel capacitive ecg system in patients with acute myocardial infarction. Clin Res Cardiol 101(3):165–174
168. Willemen T, Van Deun D, Verhaert V, Vandekerckhove M, Exadaktylos V, Verbraecken J, Van Huffel S, Haex B, Vander Sloten J (2014) An evaluation of cardiorespiratory and movement features with respect to sleep-stage classification. IEEE J Biomed Health Inform 18(2):661–669
169. Willemen T, Varon C, Dorado AC, Haex B, Vander Sloten J, Van Huffel S (2015) Probabilistic cardiac and respiratory based classification of sleep and apneic events in subjects with sleep apnea. Physiol Meas 36(10):2103
170. Withings: Sleep mat. https://www.beddit.com/. [Accessed 3 Feb 2020]
171. Wu KF, Zhang YT (2008) Contactless and continuous monitoring of heart electric activities through clothes on a sleeping bed. In: 2008 international conference on information technology and applications in biomedicine. IEEE, pp 282–285
172. X, P.: Potential x smart earbuds. https://www.potential-x.com/. [Accessed 3 Feb 2020]
173. Xiao M, Yan H, Song J, Yang Y, Yang X (2013) Sleep stages classification based on heart rate variability and random forest. Biomed Signal Process Control 8(6):624–633
174. Xie B, Minn H (2012) Real-time sleep apnea detection by classifier combination. IEEE Trans Inf Technol Biomed 16(3):469–477
175. Zaunseder S, Henning A, Wedekind D, Trumpp A, Malberg H (2017) Unobtrusive acquisition of cardiorespiratory signals. Somnologie 21(2):93–100
176. Zhu K, Li M, Akbarian S, Hafezi M, Yadollahi A, Taati B (2019) Vision-based heart and respiratory rate monitoring during sleep-a validation study for the population at risk of sleep apnea. IEEE J Transl Eng Health Med 7:1–8
177. Ziegler D, Laux G, Dannehl K, Spüler M, Mühlen H, Mayer P, Gries F (1992) Assessment of cardiovascular autonomic function: age-related normal ranges and reproducibility of spectral analysis, vector analysis, and standard tests of heart rate variation and blood pressure responses. Diabet Med 9(2):166–175
178. Zink MD, Brüser C, Stüben BO, Napp A, Stöhr R, Leonhardt S, Marx N, Mischke K, Schulz JB, Schiefer J (2017) Unobtrusive nocturnal heartbeat monitoring by a ballistocardiographic sensor in patients with sleep disordered breathing. Sci Rep 7(1):13175

Part III
Beyond Wearables

Chapter 9
The CueMinder Project: Patient-Driven Wearable Technology to Improve Quality of Life

Danielle Kempner and Martha L. Hall

Abstract CueMinder is a wearable device aimed at promoting independence for adults with Dementia living in assisted-living memory care units. Paired with a uniquely tailored alerting system, the product assists in cueing and reminding the individual to perform Activities of Daily Living (ADLs) with hopes of increasing user independence which would ultimately aid in relieving caregiver burden. The device was created to serve this patient population through image and vocal cues, rather than text. Using images and vocal cues will help overcome the current communication barrier that exists between a caregiver and a patient, as well as other associated communication issues that are extremely common for adults living with Dementia. Problems that may arise from communication are frustration, combativeness, aggression, as well as other issues that lead to increased caregiver burden and preventable patient-related accidents, in many assisted-living memory care units. This device not only aims at reducing these types of accidents but also increasing patient independence and therefore overall patient quality of life and longevity.

9.1 Introduction

Dementia is now being classified as a global epidemic [1]. With the current estimate of 47.5 million people with dementia set to triple by 2050, there is a need for resources to care for these individuals and to support caregivers [1]. Caregiver burden and burnout is a potentially increasing risk. The goal of this project is to design and develop a device that allows for a bettered quality of life not only for individuals with Dementia but also for caregivers.

Our target population is individuals with Dementia living in nursing care/assisted-living memory care units. There is an increased need for this type of technology in these units due to understaffing and reduction in quality care [2]. Since 2014, 1 in 8 nursing homes were cited for having too few nurses [2]. Even though the device has been tailored with this a particular population in mind, it has great potential

D. Kempner · M. L. Hall (✉)
University of Delaware, Newark, DE, USA
e-mail: mlucinda@udel.edu

G. D. Gargiulo and G. R. Naik (eds.), *Wearable/Personal Monitoring Devices Present to Future*, https://doi.org/10.1007/978-981-16-5324-7_9

for widespread application and could be additionally helpful for those providing at-home, independent care to persons with Dementia. This technology as a whole may help decrease patient reliance on the caregiver, and ultimately increase independence and longevity. For this population, decline often occurs when people lose independence from homogenous treatment, rather than individualized care. The aim of our device is to uniquely tailor each user's device to fit with their own agenda/schedule, thus treating them as an individual on their own care path.

9.2 Design Process

9.2.1 Empathy

Our design process starts with empathy and is based on design thinking (Fig. 9.1). We believe it is key to understand patient and caregiver needs prior to starting to design. We often use interviews and patient observation to more deeply understand and ultimately define the "problem". We begin by identifying daily challenges for both the patient and caregiver. By engaging the interviewee with discussion prompts, rather than yes or no questions, we can explore the challenges in depth and breadth.

For this project, we interviewed patients, caregivers, and additional stakeholders, such as therapists and specialists. However, it should be noted that individuals with Dementia are often not active subjects for interviews due to ethical considerations [3]. Often, it is difficult to interview people suffering from Dementia because of the communication barrier that exists in relation to any aphasia, speech impediments, lack of understanding, or other similar issues. In order to compensate, patients with similar

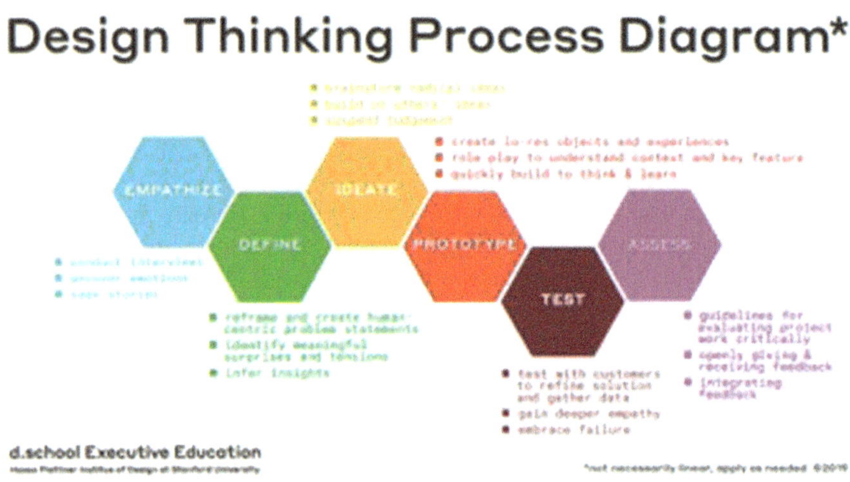

Fig. 9.1 Design thinking process adopted for device design

challenges served as interview subjects, specifically patients recovering from a stroke. There are similarities when it comes to communication barriers and memory-related issues between stroke patients and people suffering from Dementia. For patients recovering from a stroke, some are able to recall, in great detail, what types of communication strategies worked best, what type of care was best responded to, the frustrations they felt, and more [4]. These interviews were insightful when it came to the functionality and design of our device.

9.2.2 Define

The next step is to define the problem. The main needs for this device are to increase patient independence, quality, and longevity of life, relieve caregiver burden, prevent patient accidents, and increase care efficiency. For individuals with Dementia, loss of independence can stem not only from neurological decline but also due to a cultural perception of disability [5]. Increasing patient independence may lead to increased confidence and overall quality of life. If a patient is able to be more independent, they will be more active as well as have more confidence in their capabilities. With this increase in independence and life quality, health outcomes may improve, which may slow the decline and progression of Dementia.

Individuals with Dementia could be at a higher risk of accidents. Accidents can be related to falls, dehydration, malnutrition, over-medication, bedsores, UTI/bowel issues, and much more. The idea of our device is to target each issue individually and have tasks and cues set in place to try and prevent certain accidents from occurring. In addition to accident prevention, these cues are set to a cyclic schedule in order to improve overall efficiency in the units. If users themselves can be kept on schedule without having to be given constant assistance and reminders to do their daily activities, then operations in these units will move much more smoothly and will carry much less burden. Since the device aims to have a schedule set to highlight the user's greatest moments of lucidity and activity, the hope is that it will aid in giving users a sense of rhythm and keep them more independent and efficient while also making them less reliant on caregivers.

9.2.3 Ideate

People suffering from Dementia face a harsh reality when it comes to suffering deficits. In more ways than one, they received the short end of the stick. Not only is memory decline a large factor but the people also suffer when it comes to communication, as well as taking constant blows to their self-pride as they see their independence dwindle down to next to nothing.

The first thing that tends to come to a person's mind when Dementia is mentioned is the toll it takes on one's memory. Memory takes many forms. It can be in reference to one's past, one's ability to recall facts or particulars, remembering what you had for dinner last night, or even remembering what you need to get done in a day. All of these forms of memory are severely impacted by Dementia, and since many of these factors guide our lives and our decisions, it can be easy to see how someone with Dementia may easily feel lost and look for a sense of guidance.

Communication issues also commonly arise for people with Dementia. Depending on the person, they may have difficulty in perceiving what is communicated to them, responding to what they are perceiving, or in many cases, both. Understandably, a large amount of frustration accompanies these communication issues: frustration with not being understood, frustration with not understanding, and frustration of not being able to engage as they once could. Communication is a two-way street, and thus if one of those streets is blocked, the connection is severed. In addition to frustration, problems that consistently arise from communication are combativeness, aggression, and a whole slew of other issues that lead to increased caregiver burden, as well as accidents, in many care units. The communication barrier is a problem in need of a comprehensive solution, one that can help bridge the gap that currently exists.

In addition to the memory deficits and communication issues, a person with Dementia loses one thing that each and every person holds near and dear, their pride. It is hard to watch as your independence is slowly stripped away. Much of our pride comes from what we are able to do and make of ourselves. If we are limited in our abilities by the amount we come to depend on others, then our self-confidence and our sense of self-pride are greatly diminished. "Studies focused on enhancing the lives of people with dementia suggest that a supportive and inclusive environment is crucial in moving forward post diagnosis, sustaining identity, and continuing to live a life with meaning and value" (Yates). Keeping a person with Dementia independent and reliant on their own abilities can go a long way in improving life quality and ultimately, longevity.

9.2.4 Prototype

The three primary tasks being focused on during the ideation phase of the device functionality include independently going to the dining room at mealtime, getting and drinking water independently, and independent toileting.

Within memory care units, it can take immense cueing/prompting to get people to come to the table. It takes much time, patience, and corralling to make mealtime happen. The only way to do so is to individually go to each resident and tell them it is time to get ready for a meal, followed by revisiting each resident to get them to actually come to the table. The device sends an alert in the form of an image (i.e. a place setting) and a vocalized reminder (i.e. "Time to eat"), reminding them to get ready for a meal, followed by an aide or caretaker cueing them to actually make their way to the table to eat. This would eliminate added caregiver burden of having to make rounds to each resident repeatedly, thus saving time and allowing for caretakers to focus on those who require greater assistance such as someone who may suffer greater impairments or lack mobility. In turn, this will ultimately increase efficiency as well as overall independence for residents.

As said above, the idea would be that the reminder is sent to one's device, thus leading the resident to get or drink water independently. This is an important aspect to focus on simply because it is a task we easily take for granted. When we feel thirsty, we get ourselves a drink. However, being that dementia is impacting one's brain, the thirst sensor in the brain, as well as the cognition that goes along with drinking and thirst, has greatly declined. Without a 1-on-1 caretaker there to cue and monitor water intake, there is no way to measure the resident's hydration if they are simply incapable of telling you the last time they had any water. Without a constant intake of fluids, one's body can easily become dehydrated which can lead to a whole slew of medical issues.

The idea of getting an individual to drink water following a reminder to their device can be done through the use of classical conditioning. To picture this, we can make an analogy to that of Pavlov's dogs. Where the resident would be the "dog", the device would be the "bell" and the water would be the "treat". After a primary introduction to a memory care unit, it may take initial prompting to formulate associations between the task reminder and the action, but the end goal would be to create a subconscious habit so that the resident is eventually getting and drinking water independently without really thinking about it.

As is with drinking water, going to the bathroom is likely a task we do without even thinking about and likely take for granted. Similar to the deterioration that affects one's ability to sense thirst, the brain's sensing for the need to use the bathroom, such as feelings/warnings that come with bowel movements, are not registered in the manner they once were. It is because of this that residents forget to use the restroom, can't recall the last time they did so, or are incapable of even sensing their body telling them that they need to use a restroom. As one can predict, this leads to a multitude of problems such as constipation, UTIs, stomach aches, infections, accidents, and much more. All of these are at risk of occurring because a resident cannot tell what their body is trying to tell them, or cannot put into words how they are feeling.

Ideally, a resident would receive the cue/reminder in the form of an image (i.e. a toilet) as well as the vocalized cue (i.e. "Go to the bathroom"), make their way to the bathroom, and ultimately use the bathroom. The goal would be to subconsciously stimulate the bowel movement through an association between the alert and the action, aided by repetition. Once the reminder is received, one may feel an urge to use the bathroom, and because of past experience, one should subconsciously and automatically make their way to the toilet. This will hopefully serve as a preventative measure for accidents as well as gastrointestinal issues, bowel problems, and even UTIs.

Secondary alerts that the device includes are exercise, medication, and group activity alerts. These additional alerts are pertinent to a resident's schedule, and therefore should be cyclically included as task reminders given through the device. As with all of the other reminders, these alerts are aimed to keep the user more independent. The more control a user has over their life, the better their quality of life will be, as well as making them feel more confident in their own abilities.

For reminders, it was important to incorporate as many different forms of stimuli as possible. The reason for doing this is that if one pathway of communication fails, there would be another one occurring simultaneously. Each form of stimuli aims to stimulate something different in one's brain. The forms of stimuli include written word, pictures, pre-recorded commands, songs/tunes, and LED Color-Changing Lights.

The device was created to serve the dementia population and is catered to do so with image and vocal cues, rather than solely written ones. Using images and vocal cues will help overcome the current communication barrier that exists between a caregiver and a patient, as well as other associated communication issues that are extremely common for people with Dementia. The inclusion of an LED light that gives off a different color to match each alert is also a form of communication and increases the stimuli that are connected to each task/alert being sent with hopes of greater reaching the user's subconscious. Music has also been a powerful tool in bringing about moments of lucidity, and at times can also lead to a period of awakening. It is for this reasoning that songs/tunes are also included as forms of stimuli.

9.2.5 Test and Assessment

All of these things are needed to factor into the device's design. By focusing on what daily struggles exist, we stand a better chance of user acceptance. This is because if they are able to see a real change, a real improvement in themselves, they will want to hold onto that as tightly as they can, and will do whatever is necessary to do so. This is why the device has been designed the way it was. Each of its adaptations was made with a specific purpose in mind. The alert features aim to both provide a clearer sense of communication, as well as become associated with the actions they depict in hopes of triggering the user to complete actions without additional aid from caregivers. The personalization aspect aims to tap into one's circadian rhythm in order to try to help the user get into a routine that can hopefully be ingrained in their brain so that they have a sense of rhythmicity in their life. Additionally, making sure the device itself requires as little power as possible was important to help decrease the need for charging and keep things flowing as smoothly as possible.

When trying to tackle a problem such as Dementia, not much can truly be learned from a book or a website, rather the best way to gain insight into a problem is seeing its effects face-to-face. Speaking to both people who suffer from Dementia and the people who see the toll it takes day in and day out provides much more valuable information than anything one could read on paper or on some website. When you go directly to the source of a problem, you get to see the inner workings of the entire system and start to notice areas of weakness and possible holes in the system itself. While simply observing the effects of dementia, adaptations become clear. Being in the field, one is able to think on their feet and brainstorm real-time solutions to real-time problems. Hearing about what barriers lead to frustration, what forms of communication lead to the greatest success, and what types of problems commonly arise allows for previous ideas and designs to be adapted to fit with what the population actually needs. Who are we, as researchers, to say what people with Dementia need? It is up to us to listen to them, and those closest to them, to guide our designs and help us choose ways to adapt a device that will best cater to the population we are trying to serve.

9.3 Future Directions

In the future, we hope to see widespread implementation of this device and ultimately an improvement of care and life quality for the Dementia population. Although this tends to be a difficult population to work with, by keeping them in focus throughout the designing and creation of this device, the feasibility and likelihood of application are much greater. The design itself was created using feedback from not only people with Dementia and their friends and families but also a wide range of professionals who work with the population. It was from these sources that we were provided our metrics and constraints that dictated our design phase.

9.4 Conclusion

CueMinder has the potential to revolutionize and greatly improve the standard of care for people suffering from Dementia. Its implementation into memory care units will help streamline care by keeping everyone on schedule, while also aiding in users taking an active role in their own lives. With the potential to greatly improve efficiency, CueMinder can help keep things in memory care units running much more smoothly and in turn has the potential to decrease accidents that are easily preventable. Certain factors like malnutrition, dehydration, and bowel/bathroom issues are of high priority to prevent. By having reminders specifically tailored to keep these things from happening, CueMinder has the ability to truly better the lives of people suffering from Dementia.

References

1. Cuff A (2015, May 27) Dementia - a global epidemic. Retrieved from https://blogs.biomedcentral.com/on-medicine/2015/05/18/dementia-global-epidemic/
2. Rau J (2018, July 13) Most nursing homes are not adequately staffed, new federal data says. Retrieved from https://www.pbs.org/newshour/health/most-nursing-homes-are-not-adequately-staffed-new-federal-data-says

3. Hellström I, Nolan M, Nordenfelt L, Lundh U (2007) Ethical and methodological issues in interviewing persons with dementia. Nurs Ethics 14(5):608–619. https://doi.org/10.1177/0969733007080206
4. Mijajlović MD, Pavlović A, Brainin M, Heiss WD, Quinn TJ, Ihle-Hansen HB, Hermann DM, Assayag EB, Richard E, Thiel A, Kliper E, Shin YI, Kim YH, Choi S, Jung S, Lee YB, Sinanović O, Levine DA, Schlesinger I, Mead G, ... Bornstein NM (2017). Post-stroke dementia-a comprehensive review. BMC Med 15(1):11. https://doi.org/10.1186/s12916-017-0779-7
5. Yates L, Csipke E, Moniz-Cook E, Leung P, Walton H, Charlesworth G, Spector A, Hogervorst E, Mountain G, Orrell M (2019) The development of the Promoting Independence in Dementia (PRIDE) intervention to enhance independence in dementia. Clin Interv Ag 14:1615–1630. https://doi.org/10.2147/CIA.S214367

Chapter 10
IoT and Machine Learning Algorithms for Fall Detection

Pooja, S. K. Pahuja, and Karan Veer

Abstract **Context**—Fall is a vital health risk for the elderly worldwide. Every year, approximately 37 million people need medical attention due to falls in which the mostly affected are the elderly. Fall-related injuries are found to be the leading cause for them to go to an emergency room in hospitals. It increases the dependency on caregivers and reduces the quality of life. So automatic and pre-detection of falls is necessary to resolve the issues. **Methodology**—The studies included for analysis in the chapter are from the period 2013–2020. Approximately, 1000 book chapters, research, and review articles are extracted using the relevant search strings. Among them, 25 studies from reputed journals related to IoT and machine learning algorithms are used here. **Conclusion**—The study finds the task of the Internet of Things (IoT) and machine learning in fall detection, and how IoT works as a mediator for sharing the situation of patients or the elderly with medical staff, ambulance, and caretakers. The study focused on the pros and cons of IoT for various fall detection algorithms. It also described the security threats in IoT devices to make the individual's health record safe from unauthorized people but to be able to share with medical staff. So the safety of IoT becomes a significant parameter. **Significance**—Wearable devices and sensors using machine learning tools play a significant role in fall detection. Cost and time can be reduced by connecting it with Internet of Things (IoT). IoT provides a comfortable environment for the elderly at home than in hospitals, where they can engage in meetings with medical staff happily.

Keywords Fall detection · IoT (Internet of Things) · Wearable and non-wearable devices · Machine learning

Pooja · S. K. Pahuja · K. Veer (✉)
Department of Instrumentation and Control Engineering, DR BR Ambedkar National Institute of Technology, Jalan dhar, India
e-mail: veerk@nitj.ac.in

G. D. Gargiulo and G. R. Naik (eds.), *Wearable/Personal Monitoring Devices Present to Future*, https://doi.org/10.1007/978-981-16-5324-7_10

10.1 Introduction

The advancement in the diagnosis of diseases has escorted life expectations. Throughout the world, the number of elderly people is increasing in society. World Health Organization (WHO) estimated that the number of people having age more than 60 years might increase to two million by 2050 [47]. With age, people suffering from falls are increasing. A WHO report says fall-related injuries are escalating from 28 to 42% when they cross 65 years [48]. After 2013, these injuries are becoming a leading cause of death in elderlies [8]. Fall injuries are increasing social costs also, which makes it a vital public health issue. The number of deaths due to falling is 42,000 per year worldwide [11]. After a fall occurs, immediate medical care becomes significant resulting in a high survival rate. So, fall detection as soon as possible is necessary.

In recent years, fall detection devices have become popular. There are two categories of fall detection devices: wearable devices and context-aware systems [21]. Wearable devices are those that are worn by a suffering person to detect falls. These devices have an accelerometer and gyroscope, which measure velocity and acceleration. The results are verified in terms of variation in different activities, velocity, and acceleration. By analyzing the recorded data, the type of movement and activity carried out by the user can be determined. Wearable devices are helpful as they record data without disturbing the privacy of users. Many smartphones are available in the market with the features of gyroscope and accelerometer and they can measure velocity and acceleration as the user performs any activity. This data is useful in real-time analysis of fall detection. The idea of fall detection is alluring because of its cost-effective equipment. When the weight of the device is heavy, or a suffering person forgets to wear the device, it becomes difficult to monitor the performance. It depicts the prime limitation of the wearable device.

The second category of fall detection devices is the context-aware system. These systems depend upon sensors located in the nearby user area to monitor the activity. The sensors may be a pressure sensor, floor sensor, camera, and microphone, and there can be one or more types of sensors imposed on either one or more places. This category of devices can be advantageous in monitoring areas only. A camera is used to continuously acquire the images or video surveillance controlled by a classification algorithm to detect a fall [38]. The major disadvantage of video surveillance is the compromise with patient privacy. These systems are not too effective due to camera effects like illuminance [27].

Real data that perform the task should be recorded first, although it is challenging to take records of the elderly. Usually, the recording of real-world data needs monitoring of the patient for several weeks. Sometimes, this several-week data may become insufficient for getting the required results [10, 28, 3]. In place of real-world data, simulated data recorded by volunteers can be used for analysis. Maximum time fall consequence is long-lie, and persons remain on the ground till help arrives [7]. This long-lie itself becomes a problem as it leads to pneumonia, hyperthermia, and dehydration and if the lie increases more than 6 months, it may also turn into death. Fall affects the quality of life and independence of the elderly. So, the evolution of machine learning algorithms and IoT-based systems that can detect falls early, accurately, and alert in an emergency is a society need. Various solutions have been investigated for fall detection: sensor-based wearable, non-wearable, and hybrid systems. IoT-based automatic systems that can send messages using mobile or any platform during an emergency have been increased with technology advancement. Different approaches are designed to detect a fall that alerts medical staff or relatives for support if required using IoT systems [43]. The significant parts of IoT-based fall detection systems are a camera to capture the images [35], environmental [30], wearable sensors [26, 44], and IoT. IoT is a combined system of computing devices, digital or mechanical machines, and people with unique identities to access a network without human–computer interaction. Data is recorded either by a portable or wearable device and sent to the IoT network through Wi-Fi. The big data is generated and transmitted to health experts and caretakers for analysis.

10.2 Fall Detection

10.2.1 Practical Factors for Fall

There are two types of factors that affect the fall: intrinsic and extrinsic. Intrinsic factors include age, lack of balance, muscle weakness, surgery, visual impairment, medication, disease, and sometimes intentional needs like jumping. All of them are personal things with the patient, which demands attention. A few extrinsic factors are low/high lighting, less friction floor, slip, accident, stairs, uncomfortable shoes, obstacle/head collision with obstacles, free fall, lack of apparatus like handle/walker, etc. Falling consequences described by Centers for Disease Control and Prevention (CDC) are permanent care, disability, cost increment, fracture, and mental damage, reduced daily living activity, and loss of self-confidence with more dependency on others [12]. In case of severe fall, permanent bed rest and death of patient may occur. CDC reported the cost of fall injuries in hospitals is more than $30,000. So, early fall detection becomes significant to improve the quality of life.

10.2.2 Wearable Devices

With the age of the elderly, fall risk increases; however, fall detection systems cannot pause the fall but can reduce the effects after fall [18]. These devices provide a secure life to independent elderlies. Smartphones assist in these systems (see Table 10.1).

Table 10.1 Description of wearable devices used to classify fall detection and daily living activities

Article	Platform/sensor	Data collection	Machine learning Tool	Outcomes
Anguita et al. [1]	Waist-mounted tri-axial accelerometer	6 volunteers	SVM	12 activities of daily living (ADL) are classified with accuracy 90.8%
Billis et al. [9]	NA	Lone-living Seniors	Decision support system, fuzzy cognitive maps	Sensitivity—92.65%, Specificity—100%. Geriatric depression is classified as low, medium, and high
Hegde et al. [20]	Insole-based wearable sensors; gyroscope, accelerometer, pressure sensor. CC2540 microcontroller	One healthy subject, 20 min of data is recorded for 4 weeks	NA	Insole-based device scored the best result and can be used comfortably than a wrist-based device
Yuan et al. [53]	Wrist-worn device, CC2430 chip	3 elderly, and 4 young persons	ANN, SVM, C4.5, decision tree	Power-efficient algorithm, implementable on 8-bit microcontroller unit, an interrupt-driven approach which processes only accelerometer or time-interrupts
Pierleoni et al. [36, 37]	Waist, accelerometer, gyroscope, magnetometer, MCU-ATmega328	540 tests done	SVM	Accuracy—99.6%, Specificity—98.89%. Three-dimensional orientation measurement of the human body in terms of yaw, pitch, and roll
Wertner et al. [50]	Five different types of Android phone models are used	Labeled dataset of accelerometer and gyroscope is created	NA	Four types of fall and ten types of non-fall activities
Ojetola et al. [34]	SHIMMER sensor implemented at chest and thigh	42 volunteers	J48 decision tree	Discriminate near-fall, falls, and ADL with F-measures of 94%

(continued)

Table 10.1 (continued)

Article	Platform/sensor	Data collection	Machine learning Tool	Outcomes
Sucerquia et al. [45]	Waist, MKL25Z128VLK4 microcontroller, ADXL345 accelerometer, ITG3200 gyroscope	23 young adults, and 14 healthy elderly	Threshold-based classifier	Threshold-based classification accuracy up to 96% in fall detection
Hegde et al. [19]	Wrist and insole	15 healthy adults	Max relevance minimum redundancy	Comparison of SmartStep and insole-based sensor with a wrist-based sensor where SmartStep gives a better result. In 13 activities, lying down activity attained a maximum accuracy of 98%
Baig et al. [4]	ADXL345 accelerometer, ITG3200 gyroscope, and MKL25Z128VLK4 microcontroller	23 young adults, 41 healthy independent elders with age more than 62	Threshold-based classifier	96% accuracy in fall detection
Kheirkhahan et al. [23]	Wrist, smartwatch sensors	5 volunteers	Application Program Interface	Real-time and Online Assessment and Mobility Monitoring (ROAMM)
Khojasteh et al. [24]	Wrist, chest, thigh, or lumbar area	A group of the healthy young participants	Threshold-based solution using SVM	Fall detection classification, UMA fall, simulated fall, and ADL
Santoyo-Ramon et al. [41]	Chest, waist, wrist, ankle, thigh, CC2650 ARM microcontroller, and MEMS sensors	19 volunteer	SVM, KNN, NB, decision tree	ADLs classification is verified using ANOVA
Yacchirema et al. [52]	3D accelerometer, 6 LowPAN wearable devices	38 participants who performed 19 types of ADL and 15 types of falls	Decision trees, ensemble logistic regression, and deepnets	Accuracy, specificity, sensitivity, and precision are more than 94% for fall detection
Farsi [17]	Wearable smartwatch devices	14 healthy individuals	Ensemble methods	Recall, specificity, accuracy

10.2.3 Context-Aware Systems

See Table 10.2.

10.3 How IoT Helps in Fall Detection

The block diagram of the IoT-based fall detection system is shown in Fig. 10.1. It contains sensors to record the data like a gyroscope, accelerometer, Wi-Fi module, cloud services, and an alarm system with a snooze button, IoT gateway, data center, and the elderly's connectivity with medical staff, caregivers, and health experts. Accelerometer and gyroscope are used to find the acceleration and angular velocity, and buzzer/alarm will buzz when the value of parameters exceeds the threshold value. The other major part of it is the Wi-Fi module, which transfers the message via the internet. GPS shared the location of the suffering person to already saved contacts.

The E-health trend of detection using IoT starts from sensing. Sensors collect the data from the physical environment that is further processed for prediction, prevention, and fall detection. This trend is applicable in smart homes and smart towns, cities, etc., which carries an edge, cloud, fog layers, and users (see Fig. 10.2).

10.3.1 Edge Section

It is the first layer of the system where smart sensors and actuators record the physical environment data. These data recording items are either wall-mounted, static devices like thermometers or wearable devices such as accelerometers applied on the body. These devices work in two ways: preventive and detective modes. Action management works to determine the current position of the device either in the working mode or sleeping mode to deduct the energy consumption and identify that the device is in preventive mode or detective mode. In detective mode, sensors collect the filtered data, then processing and edge computing will be done. The individual signals are compared with a threshold value, and then collected data is sent to the fog layer for fall detection. Preventive devices give comfortable zones for the elderly before and after the fall. For example, IoT-based handrail assists in walking before a fall happens, and a wearable air-bag avoids severe injury after a fall (see Fig. 10.3).

10.3.2 Fog Layer

Fog or gateway performs like a bridge between network and cloud service, making possible wireless communication between the system and gateway. Fog layer benefits

Table 10.2 Description of context-based systems used for fall detection

Article	Platform/sensor	Data collection	Machine learning tool	Outcomes
Rougier et al. [39]	4 cameras	75 different events for more than 12 min	Shape-based technique to track a person's silhouette. Gaussian Mixture Model	DALs and fall detection with 0% error
Campilho and Kamel [13]	Two monocular cameras, Kinect, and JVC camcorder	2 subjects	A Kinect in front of a treadmill and a set of two camcorders on the sides of the treadmill	Comparisons to the Kinect and a marker-based method where Kinect gives more accurate result
Baldewijns et al. [6]	72 video segments at 1 location with five cameras	10 participants	NA	Comparison of simulated dataset and recorded one to identify the fall. Precision, recall, and accuracy are extracted for 5 all cameras
Charfi et al. [14]	Single camera for silhouette tracking of humans	250 video sequences in 4 different locations	SVM	Global error rate <1%, and recall, specificity, and precision are 0.98, 0.996, and 0.942, respectively
Sehairi et al. [42]	Mono static camera for video recording	CDnet video dataset	Simp-SOBS, RGA, GMM, Eig-Bg, RAF, FMTG, DSTEI, FD, 3FD, MRFMD, $\sum\Delta$, STEI	Recall, specificity, FPR, FNR, PWC, precision, F-Measure, R, R C
Tan et al. [46]	20 sensors are applied in the kitchen, bathroom, bedroom, etc.	14 elders with the age 73 and above	Algorithms are generated for automatic classification of activities	New classification is evaluated "Brief-Return-and-Exit"
Min et al. [31]	For fall events, 2 Microsoft Kinect cameras; for ADL events, one device and accelerometer	70 video sequences (30 falls +40 activities of daily living)	R-CNN, Fast R-CNN, and Faster R-CNN	94.44% precision, 94.95% recall, and 95.50% accuracy
Espinosa et al. [16]	3-axis accelerometer, a 3-axis gyroscope, and an ambient light sensor	9 males and 8 females	Convolutional neural networks (CNNs)	Accuracy of 95.64%

(continued)

Table 10.2 (continued)

Article	Platform/sensor	Data collection	Machine learning tool	Outcomes
Zhong et al. [54]	Thermal vision sensor on the ceiling of the room	9 males and 8 females	MoT-LoGNN, MOD-LoGNN, T-LoGNN	Accuracy of 98.39%

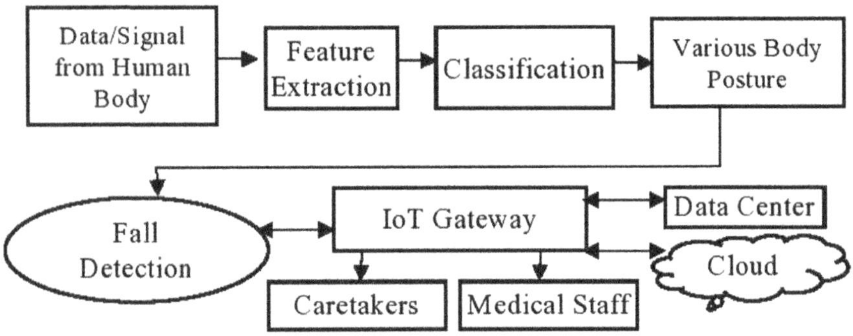

Fig. 10.1 How IoT helps in gait detection

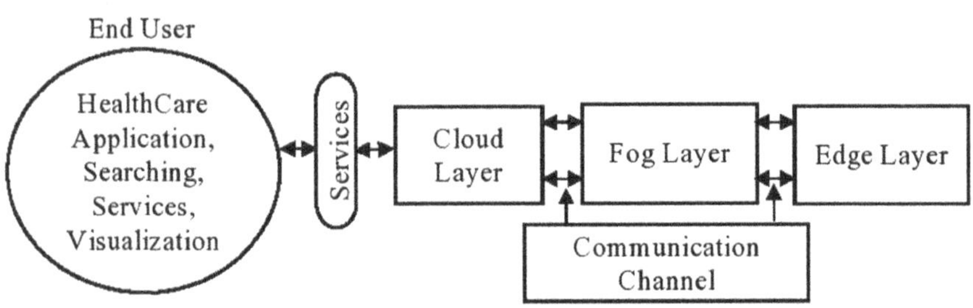

Fig. 10.2 Schematic diagram of fall across IoT layers

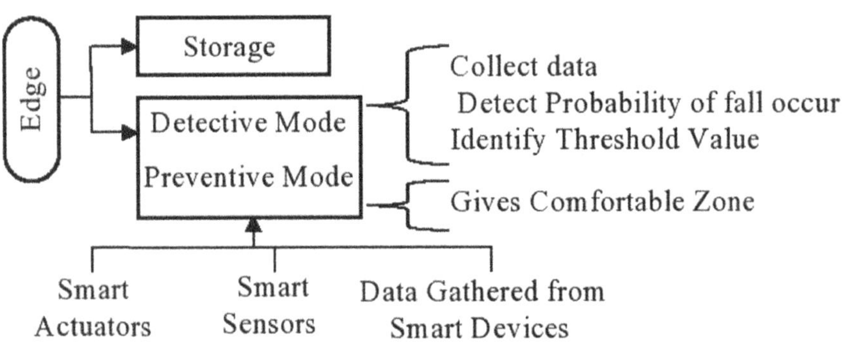

Fig. 10.3 Components of Edge Layer of IoT-based system

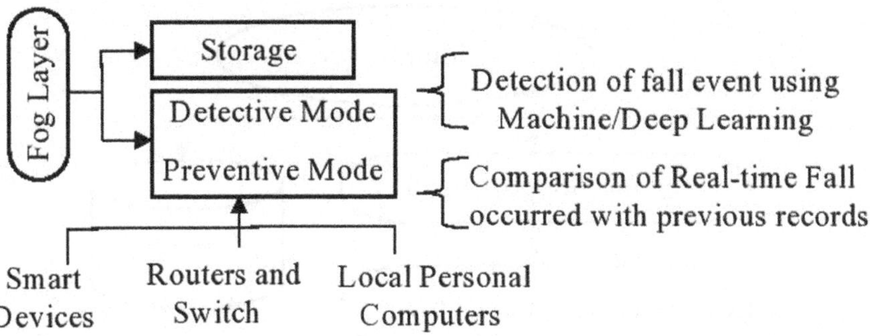

Fig. 10.4 Components of fog layer of IoT-based system

are mobility support, flexibility, low latency, scalability, and heterogeneity [44, 51]. The main components of the fog layer are personal computers, switches, routers, and Arduino, which works as a gateway mounted either in the room or in a PC, tablet, and smartphone as a wearable part. These devices have two modes, Detective and Predictive modes, and then signals are collected to detect the fall. Machine learning algorithms are applied to classify falls and different activities of daily living (ADLs). The two management actions, local and general, are applied after fall detection. Local is to give orders and information to the sensors in the edge layer for the required environment. General transmits the information to caretakers, health experts, and data analysis centers. Results will stored in the storage section and the information will be sent to cloud via channel (see Fig. 10.4).

10.3.3 Cloud Layer

It contains the action management system, health regarding decision support system, data management, and computation center. Cloud can be installed in data analytics and hospitals. Data are analyzed in two ways: one compares an individual's data with previous records to detect fall risk, unusual gait patterns in terms of fall, etc. The second one compares individuals with each other to identify the standard fall, uncommon patterns, fall risks, etc. The action management system controls all activities in fog and edge layers. Monitor system controls the active or sleep devices' energy consumption, active or inactive condition of sensors, gathering, and alliance of devices in all layers. The data management system processes the data; after analysis, a storage notification will be sent to the system (see Fig. 10.5).

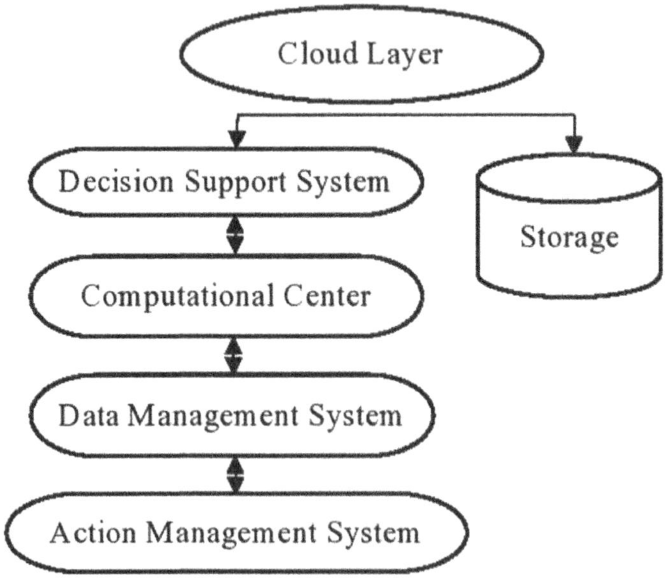

Fig. 10.5 Components of cloud layer of IoT-based system

10.4 Phases During Fall Detection

When a fall occurs, the body balance and acceleration of the suffering person change suddenly. To understand the mechanism of fall detection, pre-fall and post-fall phases have a vital role.

10.4.1 Pre-impact Phase

This phase takes a few milliseconds and happens just before the fall, called lead time. It consequences a decrease in balance, weight, and acceleration of the body. Researchers noted 700 ms as the maximum time for pre-impact before the fall [33].

10.4.2 Impact Phase

This is when the collision occurs and changes the body balance, movements, and acceleration rapidly. It can become the cause of severe damage.

10.4.3 Post-impact Phase

After a fall, small changes happen in body acceleration, and the patient remains in the shock stage which is called the post-impact phase. Fall detection is possible in all three steps; the system can inform doctors, caretakers, and data analysis centers before severe injury.

10.5 Advantage and Disadvantage of IoT in Fall Detection

IoT has several pros and cons like other systems.

10.5.1 Advantages

There are several benefits of an IoT-based system as social impact, low cost, accuracy, monitoring, time, and transparency [5].

- **Communication and Always Connected**—IoT supports man-to-machine and machine-to-machine communication and sets up a connection to make them always connected. For example, in an IoT-based production line connected with various machinery, all factors and operation information can be sent to all entities (field engineers, employees, managers, etc.), and factory control becomes easy and transparent.
- **Monitoring**—Control of machines is a significant advantage of IoT. For example, knowing the home devices are on/off or room temperature level in the home, a precise amount of supplies in the industry, etc. It can be said that control rooms are better informers than human beings. This continuous control can be recorded for analysis. Fitness bands, wristwatches, and different IoT-based wearable devices can monitor the body parts for sugar levels, pulse rate, blood pressure, etc.
- **Social Impact**—It affects individuals, such as the elderly, kids, youngsters, business holders, stock market people, and industry persons. IoT is contributing to society's advancement by improving the quality of life.
- **Low Cost**—IoT connects machine to machine and human to machine, reducing the labor cost of industry, enterprises, or companies. Small and middle-level sectors are also increasing because of cost-effective systems.
- **Time**—IoT saves a lot of time. Especially in health care for preliminary tests, elderlies do not need to go to hospitals. They are simultaneously connected with doctors and healthcare experts, and caregivers also [2].
- **Accuracy**–IoT systems analyze the big data. A considerable amount of data permits someone to make the right decision accurately. When the data is more, the decision will be more accurate. For example, as information about things is

more in the supermarket, the customer can buy with the right choice. If enterprises have satisfactory stock and supply, results can be more accurate [22].

- **Enhancement of Customer Engagement**—In general information applications, customer engagement is significantly less. In IoT, it is just the opposite; users/customers are engaging actively [2].

10.5.2 Disadvantage

Along with advantages, IoT-based devices/systems have a few disadvantages as described below.

- **Security and Privacy**—As depicted in the schematic diagram of IoT, many technologies complete the system. Several engineers and developers are involved in it, so privacy becomes the major challenge. It generates a threat to security also. The IoT-based devices are getting attention from cybercriminals and hackers. The criminal attack can destroy the critical deployment of IoT and results in loss of public privacy [25, 40]. Users can take the help of security algorithms to avoid any threat. System or relevant passwords must not be saved in the cloud, and it should be changed regularly to enhance security. The authenticity of the software of a third party should be checked. If the IoT-based devices are not in use, they should be keep off to refrain from a hacking attack.
- **Complex System**—The design, formation, and maintenance of IoT-based systems is complex and crucial. Generally, as the complexity of the system increases, the possibility of failure also increases. For example, the elderly living alone cannot understand the system very fast if it gets any update or changes from the server. If the design gets some fault due to developer error, it can damage the system's component. In an IoT-based, automatically operated dam, an error in measuring the water level and opening a wrong gate can bring the flood [52].
- **Safety Hazard**—In IoT-based systems, safety is in the hands of customers [15]. If a cybercrime event occurs, hackers can change the order's details, and existing orders may reach the wrong address. On the other hand, due to the customer's open information, there is always a chance to access privacy by unauthorized people. Data encryption techniques should be safer.
- **Social Issues**—Either in industry or business, everything is becoming automatic, reducing the labor work by human beings and increasing unemployment [29]. Human life is going in control of machines. The reliability of machine is increased at enormous scale and consequenently the possibility of harmness from the device also gets increased. The question of dependency and management of human life on devices is getting researchers' and scientists' attention.

10.6 Challenges for IoT Devices

In the market, many wearable and non-wearable devices are available for either fall detection or any other healthcare activity; still, some issues exist. Here, some problems are described on behalf of fall-related applications, customer privacy, and future work scope.

- **How to Reach on Prediction from Detection**—Fall detection is the first step toward the quality of life of the elderly but not the final. Detection can help them rescue themselves and reduce the falling interval, but they cannot refrain from the injuries. So the early prediction of fall becomes necessary, which can control the fall damages. Various studies achieved an accuracy of fall detection, but early prediction still requires quality research.
- **Way to Privacy**—For users, privacy and security from unauthorized users is a big challenge, particularly in vision-based algorithms. The camera in the bedroom, kitchen, and hall may leak confidentiality. Some researchers use Kinect to remove the camera problem, but it blurs pictures of human skelton and deep body parts to get information about human activities. Wi-Fi-based techniques are giving a better response from users than wearable devices for fall detection. To detect the fall from a group of a persons simultaneously is far from research.
- **Practical Implementation of Theoretical Research**—Maximum number of fall detection algorithms give good accuracy either in the laboratory or on simulated data. However, some ideas are doing well in practical life also. In the published research, machine learning algorithms and sensors are not working accurately. Kinect is a device that gives 3D information by tracing the human being, and several studies are using it for fall detection [32]. But Kinect has the limitation of distance (0.4–3 m). Sensors installed in the room may record varying information due to environmental or man-oriented changes. Most researchers are using the simulated database; it doesn't give reliable results practically.
- **Standard Features for Fall**—Before the solution of any issue, first, it should be defined to understand it deeply. Fall detection features can be described as how designed algorithms detect the fall and how sensors are mounted, and the system's working style. A few studies defined fall detection in terms of threshold-based, rule-based, and shape-based algorithms. Researchers do not involve the dynamic kinetic model. Most of the work on fall detection has been done using machine learning algorithms, and extracted features classify the fall and different activities. It represents a particular condition that may change with the environment, so it makes the system sensitive. After performing the experiment, condition change and reuse of system become difficult. So, there should be some specified standard features either by categorization or by mathematical modeling.

10.7 Discussion

It is true that the fall rate increases with age in the elderly. Accelerometer, gyroscope, and magnetometer are the majorly used devices for fall detection. These devices record real-time data in the indoor and outdoor environments and can transmit the information to health experts in any emergency. In wearable devices, different sensors on the waist, wrist, thigh, chest, and ankle are also used. Sensors mounted on the chest give more accurate results for fall detection but wearing wrist, ankle, and thigh sensors brings the person to uncomfortable mode [29, 54]. In context-based detection, the camera plays a vital role either by capturing the pictures or video segmentation. When these techniques connect with IoT, they give tremendous results as it increases elderly's connectivity with doctors, medical staff, data center, and caregivers, as shown in Fig. 10.1. It can be said that IoT improves the elderly's quality of life and makes them more independent. The E-health monitoring starts from sensor data and contains cloud, fog, and edge layers. The edge layer records data using smart sensors, smart actuators, and smart devices and makes them comfortable with these devices. The fog layer behaves as a bridge between cloud and the network. It analyzes the data by comparing it with previous real-time data. All the activities that occur in edge and fog layers are controlled and managed by the cloud layer. To understand fall fully, it is described in three phases: pre-impact, impact phase, and post-impact phase. Machine learning tools are used to find the gait pattern of individuals. In the literature, the most used technique is support vector machine (SVM) [5], but in one study, fuzzy also classifies the data with 100% specificity and 92.65% accuracy, as shown in Table 10.1. There are various advantages of using IoT in fall detection; a few disadvantages are also shown in the study. Still, there are some challenges in fall detection using IoT: how to maintain patients' privacy and attain the prediction from detection with more accuracy, sensitivity, and specificity.

10.8 Conclusion and Future Scope

The fall detection field is growing with machine learning, and IoT adds to it to improve the quality of life of the elderly. The detection of fall is categorized using different body postures like sitting, standing, walking, walking up-down stairs, fall posture, and lying. This whole system decreases the dependency of suffering people on caretakers. Accuracy, sensitivity, precision, and specificity are the generally used parameters in machine learning tools to show the effectiveness of the algorithms. Various studies are compared in tabular format for analyzing their results. Most of the studies show the support vector machine (SVM) and the ensemble algorithm as a useful fall detection tool. IoT maintains the connectivity between human beings, appliances, and devices, making monitoring an easy task. IoT allows a large amount of data handling but its high accuracy also made it an exciting research field. It saves lots of money and time. With all benefits, IoT has pitfalls also like privacy, complexity, less compatibility, and safety. With the use of smartphones and the Internet, IoT is making devices smart throughout the world. Cost, convenience, and safety are essential parameters in IoT. Although hackers remain attentive to vulnerability as it may be applicable in many other devices also, practitioners need to be careful about insecure components, doubtful network services, uncertain ecosystem interface, privacy, unresolved data transfer, unsure device management, and default settings. In the future, the role of edge computing in IoT can be increased as it analyzes the data traffic at the local server before going to the cloud. The simultaneous recording of data for a group of people can reduce the recording time if gait shuffling is managed.

References

1. Anguita D, Ghio A, Oneto L, Parra X, Reyes-Ortiz JL (2013) A public domain dataset for human activity recognition using smartphones. In: ESANN 2013 proceedings, 21st European symposium on artificial neural networks, computational intelligence and machine learning, pp 437–442
2. Applications IH (2016) IoT for healthcare. Int J Sci Res 5:322–326
3. Automatic Fall Detection (Oct 2013) Complementary devices for a better fall monitoring coverage. In: Proceedings of the IEEE 15th international conference on E-health networking, application & services (Healthcom), Lisbon, Portugal, pp 590–593
4. Baig MM, Afifi S, GholamHosseini H, Mirza F (2019) A systematic review of wearable sensors and IoT-based monitoring applications for older adults–a focus on ageing population and independent living. J Med Syst 43
5. Balas VE (2020) A handbook of internet of things in biomedical and cyber physical system (Book). Springer International Publishing
6. Baldewijns G, Debard G, Mertes G, Vanrumste B, Croonenborghs T (2016) Bridging the gap between real-life data and simulated data by providing a highly realistic Fall dataset for evaluating camera-based fall detection algorithms. Healthc Technol Lett 3:6–11
7. Bathon JM, Pongprutthipan M, Wrone DA, Brunswick N, Lord BSR, Sherrington C, Menz HB, Close JCT (2015) F o p: r f s, pp 16–17
8. Bergen G, Stevens MR, Burns ER (2016) Falls and fall injuries among adults aged ≥65 years—United States, 2014. MMWR Morb Mortal Wkly Rep 65:993–998
9. Billis AS, Papageorgiou EI, Frantzidis CA, Tsatali MS, Tsolaki AC, Bamidis PD (2015) A decision-support framework for promoting independent living and ageing well. IEEE J Biomed Heal Informatics 19:199–209
10. Bourke AK, Van De Ven PWJ, Chaya AE, ÓLaighin GM, Nelson J (2008) Testing of a long-term fall detection system incorporated into a custom vest for the elderly. In: Proceedings of the 30th annual international conference of the IEEE engineering in medicine and biology society, EMBS'08- Personalized Healthcare through Technology." IEEE Computer Society, pp 2844–2847
11. Burns ER, Stevens JA, Lee R (2016) The direct costs of fatal and non-fatal falls among older adults—United States. J Safety Res 58:99–103
12. CDC (2020) https://www.cdc.gov/, https://www.google.com/search?q=CDC, https://www.cdc.gov%2F)&oq=CDC, https://www.cdc.gov%2F)&aqs=chrome..69i57j0l2.586j0j4&sourceid=chrome&ie=UTF-8. Accessed August 9, 2020
13. Campilho A, Kamel M (2014) Image analysis and recognition: 11th International Conference, ICIAR 2014 Vilamoura, Portugal, October 22–24, 2014 proceedings, part II. Lecture notes computer science (including Subser Lect Notes Artif Intell Lect Notes Bioinformatics), vol 8815, pp 66–73

14. Charfi I, Miteran J, Dubois J, Atri M, Tourki R (2012) Definition and performance evaluation of a robust SVM based fall detection solution. In: 8th international conference on signal image technology & internet based systems SITIS 2012, pp 218–224
15. Chelli A, Patzold M (2019) A machine learning approach for fall detection and daily living activity recognition. IEEE Access 7:38670–38687
16. Espinosa R, Ponce H, Gutiérrez S, Martínez-Villaseñor L, Brieva J, Moya-Albor E (2019) A vision-based approach for fall detection using multiple cameras and convolutional neural networks: a case study using the UP-Fall detection dataset. Comput Biol Med 115:103520
17. Farsi M (2020) Application of ensemble RNN deep neural network to the fall detection through IoT environment. Alexandria Eng J
18. Frailty T (2001) Age Ageing-2001-Masud-3-7.pdf. 3–7
19. Hegde N, Bries M, Swibas T, Melanson E, Sazonov E (2018) Automatic recognition of activities of daily living utilizing insole-based and wrist-worn wearable sensors. IEEE J Biomed Heal Informatics 22:979–988
20. Hegde N, Sazonov E (2014) Smartstep: A fully integrated, low-power insole monitor. Electronics 3:381–397
21. Igual R, Medrano C, Plaza I (2013) Challenges, issues and trends in fall detection systems. Biomed Eng Online 12:66
22. Indrakumari R, Poongodi T, Suresh P, Balamurugan B (2020) The growing role of Internet of Things in healthcare wearables. Elsevier Inc
23. Kheirkhahan M, Nair S, Davoudi A, Rashidi P, Wanigatunga AA, Corbett DB, Mendoza T, Manini TM, Ranka S (2019) A smartwatch-based framework for real-time and online assessment and mobility monitoring. J Biomed Inform 89:29–40
24. Khojasteh SB, Villar JR, Chira C, González VM, de la Cal E (2018) Improving fall detection using an on-wrist wearable accelerometer. Sensors (Switzerland) 18:1–28
25. Kim TH, Choi A, Heo HM, Kim K, Lee K, Mun JH (2019) Machine learning-based pre-impact fall detection model to discriminate various types of fall. J Biomech Eng 141:1–10
26. Kim JH (2013) Advances in intelligent systems and computing: preface. Adv Intell Syst Comput 208 AISC, 623–630
27. Klonovs J, Haque MA, Krueger V, Nasrollahi K, Andersen-Ranberg K, Moeslund TB, Spaich EG (2016) Introduction. Springer Briefs Comput 1–21
28. Kostopoulos P, Nunes T, Salvi K, Deriaz M, Torrent J (2015) F2D: A fall detection system tested with real data from daily life of elderly people. 2015 17th Int Conf E-Health Networking. Appl Serv Heal 2015:397–403
29. Kwon SB, Park JH, Kwon C, Kong HJ, Hwang JY, Kim HC (2019) An energy-efficient algorithm for classification of fall types using a wearable sensor. IEEE Access 7:31321–31329
30. Lee RYW, Carlisle AJ (2011) Detection of falls using accelerometers and mobile phone technology. Age Ageing 40:690–696
31. Min W, Cui H, Rao H, Li Z, Yao L (2018) detection of human falls on furniture using scene analysis based on deep learning and activity characteristics. IEEE Access 6:9324–9335
32. Mozaffari N, Rezazadeh J, Farahbakhsh R, Yazdani S, Sandrasegaran K (2019) Practical fall detection based on IoT technologies: a survey. Int Things 8:100124
33. Nyan MN, Tay FEH, Murugasu E (2008) A wearable system for pre-impact fall detection. J Biomech 41:3475–3481
34. Ojetola O, Gaura E, Brusey J (2015) Data set for fall events and daily activities from inertial sensors. In: Proceedings of the 6th ACM multimedia systems conference MMSys 2015, pp 243–248
35. Pannurat N, Thiemjarus S, Nantajeewarawat E (2014) Automatic fall monitoring: a review. Sensors 14:12900–12936
36. Pierleoni P, Belli A, Maurizi L, Palma L, Pernini L, Paniccia M, Valenti S (2016) A wearable fall detector for elderly people based on AHRS and barometric sensor. IEEE Sens J 16:6733–6744
37. Pierleoni P, Belli A, Palma L, Pellegrini M, Pernini L, Valenti S (2015) A high reliability wearable device for elderly fall detection. IEEE Sens J 15:4544–4553

38. Popoola OP, Wang K (2012) Video-based abnormal human behavior recognitiona review. IEEE Trans Syst Man Cybern Part C Appl Rev 42:865–878
39. Rougier C, Meunier J, St-Arnaud A, Rousseau J (2011) Robust video surveillance for fall detection based on human shape deformation. IEEE Trans Circuits Syst Video Technol 21:611–622
40. Saleh M, Jeannes RLB (2019) Elderly fall detection using wearable sensors: a low cost highly accurate algorithm. IEEE Sens J 19:3156–3164
41. Santoyo-Ramón JA, Casilari E, Cano-García JM (2018) Analysis of a smartphone-based architecture with multiple mobility sensors for fall detection with supervised learning. Sensors (Switzerland) 18
42. Sehairi K, Chouireb F, Meunier J (2017) Comparative study of motion detection methods for video surveillance systems. J Electron Imaging 26:023025
43. Seniors' Falls in Canada: Second Report-Canada.ca (nd) https://www.canada.ca/en/public-health/services/health-promotion/aging-seniors/publications/publications-general-public/seniors-falls-canada-second-report.html. Accessed August 9, 2020
44. Stone EE, Skubic M (2015) Fall detection in homes of older adults using the microsoft kinect. IEEE J Biomed Heal Informatics 19:290–301
45. Sucerquia A, López JD, Vargas-Bonilla JF (2017) SisFall: A fall and movement dataset. Sensors (Switzerland) 2017
46. Tan TH, Gochoo M, Jean FR, Huang SC, Kuo SY (2017) Front-door event classification algorithm for elderly people living alone in smart house using wireless binary sensors. IEEE Access 5:10734–10743
47. WHO (Jun 2018) World report on ageing and health. http://apps.who.int/iris/bitstream/handle/10665/186463/9789240694811_eng.pdf. Google Search (n.d.). Accessed August 9, 2020
48. WHO (Jun 2018) World Health Organization: global report on falls prevention in older age. https://extranet.who.int/agefriendlyworld/wp-content/uploads/2014/06/WHo-Globalreport-on-falls-prevention-in-older-age.pdf. Google Search. Accessed August 9, 2020
49. WHO | Falls Prevention in Older Age. https://www.who.int/ageing/projects/falls_prevention_older_age/en/. Accessed 10 Aug 2020
50. Wertner A, Czech P, Pammer-Schindler V (2015) An open labelled dataset for mobile phone sensing based fall detection
51. Yacchirema D, De Puga JS, Palau C, Esteve M (2018) Fall detection system for elderly people using IoT and Big Data. Procedia Comput Sci 130:603–610
52. Yacchirema D, de Puga JS, Palau C, Esteve M (2019) Fall detection system for elderly people using IoT and ensemble machine learning algorithm. Pers Ubiquitous Comput 23:801–817
53. Yuan J, Tan KK, Lee TH, Koh GCH (2015) Power-efficient interrupt-driven algorithms for fall detection and classification of activities of daily living. IEEE Sens J 15:1377–1387
54. Zhong C, Ng WWY, Zhang S, Nugent C, Shewell C, Medina-Quero J (2020) Multi-occupancy fall detection using non-invasive thermal vision sensor. IEEE Sens J XX:1–1. https://doi.org/10.1109/jsen.2020.3032728

Part IV
Signal Processing for Wearable/IOT Applications

Chapter 11
Smart Home Automation Using Wearable Technology

Rohit Volety and P. Geethanjali

Abstract Technological advancements have made people lead a sophisticated, comfortable, and luxurious life. The products available in the market are not designed to make difference to the life of the disabled and the old age population of the world. The disabled and aged people have become grounded in life to perform everyday tasks even after technological advancements. This affects physical health and hygiene in addition to mental depression. In the market, few products are available targeting the aged and the disabled population. In smart home automation, using wearables made a major impact on the quality of life of the disabled and aged population. The smart home automation features automatic opening/closing of the doors, object/face recognition, and switching on/off lights, without moving from a place to the target location. Various technologies are available to build automation features to manage everyday tasks independently and to improve the quality of life of disabled and aged people. This chapter focuses on various technologies available for smart automation to support the needy population of the world, from the perspective of wearables. This chapter also discusses the development of gesture-based home automation and security for visually impaired people and old age people.

11.1 Introduction

In the world, there were 703 million people aged 65 years or above in the year 2019. The number of the aged population would reach 1.5 billion in 2050. Globally, the share of the population aged 65 years or over increased from 6% in 1990 to 9% in 2019. It's difficult for aged people to manage activities of daily living. In recent years, advances in technology enabled the aged people, chronically diseased people, and people with different disabilities to use various types of human–machine interfaces using biomedical signals to communicate and or control the external environment and or devices. However, there has been a remarkable effort to aid the aged community based on wearable technology for smart home automation, nursing, and telehealth

R. Volety · P. Geethanjali (✉)
School of Electrical Engineering, Vellore Institute of Technology, Vellore 632014, India
e-mail: pgeethanjali@vit.ac.in

G. D. Gargiulo and G. R. Naik (eds.), *Wearable/Personal Monitoring Devices Present to Future*, https://doi.org/10.1007/978-981-16-5324-7_11

services. Most of this wearable technology received support from caretakers as well as from the adults seeking support for residential monitoring. The wearable devices are connected mostly through wireless sensor networks. The devices are equipped with sensors to monitor physiological parameters, activity, and fall, environmental conditions, etc., and communicate with the network to alert the caregiver.

Wearable technology has made a mark in history because of its many popular features like accessibility, convenience, fitness tracking, health and safety, and Internet of Things (IoT) [1]. Wearable technology can be used to send an alarm quickly and discreetly. With a simple press of a button in the case of an accident or injury, alerts can be sent to caretakers or even emergency services. This is very ideal for employees or students working alone. Television (TV), radio, speakers, lights, and even heating in a home can be controlled from one place which links all these devices together through the Internet. Brands like Fitbit and Garmin have integrated fitness trackers that monitor an individual's every step, heart rate, temperature, etc., and compiles multiple data to provide custom fitness solutions. Calls, emails, texts, and other things can be easily dealt with, without smartphones.

Smart home automation has been a boon for the general population, and it has been a blessing for the old and disabled people. Hence, they need a security system that can be operated with the flick of a hand or any gesture. More user-friendly techniques are needed to help the old and disabled community in smart home automation. Controlling basic applications like lights, fans, etc., required triggers like walking inside a room or gestures; making these gestures are very difficult for people who are old and disabled. Assistive devices lack user-friendliness to meet the needs of aged and disabled people. One such solution is the use of wearable technology in the field of smart home automation. Over the years, wearable technology-based smart home-based automation has evolved from smart auto-lacing shoes that can charge mobile phones to a small device placed on the neck to control your entire home with just a flick of your neck. The following section discusses various home automation systems developed for abled, disabled as well as aged people.

11.2 Related Works

Smart home automation projects are used to control the alarm systems, lights on/off/dim, heating, ventilation, and air conditioning (HVAC) on/off, door lock, window shutdown, on/off of different appliances, and monitor temperature and humidity, motion detection, fire and smoke detection, door status, light level, and video. These automations can be implemented using a local area network (LAN), wireless sensor network (WSN), and personal area network (PAN).

ElShafee, A. and Hamed, K. A. have implemented a Wireless Fidelity (Wi-Fi)-based home automation system [2] that consists of a server and hardware components. The server controls the hardware modules and if the situation demands more hardware modules, they can be easily added. There is a server Internet Protocol (IP) that can be used to control any of the hardware from the web browser of any personal computer

(PC) in the same network in the LAN. The system has actuators and sensors. Sensors can be used to take in the data from the environment and are sometimes used as triggers for the actuators: e.g., watering of plants automatically when the moisture content is low in the soil. Kodali, R. K. and Soratkal, S. discussed the Wi-Fi-based home automation, where the user has to log in through the web browser using his/her credentials. Data is passed through a security module to be transferred to a 128-bit hex key and is checked for availability in the data source represented in Extensible Markup Language (XML) files. The system writes this action into the logger and sends it to the Log interface. The status of the sensor and the actuator are given directly by the communication module. If there is any change in sensor data, then the data is logged into the logger. Change of the status of the actuator can be done directly by sending the signal to the hardware interface by the communication module. These changes are written in the data source represented in XML files for later checks. The automation interface is responsible for setting up new automation and saves it into the data source XML files. The Automator is in charge of carrying out the automations that have already been saved. To do this, it gathers data from the data source XML files, status, and control interfaces and compares it with data in the XML files. It sends the signals to the communication models to be given to the hardware interface. Changes are written into the logger and appear in the log interface [2, 3].

Gesture-based control of home automation has been developed for people with loss of vision, motor skills, and mobility [4]. This system also has the capacity to analyze tremors from the movements of the user. This information can then be used for medical diagnosis, therapy, and other emergency services. This pendant can identify control gestures with an accuracy of 95% and user-defined gestures with an accuracy of 97%. The other useful feature is that the camera is ringed by infrared (IR) light-emitting diode (LED) and has an IR pass filter due to which the gestures made by the hand can be recognized even in the dark. With the press of the button, gesture recognition is started. A system presents Inviz, a wearable capacitive sensor-based gesture recognition system for environmental control in people with issues related to mobility [5]. It uses textile-based capacitive plates woven into the cloth as proximity sensors. This system uses a novel hierarchical signal processing algorithm where the computation is broken down into several low-power and high-power tiers. The low-power tiers maintain the high availability of the system at low power and the higher power tiers are woken up only when required to perform advanced signal processing. A prototype of a fully functional Inviz system was developed and evaluated in the context of a home automation system. It showed that gestures can be recognized with high accuracy and low energy consumption.

Neuromuscular disorders like cerebral palsy and injuries to the spinal cord result in disorders like quadriplegia and paraplegia [6]. Activities of daily living include control of home appliances which become difficult due to impaired motor movements. Using hand and eye movements can be a solution to this problem, but some of these movements require a lot of limb movement which just increases the discomfort of the user. Hence, these kinds of systems cannot be used as a long-time solution and certainly not for quadriplegics with a varied range of motor impairments. As a solution, a multi-sensor gesture recognition system is designed with the advantage being it uses comfortable and low-powered wearable sensors. An electrooculography (EOG)-based headband using textile electrodes and a glove that uses flex sensors coupled with an accelerometer for the detection of eye and hand gestures is designed. The gestures are used to control appliances. The novel feature of this system is a set of robust gesture recognition algorithms that can be implemented in hardware and on computationally weak microcontroller platforms. The gestures from the glove and headband device are fused on a smartphone that controls home appliances using a web service and a custom home automation system. Eye movement can be used as gesture recognition for a person with limited limb mobility, while hand gestures can be used for gesture recognition for people with limited eye movements.

Automation is used for monitoring patients with conditions like arrhythmia seizures, activity, fall down, etc. These conditions when detected can be used as a trigger to automatically alert emergency services like ambulances and also their family circles. A system presents an electrocardiogram (ECG) sensor [7]. It is based on the wireless and continuous event recorder of ECG signals. The ECG sensor has two electrical contact points that contact the surface of the subject using conducting hydrogel, is directly connected to the electronic circuits for boosting the signal, and with wireless transmission of the recordings to a receiver integrated as a component within a handheld device (HHD). The HHD is said to be an intelligent unit that analyzes and records all the signals. This unit uses General Packet Radio Services (GPRS) which is a standard telecommunication facility for the sending of the alerts along with ECG recordings to a remote server. The doctors at the hospitals use a special remote client installed on a standard PC as a clinical diagnostic station (CDS). Trained staff will be able to evaluate the ECG recordings, diagnose the patients, and then follow up on the medical practices required [8].

The wear your world (WUW) system has many interesting features in the gesture recognition system with multi-touch systems, freehand gestures, and iconic gestures [9]. By augmenting surfaces, walls, and physical objects and turning them into interfaces, it frees digital information. The prototype uses a hat-mounted projector–camera configuration to identify the gestures of the users. The localization of a resident inside the house is useful knowledge. Many triggers can be derived from this knowledge. The main goal here is to localize a resident in an indoor environment by combining the information gathered from the distributed environmental sensors and body activity information provided by the wearable sensors [10]. Binary information about human motion in their region is provided by the passive infrared sensor. Estimates such as activity recognition, walking velocity, and heading estimation can be an inference from the motion information inertial measurement unit sensor. For the purpose of increasing the localization accuracy, a particle filter-based sensor fusion algorithm was proposed. This data can be used to trigger applications like switching off lights in a particular room based on where a person is in real time, or making particular adjustments to the room when a person is sleeping or doing a particular activity. Most of the systems that are used for the purpose of body motion tracking consist of a lot of wires and hence hinder movements [11]. This affects the natural limb movement of the user which in turn hinders the process of collecting sensitive data. Researchers developed a wearable wireless sensor network using accelerometers developed for the purpose of monitoring human motion. This system has the major advantage of being wireless which allows for unrestrained movement and hence improves the quality of data being gathered. The system has the added advantage of being portable and also low in cost when compared to the camera-based method of motion tracking. The low power feature of the system makes this system viable for long-term monitoring.

In addition to human monitoring, energy monitoring was also attempted. Two telemetry units with multiple sensor channels that are used to sense resistance and voltage [12] are used for the purpose of home energy monitoring. The signals from the sensors are mixed, modulated, and transmitted using a radio transceiver to a remote computer. Another important feature of this system is that it can be wireless battery charging. A wireless sensor Zigbee network is designed and tested for the purpose of monitoring and controlling home appliances. In the Zigbee network parameters, a central control unit is used to set up the network and configure it. Sensors report the sensor data like light and temperature periodically. From the central control unit, any home appliance limited to less than 10 amps can be turned on or off. Home automation is achieved with this bidirectional communication that allows wearable handsets to access the consumer services offered through the server [13]. Services can be accessed by the universal home services interface (UHSI) which is shared by all the access devices. Another feature is that the UHSI can also include a voice-activated interface to simplify the usage from non-graphical terminals. This infrastructure resides on a home gateway and makes it possible for the users to access the local subsystems or external wide area network (WAN) services in a home network.

11.3 Communication Protocols

A communication protocol which is essential in all kinds of smart automation technology is a set of rules that permit two or more parties of a system to transmit data. Some of the most popularly used wireless protocols are Bluetooth low energy (BLE), Z-wave, and Zigbee [14].

11.3.1 Zigbee

Zigbee is a wireless communication protocol that is used for short-range and low transmission rate between electronic components [15]. Zigbee has two modes in which the Zigbee node works. In the working mode, the rate of transmission is low and the amount of data transmitted is small. Due to these factors, the time of receiving and sending signals is reduced. In the working mode, the Zigbee node lies dormant, and hence these two modes are combined to reduce the power consumption significantly [15]. This protocol is very reliable because of the talk-when-ready mechanism using which the node sends packets and waits for a confirmation reply from the recipient. The packet will be transmitted again if there is no confirmation. Another important feature of a Zigbee network is that it can contain 255 Zigbee nodes. Zigbee provides security for the network and application support layers [16]. Security services include methods for key establishment, key transport, and frame protection, and device management gives an implementation guide for the usage of Zigbee devices in wireless sensor network (WSN) applications [17].

Zigbee consists of three types of devices namely coordinator, router, and end device [18]. With reference to the device types in an IEEE 802.15.4 network, the Zigbee coordinator corresponds to the personal area network (PAN) coordinator, a router corresponds to a coordinator, and an end device corresponds to an reduced function device (RFD) or a full-function device (FFD) which is neither a coordinator nor the PAN coordinator [18].

The Zigbee network layer can support the star, tree, and mesh topologies. In the star topology, the network is controlled by the coordinator, and the end devices directly communicate with the coordinator. The coordinator initiates and maintains the devices on the network. In mesh and tree topologies, the coordinator starts the network and chooses certain key network parameters, but the network can be extended with the use of Zigbee Routers [18]. In tree networks, using a hierarchical routing strategy, routers move data and control messages through the network. Zigbee provides security services and attack vectors, hence a tool was developed so that researchers, developers, and testers can check the implementation of the security services of their product [19]. This tool can be used to check encryption networks and perform tests like network leaves/joins, resetting to factory defaults, or searching for insecure key transport. The practical security analysis of major Zigbee implementations like smart lighting solutions and a Zigbee-enabled door lock showed that solutions are designed such that their implementation and setup is easy [19]. But they don't have the configuration possibilities and undergo the vulnerable device pairing procedure which allows any person to sniff the exchanged network key. Zigbee communication can easily be jammed. Zigbee is preferred due to the fact that it has low power communication and energy saving. If we send noise to the target Zigbee channel, we can prevent communication. A user would identify a lost connection and try to do a re-pairing. Now the transmitted network key can be sniffed. Hence, the entire security is compromised and the attacker gets full control of the system due to the fact that the secrecy of the key is compromised [19].

Zigbee has two security modes: standard security mode and high-security mode [20]. Standard security mode is designed for lower security residential applications while the high-security mode is used for high-security commercial applications. Dini and Tiloca [20] state that the security model of the Smart Energy Profile has been widely used as a reference security model for Zigbee applications. This security model has two main issues: (i) the requirement of the forward address is not addressed to the full extent. When a node leaves the network, it is still able to access the communication because of the fact that the onboard keying material is not properly revoked [20]. Hence, these keys are compromised and if not revoked properly they can be used to mount many attacks on the network and application level. (ii) the model uses a public key protocol for device authentication and key establishment [20]. This model allows many subjects to issue certificates, namely manufacturers, distributors, and even end-users. Hence, the device should have certificates of all potential certification subjects. But this creates the problem of scalability. This is because it clashes with the fact that Zigbee end devices have limited storage resources. IEEE 802.15.4 defines the physical and MAC layers, while Zigbee defines the network and application layers [20].

A Zigbee-based sensor network has been attempted for automatic temperature control, automatic light control, intrusion detection, and real-time location tracking [21]. Users can easily control their house application by sending the command through a short message service (SMS). When the main controller receives the command, it sends the command to the destination node using the intermediate nodes. The destination node can then carry out the task. The sensor nodes can be added anywhere in the house. The main controller can detect new nodes automatically. Apart from SMS-based commands, Zigbee networks can be used for the development of voice-controlled home automation. The speech recognition algorithm is implemented in the matrix laboratory (MATLAB) [22]. After the speech recognition commands are recognized, the Arduino pin is changed using the universal serial bus (USB). The voltage change in the pin is detected by the same pin connected to the transmitter Zigbee module. The 5 V relay is triggered when the Zigbee module detects the message. This relay can then be used to control any other application. Hence, any device can be remotely controlled with speech commands using a wireless network. A system can be implemented with a Zigbee security system with 2 modes: SMS call mode and alarm mode [23]. A system can be implemented with a home automation system with recognition of voice commands [24]. A system can be implemented with a Zigbee-based home automation system and Wireless Fidelity (Wi-Fi) network which is put together with the use of a common home gateway [25].

The basis of the communication between devices in a Zigbee network is the application profile [19]. Application profiles can be interpreted as the agreement for the messages, processing actions, and message formats that help the developers to create an interoperable application employing application entities that reside on separate devices. A Zigbee home automation public application profile allows a chain of devices to exchange control messages to make wireless home automation applications [19]. This system can be used to automate various devices like turning a fan on or off, automating the lights, etc. [19]. This means that if a manufacturer wants its device to be compatible with the other devices from other manufacturers, its device should implement the standard interfaces and practices of this profile. To do this, all the Zigbee home automation devices should implement the start-up attribute sets. Two of the attributes are discussed below [19].

Two of the attributes are as follows:

Default Trust Center Link Key

- 0 × 5A 0 × 69 0 × 67 0 × 42 0 × 65 0 × 65 0 × 41 0 × 6C 0 × 6C 0 × 69 0 × 61 0 × 6E 0 × 63 0 × 65 0 × 30 0 × 39.
- Note: The Link Key is listed in little-endian format.

Default Link K0065xy Join

- 0 × 01 (True).
- This flag enables the use of default link key joins as a fallback case at start-up time.

Because of the use of the default trust center (TC), link key ZigbeeAlliance09 introduces a very high risk to the secrecy of the network key [19]. In the home automation public application profile, the current network key is transported using the default TC link key in the situation when the joining device is not known. This permits the case where alternative pre-configured link keys associated with a device can be also used (Zigbee Alliance 2013). Since the security of the device depends on the secrecy of the keys and hence on the secure initialization and transport of the encryption keys, this default fallback mechanism is considered to be a critical risk [19]. If an attacker is able to sniff a device join using the default TC link key, the active network key is compromised, and hence the security of the whole network communication is compromised. If systems like lighting are considered, then it's lower risk but when it comes to door locks, the impact is a lot greater. Zigbee light link (ZLL) profile mainly addresses devices and their functionality in the over-the-counter, consumer lighting application domain [19]. During Zigbee commissioning when a non-ZLL device is connecting to a ZLL network without a trust center, for the purpose of authentication, an already installed link key is used to secure the transfer of the network. The ZLL pre-installed link key is shared by all the certified ZLL devices. Just like the home automation public application profile (HAPAP), the ZLL profile also uses »ZigbeeAlliance09« as the default center link key in the SAS and requires the support of an insecure join as a fallback. This leads to the same vulnerability initial key exchange [19]. Even if the manufacturer has implemented a secure key exchange and distributed the proper key material, it is possible for an external attacker to disturb the network join by selective jamming and then waiting for the insecure join to get access to the exchanged key material. Even after the implementation of the secure key exchange and distribution of proper material by the manufacturer, the network can still be disturbed by an attacker using the technique of selective jamming and then waiting for the insecure join to get access to the exchanged key material. The information of the ZLL master key permits the attacker to intercept the key exchange and get access to the current network key. If this happens, then the attacker will have access to all the devices in the Zigbee network [19]. If the ZLL master key is leaked on the net, then the security of all the ZLL devices is considered to be compromised.

11.3.2 Z-wave

Z-wave operates in the industrial, scientific, and medical radio frequency band [26]. Z-wave uses a mesh network. The number of devices that can be connected is 232. Z-wave involves the four lower layers of the Open Systems Interconnection (OSI) model, physical, data link, network, and transport. It works on the common industrial frequency 828 MHz at EU markets and also 908 MHz as a part of the Industrial, Scientific, and Medical (ISM) bands with a maximum range of 30 m. Finally, the maximum provided data rate is up to 100 kbps. The media access control (MAC) layer controls the radio frequency (RF) medium which can be controlled by the wireless hardware and is independent of the radio frequency medium [27]. The transfer layer of the Z-wave protocol sees the connection between two sequential devices which include requisition, checksum screening, and acknowledgment (ACK). The routing layer manages to forward the frames by the Z-wave network from one single device to another. The application layer is responsible for distributing the frame payload, decoding, and performing commands and supplied parameters in the Z-wave network [28].

The transport layer of the Z-wave is responsible for tasks like retransmission, packet acknowledgment, beaming, and authentication of packet origin [28]. Each frame in the Z-wave layer has the 32-bit home identity (ID) that is responsible for the identification of the associated Z-wave network, 8-bit source node ID, frame header which is mainly responsible for defining frame type (single-cast, multi-cast, routed), and control flags such as low power transmission, 8-bit payload length, and the payload and the 8-bit frame checksum value. This layer depends on a frame checksum value to detect and discard erroneous frames. Frame retransmission happens when an acknowledgment frame has not been received from the destination node before the expiration time of the frame [28]. Beam frames controlled by the transport layer are used for the purpose of waking up the battery-powered Z-wave nodes. Many Z-wave devices that are battery-powered like door locks need to listen to the network controller for the incoming commands, however, keeping their radio switched on all the time drains the battery very quickly [28]. To preserve battery power, the device goes into sleep mode and periodically turns on its radio, to look for beam frames. Several beams are sent by the transmitting node in 100 ms intervals to make sure that the device which is in sleep mode notices one of the frames when it wakes up.

The Z-wave protocol has a mesh network with a primary controller and has 232 nodes that can behave as a packet repeater which is used to route the Z-wave data even in situations when the communicating parties are not able to establish radio links to each other. Each of the devices in the Z-wave network maintains a network topology that can indicate all the devices in its proximity. Routing issues are caused in the network in situations when the device locations change or when they are removed from the network. For the purpose of detection of new network locations and routes and also to optimize the routing tables, the Z-wave protocol has support for auto topology discovery.

The application layer oversees the parsing of the frame payload and decoding of the Z-wave commands and the supplied parameters. The decoded command and associated parameters will be sent to the controller software which runs on the home control computer or appliance in case the node was a Z-wave controller device. If not, they will be processed by the device firmware that has been developed by using the Z-wave software development kit (SDK) and is running on the Z-wave chip. The payload frame is started with a one-byte command header which specifies the command is single/multi or broadcast which is followed by the command class. Z-wave command classes are used to define the device functionalities such as door lock, alarm sensor, binary sensor, and heating thermostat. Each command class consists of multiple commands. For example, COMMAND_CLASS_ALARM (defined as 0 × 71) includes CMD_GET (0 × 04) and CMD_REPORT (0 × 05), where the first one is sent by the controller to the alarm to get the current state of the alarm. The second one is sent to the controller when the alarm is being triggered [28, 29].

Some examples of Z-waves are Z-wave door locks and Z-wave light switches. Z-wave light switches can help the users take control of lights with a click of a button from the portable controller. Emergency situations can be identified by this method. An important feature of appliances having the Z-wave door lock is that the appliance lies dormant for 1.5 s and searches for the beam node frames by regularly turning on its radio.

Messages are asynchronously exchanged over the RF medium as the MAC Protocol Data Unit (MPDU) frames by using the Z-wave physical layer [29]. An MPDU has a header, which consists of identification as well as the control fields. The MPDU payload has data concerning the application-layer command, query, or report. There are some optional layers between the MPDU header and application layer, which include the network and security layers. A very important characteristic of Z-wave is the way in which devices enter and leave the network. This is very similar to the pairing process seen in Bluetooth. A user who wishes to add any device to the network first has to put the Z-wave controller and the new device into its pairing mode. Putting a device in pairing mode is done by either pressing a button or by resetting the device manually. The controller adds any device found to also be in pairing mode during the pairing mode. Based on the implementation of the controller, it gives a list of devices that have been discovered and allows the user to select the ones that the user wants to add. Alternatively, the controller also has the option to add all discovered devices within its physical proximity. Some examples of the controller are Mi Casa Verde (Vera 2016) and Aeon Labs Z-Stick (Z-Stick 2016) [29]. Removal of a device from the Z-wave network can be done by the same technique. Z-wave devices can be identified by a 4-byte home ID. This is assigned to a device by the controller in the pairing process. All the nodes that are paired to a controller have the same home ID. This is assigned by the manufacturer to the controller in the factory. The next type of identification is the node ID. A node ID is a byte value that is assigned to a device by the controller during its pairing process. The controller node is given the lowest node ID which is 1 [29]. The first device which is paired to a controller has the node ID of 2. As more devices are added, they are given the consecutive node ID after the previous one. Z-wave networks consist of controllers, sensors, and actuators. For device compatibility, the application-layer protocol has been very well-defined (OpenZwave 2016); but, not all devices need the ability to take part in all application-layer transactions. For example, there is no need for a light switch to know how to respond to a request for reading the temperature. Hence, the application layer has been divided according to its functionality into a series of command classes. There exists a subset of the application-layer commands concerning the class within the command classes. Any Z-wave device belongs to the command classes, and the associated application-layer protocol functionality is included during the compilation of its firmware image [29]. During the pairing operation, a device announces this set of supported classes to the controller. Every command class is actually a unique byte value (OpenZwave 2016). For example, a light switch may announce that it has commands to get, set, and report the state of the switch (OpenZwave 2016). Now that it has been announced, the controller now knows the subset of commands that the light switch obeys. Commands sent to the light switch from command classes that are not supported are just ignored.

11.3.3 Bluetooth Low Energy (BLE)

The data rate is 2mbps for BLE 4.2 and BLE 5. However, the rate is much lower on account of the various protocols overheads, adaptive RF connection adjustments for maintaining robust links [30]. One of the main drawbacks of BLE is low throughput. For BLE 4.2, the highest throughput ranges from 58 to 236 MBps [31]. A system can also be implemented with a fast and low-cost smart home automation based on an Android app and an Arduino [32]. A Bluetooth chip is used. Many devices like lights and direct current (DC) servo motors are used along with Arduino as used to show the feasibility, reliability, and quick operation of the Bluetooth-based smart home automation. These smartphones can be additionally worn as wearables, thus making them a suitable wearable for the purpose of home automation [33].

The BLE protocol stack has three main layers, mainly the controller, host, and application. The controller consists of the physical layer and the link layer [34]. The host has upper layer functionalities, which also include the logical link control and adaptation protocol (L2CAP), the attribute protocol (ATT), the generic attribute profile (GATT), the Security Manager (SM), and the generic access profile (GAP). The host controller interface (HCI) is the communication between the host and the controller. On top of the host, an application-layer program can be used. The logical link control and adaptation protocol (L2CAP) layer is responsible for the task of multiplexing and performing segmentation and reassembly of the packets for the lower layers. The communication between two devices playing the roles of server and client is defined by the ATT. The tasks of device pairing, authentication, and key distribution are overseen by the SM. A database in the GATT server has the data that has to be transmitted by the BLE node and then received by a BLE node. If required, the data from the GATT server can be moved to another database that resides in another node and is called the GATT client. The device role of master and slave and the GATT role are independent of each other.

For the interaction with the GATT server and GATT client databases, the GATT layer has defined a framework. Using the ATT layer, the GATT client can modify the data which is situated in the GATT server [34]. By sending requests, the GATT client can approach the data of the GATT server, which triggers responses from the GATT server. Notifications and indications which are the 2 different types of messages are sent to the GATT client by the GATT server. Notifications differ from the identifications in the way that there is a necessity of acknowledgments for identifications whereas notifications do not need any acknowledgments. On the basis of hierarchy, the data is structured on the GATT server. The data here has been organized as a profile-service characteristics–attributes hierarch. The smallest data entities defined by the GATT are attributes. Attributes contain application data or data concerning the GATT hierarchy structure. The GATT table has a set of attributes. A GATT table can be explained as a database where each row represents a single attribute and the columns represent the different fields that make an attribute. The attributes in the GATT server have been grouped into services which are represented by a service attribute. Application data is contained by the characteristic attributes [34]. Descriptors are actually attributes that have been defined and describe the characteristic values. A service attribute represents a service that has one or more characteristics. For the purpose of breaking data up into logical entities, services are used. A defined collection of services is called a profile. The device roles, modes, and procedures for the discovery of BLE devices which are in range and services, the management of connection establishment, and security are defined by the GAP layer [34]. Advertiser, scanner or initiator, master, and slave are the 4 roles that have been defined by the BLE GAP. To find how many and which BLE devices are in range, advertising, scanning, and initiating modes are used. If a device is in advertising mode, it is referred to as the advertiser, whereas the devices that are in initiating and scanning modes are called initiators and scanners, respectively. The roles of the scanner and initiator don't differ much except the fact that the scanner only discovers the advertiser, whereas the initiator requests a connection with the advertiser after it has done receiving an advertised message [34]. The scanner wakes up periodically so that it can look out for any new advertising messages. Connections can allow application data to be transmitted in a very reliable and robust manner. After a connection has been established, the advertiser becomes the slave whereas the initiator becomes the master.

11.4 Role of Internet of Things in Smart Home Automation

Internet of Things (IoT) has played a huge role in smart home automations. IoT has made it possible for us to connect to our devices from anywhere in the world. Internet of Things, or IoT, can be explained as a system of interrelated computing devices, digital machines, etc., that are given special unique identifiers that provide us with the ability to transfer the data over a network without the need for any human-to-human or human-to-computer interaction. The component of IoT that makes IoT-based smart home automation so powerful is the cloud. Some of the cloud deployment models include private, community, and public types of clouds. Private clouds are operated and managed by a single organization [35]. There are several advantages to setting up a private cloud. The first is that it maximizes and optimizes the utility of the in-house resources that are already there. The second is that issues related to the privacy of the data, security breaches, and trust can be solved to an extent using private clouds.

The third reason is that the cost of the data transfer to the public cloud from a local IT structure is considerable. The fourth is that organizations need full control over the critical activities which are behind their firewalls. In some situations, many organizations come together to share a cloud infrastructure and also the policies, requirements, values, and concerns. These types of cloud infrastructures are called community clouds and can be hosted by a third-party organization or inside of any one of the organizations in the community. This public cloud can be used by public users, but the cloud service provider has full ownership of the public cloud along with its own policy, value, and profit, costing, and charging model. Most of the popular cloud services are actually public clouds like Amazon EC2, S3, Google AppEngine, and Force.com. Now, cloud computing can be explained as a model for enabling convenient, on-demand network access to a shared pool of configurable computing resources (e.g., networks, servers, storage, applications, and services) that can be provisioned and released with minimal management effort [35]. Visualization is very important for IoT applications because it is responsible for the interaction of the user with the environment [36]. To fully benefit from the IoT revolution, visualizations that are easy to understand and also attractive to look at are critical. As we switch from 2Dimension (2D) to 3D (3Dimension) screens, information can be provided to in more efficient ways to consumers. This enables the policymakers to convert the data into knowledge which is important in quick decision-making. Extracting information from raw data is nontrivial. This takes into account both event detection and visualization of the raw and modeled data, with the information presented to the end-user according to his needs.

Figures 11.1, 11.2, 11.3, and 11.4 show the graph of the data that is being pushed to the ThingSpeak cloud. The data is coming in from a gas sensor, ultrasonic sensor, accelerometer, and ECG sensor which are connected to an ESP8266. ThingSpeak is

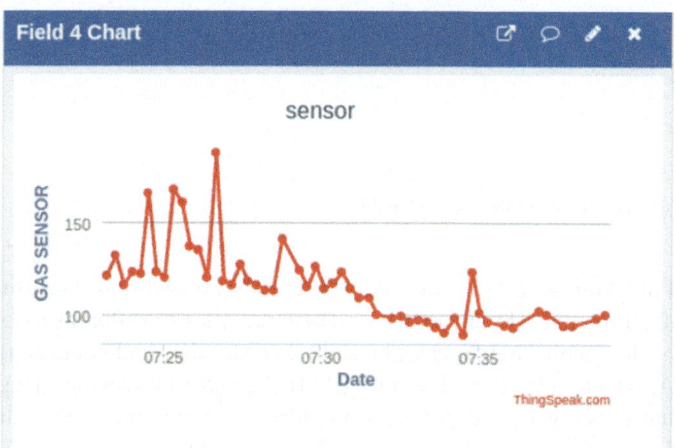

Fig. 11.1 Gas sensor recorded data

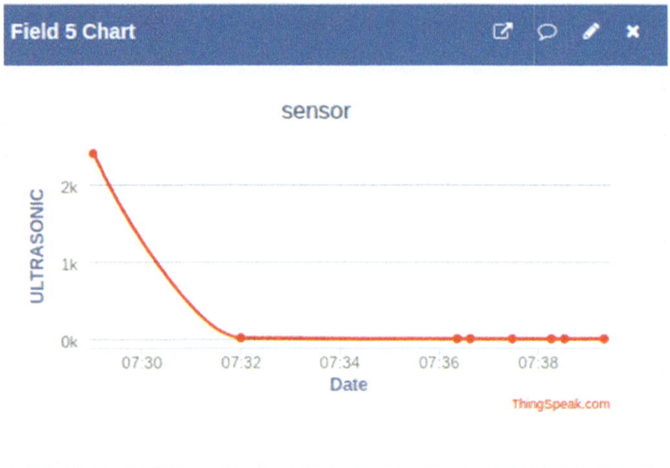

Fig. 11.2 Ultrasonic sensor recorded data

Fig. 11.3 Accelerometer sensor recorded data

also an IoT analytics service [37]. It has an API to retrieve and store data using Hypertext Transfer Protocol (HTTP) requests. There are a lot of different tasks that can be accomplished with ThingSpeak like storing sensor data and controlling projects using the Application Program Interface (API). ThingSpeak provides the graphs for the data that is being logged and also provides functions like time scaling, averaging median, summing, and rounding of the data. ThingSpeak has API read and writes keys. It has time management and Javascript-based charts. There are multiple channels that can be created in ThingSpeak with each channel supporting 8 fields.

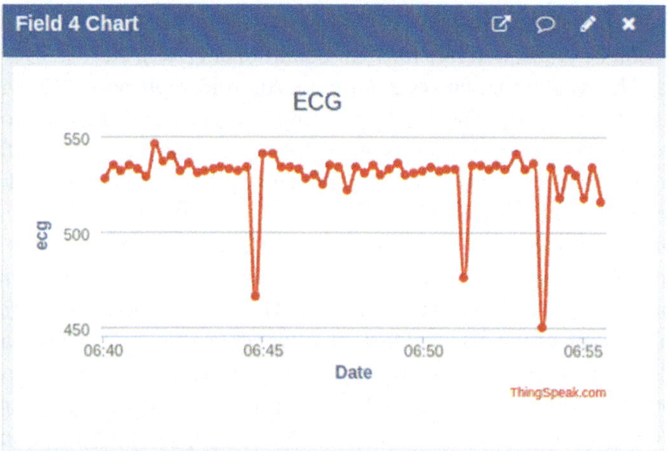

Fig. 11.4 ECG sensor recorded data

IR sensor, passive infrared sensor (PIR sensor) sensor, and fire detector sensor based on a home automation system have been implemented with an IoT-based monitoring and control system for home automation [38]. The IR sensor (Infrared sensor) is used here for knowing the status of the light inside a room. By using a mobile phone, lights can be turned off for saving power by the use of IoT concepts and Wi-Fi configuration. The PIR sensor is used to test whether a person is there in the room or not. This information will be then used to switch on/off the fans. The lights, fans can be controlled by a web server or by making an app on mobile. The fire detection sensor is for the safety of the family. In the event of a fire in the house, the fire detection sensor will be triggered and can be used to set up messages, alerts along with pictures or videos to a mobile phone followed by a phone call to the nearby fire station. Using a mobile phone instead of emails saves us from virus threats that may be included in the attachment or a situation can arise in which the mail cannot be viewed because of many spam emails. Also logging into emails to check messages wastes a lot of time.

An IoT-based smart security and home automation system was also developed with many advantages like a low-cost system, operating a home security system without data connection enabled in the phone, without security problems, and faulty triggers. The optional smartphone application was also developed to take care of the user's needs [39]. The prototype can be used as a smart security system and also as a smart home automation system. The TI-CC3200 launchpad and a Wi-Fi can be used to control all the electrical appliances inside the home. PIR sensors trigger a voice call to a predefined number and also can be used to disable the security system in case a guest arrives in the house and the user is not there to receive the guest. SMS can be sent to the concerned authority in a situation if the user feels that his house is not safe. Additional features include turning lights on to warn the intruder and turning the lights off after a fixed time delay. By pressing digits like 2, 4, and 5, many other appliances can be controlled like fans, air conditioners (AC), etc.

IoT-based home automation system uses an Android application [40] with a high-security feature to log in to the app by entering the user name and password. These passwords are user-defined. Various sensors like temperature sensors, rainfall sensor, and light-dependent resistor (LDR) are used. This system can also be used to control the fan speed, light intensity, and to check rainfall. Actuators like servo, etc., are used along with other components like buzzers, motor drivers, etc.

Some of the services that have been used a lot to make IoT-based smart home automation are Pachube, Nimbits, and iDigi [41]. These services mostly focus more on the visualization of the data and lack of secure data access. They also lack secure data access and provision for the interfaces for the linkage to the mobile or external application for further processing. iDigi is an application messaging engine that permits broadcast and receipt notification for the use of application-to-device interaction and also confirmation. iDigi platform can be described as a machine-to-machine platform-as-service [41]. This platform helps in making secure, cost-effective, and scalable solutions to tie together the device assets. It makes the process of linking remote assets very easy and provides all the tools to link, store, and move the data from one end of the enterprise to another. This platform also includes iDigi Dia which is a device connecting software that is used to simplify device connectivity and integration. It enables the management of products like Zigbee nodes. It also has permanent storage and cache options for access to historical device samples and generation-based storage. Broadcast and receipt notification for application-to-device interaction and confirmation is allowed by the application messaging engine. Nimbits is basically a data processing service. It is used to record and share sensor data on the cloud [41]. It is an open-source platform for IoT and is also free. Users can generate data points and also drop changing numeric, text-based, Global Positioning System (GPS), Javascript Object Notation (JSON), or XML values into the data points. These points can be further configured to do calculations, generate alerts, and be linked to process control diagrams, spreadsheets, websites, etc. By the use of simple mathematical relations, it offers an alert management mechanism and data calculation on the received sensor data. Pachube was one of the first database service providers which enabled the developers to link sensor data to the Web [41]. It has a scalable infrastructure with a real-time data closed based infrastructure for IoT. This platform allows users to build IoT projects, store and share real-time sensor data of the surroundings from anywhere in the world. This platform is used to make graphs of the data that came in from the sensors and monitor remote environments. Also, there are a lot of interfaces that are available for making sensor-based applications or mobile-based applications for the management of the data on the cloud. Apart from these features, one of the most important features that have made it a go-to cloud service is that the basic usage is free along with an easily accessible API coupled with an interactive website for management of the sensor data [41].

11.5 Gesture-Based Home Automation and Security for the Visually Impaired People and Old Age People

Authors have developed a gesture-based home automation device and security-enabled system to help the visually impaired and aged people. Various sensors are embedded on the home automation devices to assist disabled people with vision impairment as well as for aged people with their capability. In addition to this, the security system is included using face recognition with a voice-assistant feature using a camera placed on the door of the house. The door opens for registered faces in the database and any forced entry automatically sends an alert message to police. People with physical disabilities, such as paraplegia and Parkinson's disease, as well as old age people are greatly limited with movement inside the home.

11.5.1 Description of the System

The proposed gesture-controlled home automation device comprises XBee (uses the IEEE 802.15.4 network protocol) module, accelerometer (ADXL335), joystick (Dual Axis XY), and Arduino duo (microcontroller). As soon as the system is activated, the two XBee modules are connected. The gestures are sensed using accelerometers placed on the hand/neck of the person. The sensed values are given to the Arduino microcontroller that is powered by 5 V. A threshold value is fixed from the capability of a person. The values which are recorded from the accelerometer are compared with the set threshold value. If the value goes above the threshold, signals from Arduino are given to the XBee transmitter and to the receiver. The transmitted data is received by an XBee operating at the same frequency as that of the transmitter. The received signals are given to the Arduino that compares the received and stored hand gestures and then signals are given to the home appliances through relays to control them. Arduino duo controllers were used to control the system during the operation. People with various impairments can extraordinarily decrease the self-sufficiency and personal satisfaction of a patient while exhibiting a noteworthy repeating cost in home-human services. Expanded with a non-intrusive wearable sensor framework and home-mechanization gear, the patient can recapture a degree of independence without depending on other family members or caretakers for doing simple tasks. A gesture-based controlled automation system would assist aged people. The aim of this project is to help such people to live safely inside the home with home automation features like switching on/off the lights and fans or opening/closing the door. The glove is a rechargeable weightless handheld portable device.

This device includes features to assist vision-impaired people with voice-assist features using object recognition in the workspace. In addition, the device includes features to alert police and send a message to hospitals, caretakers, etc. The security system uses face recognition technology with a voice-assist feature for the purpose of security. Therefore, when a known person asks to open the door, the face will be recognized and a voice-assist feature will tell the name of the person if registered, otherwise it sends the message to the police. The same device can also be used to assist visually challenged people to identify objects/obstacles in a workspace like a cupboard, chairs, etc. The glove includes a panic button to send an immediate alarm to the hospital, police, and neighbors, respectively.

11.6 Conclusion

Automation is changing industries to make high productivity, efficient use of resources, etc. Apart from industries, health care has embraced the concept of automation, and IoT and has developed an entirely new concept of telemedicine. Robots are being used to do tasks that human beings can't with precision that is difficult to achieve. There are numerous advantages to automation, but there are also some disadvantages. The one general notion is that automation is replacing the workforce. The tedious and repeating tasks that were being carried out by a large workforce are now being replaced by technology. It is a highly controversial statement that automation is replacing the general workforce and most of the general public considered it is negative. The other major disadvantage of automation is the high capital required to invest in automation. Moreover, there is a higher level of maintenance required for autonomous machines as compared to a manually operated machine. The transitions to an autonomous industry have to overcome several challenges such as 'What to automate?', to reach the end goal of maximum productivity and efficiency with the minimum investment. The second challenge is skillset. Industry leaders are often in the deficit of a team with the right skillset to implement automation and test it. The next challenge comes in the form of a high upfront investment cost. Industries often fail to put together a capital high enough to achieve automation. But smart automation is not a dream anymore; it has been achieved, embraced, and is now one of the most important components of health care. With continuous advancements, smart automation is set to continually grow with no limits to what it can achieve to aid people with disabilities and aged people to manage day-to-day activities. The role of Internet of Things in smart home automation needs cloud services, deployment models, and some innovative notions in smart home automation using IoT. Network architectures, security concerns, and their role in smart home automation have been discussed. Furthermore, gesture-based home automation and security for visually impaired people and old age people have been discussed in an attempt to contribute to society. With more innovation in smart automation, its future can be described as interesting at the very least.

References

1. Rifon LA, Costa CR, Carballa MG, Rodriguez SV, Iglesias MF (2013, April) Improving the quality of life of dependent and disabled people through home automation and tele-assistance. In: 2013 8th international conference on computer science & education. IEEE, pp 478–483
2. ElShafee A, Hamed KA (2012) Design and implementation of a WIFI based home automation system. World Acad Sci Eng Technol 68:2177–2180
3. Kodali RK, Soratkal S (2016) MQTT based home automation system using ESP8266. In: 2016 IEEE region 10 humanitarian technology conference (R10-HTC). IEEE, pp 1–5
4. Starner T, Auxier J, Ashbrook D, Gandy M (2000, October) The gesture pendant: a self-illuminating, wearable, infrared computer vision system for home automation control and medical monitoring. In: Digest of papers. fourth international symposium on wearable computers. IEEE, pp 87–94
5. Singh G, Nelson A, Robucci R, Patel C, Banerjee N (2015, March) Inviz: low-power personalized gesture recognition using wearable textile capacitive sensor arrays. In: 2015 IEEE international conference on pervasive computing and communications (PerCom). IEEE, pp 198–206
6. Nelson A, Schmandt J, Shyamkumar P, Wilkins W, Lachut D, Banerjee N et al (2013, November). Wearable multi-sensor gesture recognition for paralysis patients. In: SENSORS, 2013 IEEE. IEEE, pp 1–4
7. Fensli R, Gunnarson E, Gundersen T (2005, June) A wearable ECG-recording system for continuous arrhythmia monitoring in a wireless tele-home-care situation. In: 18th IEEE symposium on computer-based medical systems (CBMS'05). IEEE, pp 407–412
8. Cheng MH, Chen LC, Hung YC, Yang CM (2008, August) A real-time maximum-likelihood heart-rate estimator for wearable textile sensors. In: 2008 30th annual international conference of the IEEE engineering in medicine and biology society. IEEE, pp 254–257
9. Mistry P, Maes P, Chang L (2009) WUW-wear Ur world: a wearable gestural interface. In: CHI'09 extended abstracts on human factors in computing systems, pp 4111–4116
10. Pham M, Yang D, Sheng W (2018) A sensor fusion approach to indoor human localization based on environmental and wearable sensors. IEEE Trans Autom Sci Eng 16(1):339–350
11. Low KS, Lee GX, Taher T (2009, May) A wearable wireless sensor network for human limbs monitoring. In: 2009 IEEE instrumentation and measurement technology conference. IEEE, pp 1332–1336
12. Kumar SK (2012) Design of a wireless platform for wearable and home automation applications (Doctoral dissertation, University of Missouri--Kansas City)
13. Corcoran PM, Desbonnet J (2002, June) Wireless home network infrastructure for wearable appliances. In: 2002 digest of technical papers. International conference on consumer electronics (IEEE Cat. No. 02CH37300). IEEE, pp 104–105
14. Gomez C, Paradells J (2010) Wireless home automation networks: a survey of architectures and technologies. IEEE Commun Mag 48(6):92–101
15. Li P, Li J, Nie L, Wang B (2010, March) Research and application of Zigbee protocol stack. In: 2010 International conference on measuring technology and mechatronics automation, vol 2. IEEE, pp 1031–1034
16. Somani NA, Patel Y (2012) Zigbee: a low power wireless technology for industrial applications. Int J Control Theor Comput Model (IJCTCM) 2(3):27–33
17. de Almeida Oliveira T, Godoy EP (2016) Zigbee wireless dynamic sensor networks: feasibility analysis and implementation guide. IEEE Sens J 16(11):4614–4621
18. Zigbee Document 053474r17, Zigbee Specification, Zigbee Alliance, January 2008. http://www.Zigbee.org/

19. Zillner T, Strobl S (2015) Zigbee exploited: the good, the bad and the ugly. Black Hat–2015 Электронный ресурс. .–Режим доступа: https://www.blackhat.com/docs/us-15/materials/us-15-Zillner-Zigbee-Exploited-The-Good-The-Bad-And-The-Ugly.pdf (дата обращения: 21.03. 2018)
20. Dini G, Tiloca M (2010, June) Considerations on security in Zigbee networks. In: 2010 IEEE international conference on sensor networks, ubiquitous, and trustworthy computing. IEEE, pp 58–65
21. Rathod K, Parikh N, Parikh A, Shah V (2012, September) Wireless automation using Zigbee protocols. In: 2012 ninth international conference on wireless and optical communications networks (WOCN). IEEE, pp 1–5
22. Çubukçu A, Kuncan M, Kaplan K, Ertunc HM (2015, May) Development of a voice-controlled home automation using Zigbee module. In: 2015 23nd signal processing and communications applications conference (SIU). IEEE, pp 1801–1804
23. Ahmad AW, Jan N, Iqbal S, Lee C (2011, August) Implementation of Zigbee-GSM based home security monitoring and remote control system. In: 2011 IEEE 54th international midwest symposium on circuits and systems (MWSCAS). IEEE, pp 1–4
24. Narayanan VS, Gayathri S (2013) Design of wireless home automation and security system using PIC microcontroller. Int J Comput Appl Eng Sci 3(13):140–153
25. Gill K, Yang SH, Yao F, Lu X (2009) A Zigbee-based home automation system. IEEE Trans Consum Electron 55(2):2–430
26. Marksteiner S, Jiménez VJE, Valiant H, Zeiner H (2017, November) An overview of wireless IoT protocol security in the smart home domain. In: 2017 internet of things business models, users, and networks. IEEE, pp 1–8
27. Yassein MB, Mardini W, Khalil A (2016, September) Smart homes automation using Z-wave protocol. In: 2016 international conference on engineering & MIS (ICEMIS). IEEE, pp 1–6
28. Badenhop CW, Graham SR, Ramsey BW, Mullins BE, Mailloux LO (2017) The Z-wave routing protocol and its security implications. Comput Secur 68:112–129
29. Badenhop CW, Graham SR, Ramsey BW, Mullins BE, Mailloux LO (2017). The Z-wave routing protocol and its security implications. Comput Secur 68:112–129
30. Dian FJ, Yousefi A, Lim S (2018) A practical study on Bluetooth Low Energy (BLE) throughput. In: 2018 IEEE 9th annual information technology, electronics and mobile communication conference (IEMCON), Vancouver, BC, pp 768–771. https://doi.org/10.1109/IEMCON.2018.8614763
31. Del Carpio LF, Di Marco P, Skillermark P, Chirikov R, Lagergren K (2017) Comparison of 802.11 ah, BLE and 802.15. 4 for a home automation use case. Int J Wirel Inf Netw 24(3):243–253
32. Naresh D, Chakradhar B, Krishnaveni S (2013) Bluetooth based home automation and security system using ARM9. Int J Eng Trends Technol (IJETT) 4:4052
33. Papp I, Velikic G, Lukac N, Horvat I (2015, September) Uniform representation and control of Bluetooth Low Energy devices in home automation software. In: 2015 IEEE 5th international conference on consumer electronics, Berlin (ICCE-Berlin). IEEE, pp 366–368
34. Dian FJ, Yousefi A, Lim S (2018, November) A practical study on Bluetooth Low Energy (BLE) throughput. In: 2018 IEEE 9th annual information technology, electronics and mobile communication conference (IEMCON). IEEE, pp 768–771
35. Mell P, Grance T (2009, August 21) Draft nist working definition of cloud computing-v15
36. Gubbi J, Buyya R, Marusic S, Palaniswami M (2013) Internet of Things (IoT): A vision, architectural elements, and future directions. Futur Gener Comput Syst 29(7):1645–1660
37. Doukas C, Maglogiannis I (2011, November) Managing wearable sensor data through cloud computing. In: 2011 IEEE third international conference on cloud computing technology and science. IEEE, pp 440–445
38. Pavithra D, Balakrishnan R (2015, April) IoT based monitoring and control system for home automation. In: 2015 global conference on communication technologies (GCCT). IEEE, pp 169–173

39. Kodali RK, Jain V, Bose S, Boppana L (2016, April) IoT based smart security and home automation system. In: 2016 international conference on computing, communication and automation (ICCCA). IEEE, pp 1286–1289
40. Reddy PSN, Reddy KTK, Reddy PAK, Ramaiah GK, Kishor SN (2016, October) An IoT based home automation using android application. In: 2016 International conference on signal processing, communication, power and embedded system (SCOPES). IEEE, pp 285–290
41. Doukas C, Maglogiannis I (2011, November) Managing wearable sensor data through cloud computing. In: 2011 IEEE third international conference on cloud computing technology and science, pp 440–445

Chapter 12
A Novel Architecture Design for Complex Network Measures of Brain Connectivity Aiding Diagnosis

Chandrajit Pal and Amit Acharyya

Abstract Contrary to the conventional belief that cognitive functions are dictated by activation of brain areas, recent research shows that brain-wide functional connectivity network plays more fundamental role. Due to the prevalence of neurocognitive and motor diseases understanding of the nature of such functional connectivity is paramount in devising effective treatment and management strategy. It has already been shown that functional connectivity network derived from EEG (electroencephalogram) recordings exhibit different topography for different neurocognitive conditions. This opens up the possibility of identifying effective markers for predicting cognitive decline by quantitative analysis of such networks. More importantly this may lead to a new type of non-invasive therapy where such connectivity-guided neuromodulation techniques could be initiated for managing cognitive deficit on the fly in nomadic environment. However, one pre-requisite of such approach is the ability to extract the functional connectivity from multichannel EEG in real time. In other words, it necessitates to build a novel architecture for real-time quantitative characterization of functional brain connectivity networks derived from Electroencephalogram (EEG). It consists of two main parts—calculation of Phase Lag Index (PLI) to form the functional connectivity networks and the extraction of a set of graph-theoretic parameters to quantitatively characterize these networks. Owing to the computationally intensive nature of functional connectivity extraction, this is only possible if suitable hardware accelerator is designed. In this chapter we have explored the most efficient hardware architecture for functional brain connectivity formulation from multichannel EEG data after removing the artefacts. Extracted the markers for its quantitative characterization and finally for computing the temporal variability of such markers which typically conveys the implicit nature of cognitive efficiency. Once the EEG data is received, some form of transformation needs to be carried out on the matrix. This generates various patterns which would help in generating a network topology for analysis of the brain cognition network. This analysis of various brain states for neurological anomalies and a normal state by creating a

C. Pal (✉) · A. Acharyya
IIT Hyderabad, IITH Main Road, Near NH-65, Sangareddy, Kandi, Telangana 502285, India
e-mail: ee17pdf04@iith.ac.in

A. Acharyya
e-mail: amit_acharyya@iith.ac.in

G. D. Gargiulo and G. R. Naik (eds.), *Wearable/Personal Monitoring Devices Present to Future*, https://doi.org/10.1007/978-981-16-5324-7_12

graphical model-based visualization of the transformation which is time consuming and requires hardware acceleration. Novel signal processing algorithms has been developed and holistic algorithm—architecture optimization has been carried out for efficient mapping of the algorithm into hardware which has been implemented in FPGA.

12.1 Introduction

The brain has areas of specialized neurons which are responsible for distinct functions. These assemblies influence each other through excitatory and inhibitory synaptic connections [1]. Ensembles of segregated cortical areas of the brain form a big coherent organization which dynamically interacts to steer the brain into specific cognitive states. The temporal evolution of these synchronized cortical areas is instrumental in understanding how the human brain performs certain tasks given a particular stimulus. Functional brain connectivity is defined as the time-based temporal correlations between distributed neuronal units or the synchronization of activation of brain areas when performing a particular cognitive task. In other words, FC refers to the statistical dependence between the signals stemming from two (or among many) distinct units within a nervous system (from single neurons to whole neural net works) [12]. Due to the prevalence of neurocognitive motor diseases understanding the nature of such FC network is important in devising an effective treatment strategy. It has already been shown that FC network derived from EEG (electroencephalogram) recordings exhibit different topography for varying neurocognitive conditions. In other words the graph theory (GT) parameters (coined as features) of the FC network can be used for classifying the various cognitive pathological states. This opens up the possibility of identifying effective markers for predicting cognitive decline by quantitative analysis of such network parameters, which could lead to non-invasive therapy for managing cognitive deficits. An on-the-fly therapeutic approach necessitates the extraction and analysis of network parameters from multichannel EEG in real time. This can be achieved by a suitable hardware accelerator design, given the compute intensive nature of the FC/GT measures [13]. In this study, we have designed the architecture for one brain FC measure, Phase Lag Index- (PLI), derived from the phase value of EEG data, since phase is an important component determining connectivity among multiple electrodes (nodes) [13]. PLI is an alternative measure of statistical interdependencies between time series, which reflects the strength of the coupling but is expected to be less sensitive to the influence of common sources and amplitude effects. The acquired 8-channel EEG data (headcap [4]) is generally pre-processed (e.g., artifact reduction) and the corresponding phase information is extracted using available off-the shelf (COTS) components. Hence, in this paper we have focused on the computation of (1) FC measure- PLI and (2) 8 GT parameters of PLI (which are computed in parallel). A holistic algorithm-architecture optimization has been carried out and the design has been synthesized using STMicroelectronics 130-nm technology library @ 25 MHz, consuming 10nW of dynamic power. Several studies have been performed on brain FC analysis [2, 5, 10, 11, 14, 15, 17, 18]. Authors in [10] have designed an open-source hardware platform to accelerate the determination of scale-freeness (a fundamental topological feature of biological networks), as a benchmark to evaluate time series extraction settings required for reconstruction of brain connectivity. An efficient hardware design was proposed with dedicated parallel coprocessor, implementing only the Dijkstras algorithm using field-programmable gate arrays (FPGA) as a solution to the high computational demand [11]. However, we claim that our proposed design is a first study to accelerate the computation of brain FC/GT measures, making the hardware more suitable for real-time connectivity analysis. The remainder of this paper is organized as follows. Section 12.2 describes the processing overview and the algorithm to architecture mapping, Sect. 12.3 analyses the results and conclusions are drawn in Sect. 12.4 [13]. In the subsequent section we will be discussing on a paper [1] which describes hardware design methodology for reliable and automated removal of ocular and muscular artefact from EEG signals.

12.2 Materials and Methods

12.2.1 Artefact Removal Methodology

EEG is a non-invasive tool for neuro-developmental disorder diagnosis and treatment. However, EEG signal is mixed with other biological signals including Ocular and Muscular artifacts making it difficult to extract the diagnostic features. Therefore, the contaminated EEG channels are often discarded by the medical practitioners which may result in less accurate diagnosis. Many existing methods require reference electrodes, which will create discomfort to the patient/children and cause hindrance to the diagnosis of the neuro-developmental disorder and brain-computer interface in the pervasive environment. Therefore, it would be ideal if these artifacts can be removed real time on the hardware platform in an automated fashion and then the denoised EEG can be used for online diagnosis.

The authors [1] propose a reliable, robust, and automated methodology to solve the aforementioned problem. The proposed methodology is based on the Haar function-based Wavelet decompositions with simple threshold-based wavelet domain denoising and artifacts removal schemes. Subsequently hardware implementation results are also presented. 100 EEG data from Physionet, Klinik für Epileptologie, Universität Bonn, Germany, Caltech EEG databases and 7 EEG data from 3 subjects from University of Southampton, UK have been studied and nine exhaustive case studies comprising of real and simulated data have been formulated and tested. The proposed methodology is prototyped and validated using FPGA platform. Like existing literature, the performance of the proposed methodology is also measured in terms of correlation, regression, and R-square statistics and the respective values lie above 80%, 79%, and 65% with the gain in hardware complexity of 64.28% and improve-

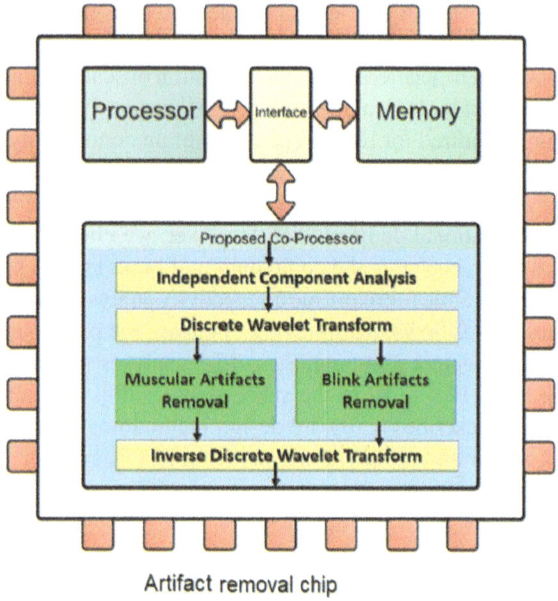

Fig. 12.1 Block Diagram along with the flowchart of the proposed system

ment in hardware delay of 53.58% compared to state-of-the art approaches. Hardware design based on the proposed methodology consumes 75 mW power.

The automated methodology proposed can remove blink and muscular artefacts real time without the need of any extra electrode. Its reliability and robustness is also established after exhaustive simulation study and analysis on both simulated and real data. This would be useful in next generation personalized pervasive health care for brain-computer interface and neuro-developmental disorder diagnosis and treatment [3]. Overview of the complete system design has been depicted in Fig. 12.1. In the next subsections we will discuss the analysis of wireless EEG-based functional connectivity measures with respect to change in environmental factors [3].

12.2.2 Analysing Wireless EEG-Based Functional Connectivity Measures with Respect to Change in Environmental Factors

Here the authors presented a systematic exploration to formulate a predictive model of the human cognitive process with the changing environmental conditions at workplace as shown in Fig. 12.2.

Here the authors aim to investigate the EEG-based features extracted from brain signals acquired during changing environmental factors (acting as a stimuli) to determine the significant features involved in cognitive processing. These features are used to model the relationship between the cognitive processes (functional units)

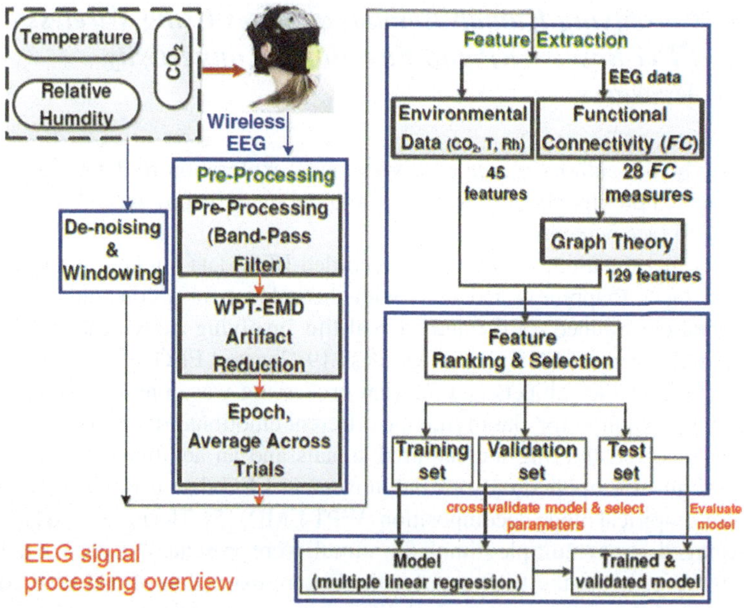

Fig. 12.2 Overview of the methodology

and the environmental variables. Identification of significant features and the model parameters can help to control the environmental factors and lead toward intelligent workspace design.

For this investigation, the designers recorded EEG data elicited by six environmental conditions from three participants in a naturally ventilated building. A wireless 19-Channel EEG system was used to collect data from the subjects and relevant processing was done to identify EEG features which were used for modeling the relationship of the cognitive process with each different condition. The acquired EEG data was denoised using wavelet packet transform-empirical mode decomposition (WPT-EMD) [13] artifact reduction algorithm. The processed EEG data were used to generate FC-based measures which were represented with a reduced dimensionality by applying the Brain Connectivity Toolbox (BCT) yielding 129 features. Similarly, we extract 15 time and frequency domain features from the environmental monitoring data acquired through carbon dioxide (CO_2), temperature (T), and relative humidity (Rh) sensors. We select five significant features that maximize the variance in the feature space, each from EEG, CO_2, T, and Rh data. These features are successfully used for building a model using multiple linear regression in conjunction with 10 runs of tenfold cross-validation (CV) to prospectively select the parameters that relate the environmental data to the EEG. Conversely, it is also important to explore on determining several EEG-based features for classifying three emotional states (happy, fearful, and neutral) pertaining to face perception.

12.2.3 Classifying Human Emotional States Using Wireless EEG-Based ERP and Functional Connectivity Measures

In this study, the authors presented a systematic exploration to determine several EEG-based features for classifying three emotional states (happy, fearful, and neutral) pertaining to face perception.

For this investigation, the designers recorded EEG data elicited by neutral and emotional faces (happiness and fear) with the subject: in a constrained position (conditional-constrained EEG) and in real-life involving body/head movements (condition2-movement EEG) [6]. A wireless 19-Channel EEG system was used to collect data from the subjects and relevant processing was done to identify robust EEG features required for classifying the different emotional states. The processing hierarchy involved filtering the captured signals and an additional artifact reduction algorithm was applied only on the movement EEG data using wavelet packet transform-empirical mode decomposition (WPT-EMD) [5]. The processed EEG data were averaged across multiple stimuli (26 stimuli were presented for each of the three emotions) to generate the ERP data for the feature extraction of: (1) ERP components and (2) FC-based measures. Each FC measure was represented with a reduced dimensionality by applying: (1) gray level co-occurrence matrices (GLCM) [2] and (2) brain connectivity toolbox (BCT). Out of the 463 features extracted, the significant features were selected from a ranked list, using sequential forward selection (SFS) algorithm in conjunction with two classifiers following a cross-validation technique. Our results show that we are able to prospectively classify the three emotional states from condition with an accuracy of 89% using only 17 features with the linear discriminant analysis (LDA) classifier. The overall picture with the complete process flow has been shown in Fig. 12.3. In the next subsection we are going to discuss on the brain connectivity analysis from EEG signals using stable phase-synchronized states during face-perception tasks.

12.2.4 Brain Connectivity Analysis from EEG Signals Using Stable Phase-Synchronized States During Face-Perception Tasks

Degree of phase synchronization between different Electroencephalogram (EEG) channels is known to be the manifestation of the underlying mechanism of information coupling between different brain regions. In this study, the authors apply a continuous wavelet transform (CWT) based analysis technique on EEG data, captured during face-perception tasks, to explore the temporal evolution of phase synchronization, from the onset of a stimulus. Explorations show that there exists a small set (typically 3–5) of unique synchronized patterns or synchrostates, each of which are stable of the order of milliseconds. Particularly, in the beta (β) band, which has been

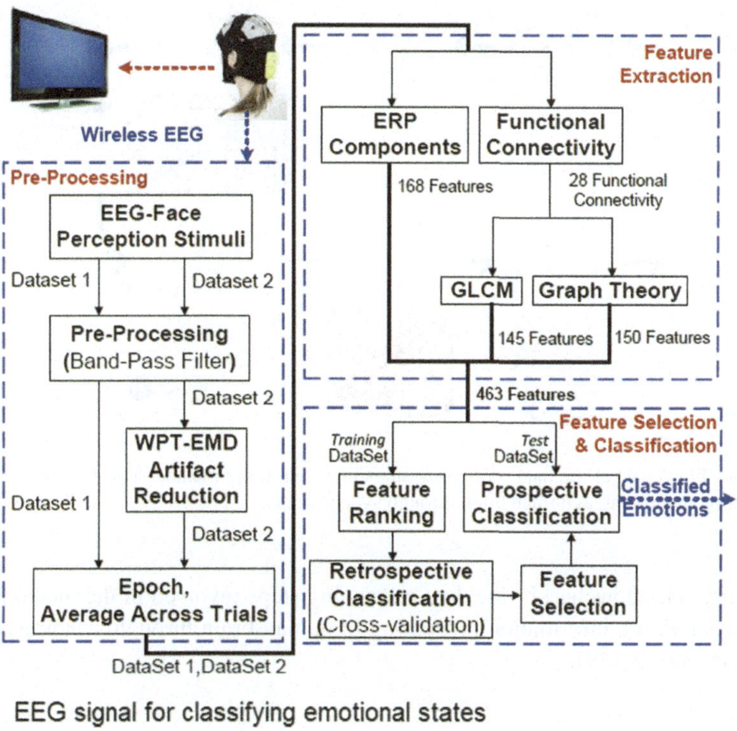

Fig. 12.3 Overview of the methodology

reported to be associated with visual processing task, the number of such stable states has been found to be three consistently. During processing of the stimulus, the switching between these states occurs abruptly but the switching characteristic follows a well-behaved and repeatable sequence. This is observed in a single subject analysis as well as a multiple-subject group-analysis in adults during face perception. It has also been reported that although these patterns remain topographically similar for the general category of face-perception task, the sequence of their occurrence and their temporal stability varies markedly between different face-perception scenarios (stimuli) indicating toward different dynamical characteristics for information processing, which is stimulus-specific in nature. Subsequently, these stable states have been translated into brain complex networks and derived informative network measures for characterizing the degree of segregated processing and information integration in those synchrostates, leading to a new methodology for characterizing information processing in human brain.

The proposed methodology of modeling the functional brain connectivity through the synchrostates may be viewed as a new way of quantitative characterization of the cognitive ability of the subject, stimuli, and information integration/segregation

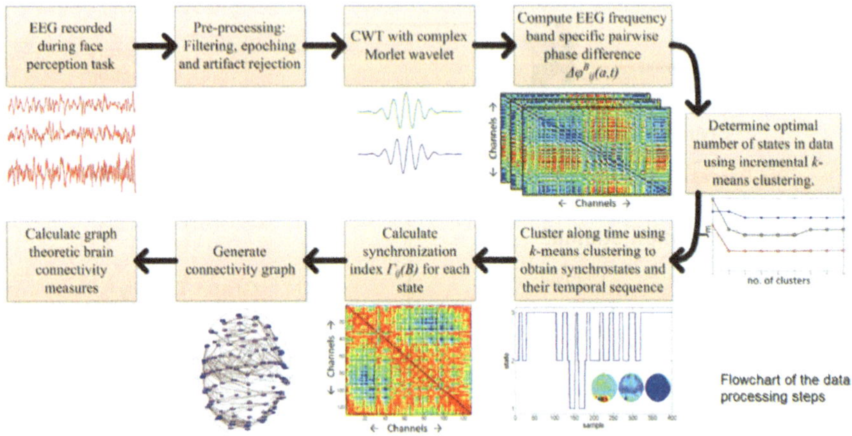

Fig. 12.4 Flowchart of the data processing steps involved in the methodology for characterizing information processing in human brain through synchrostates

capability. The flowchart of the data processing steps involved in the methodology for characterizing information processing in human brain through synchrostates is shown in Fig. 12.4 [9].

12.2.5 On the Existence of Synchrostates in Multichannel EEG Signals During Face-Perception Tasks

In the previous Sect. 12.2.4, during a visual perception task, it has been shown that at ms time scales, there exists a small set of unique phase synchronized patterns, each being stable of the order of ms, and then abruptly switching from one to another. These quasi-stable phase difference patterns are termed as synchrostates. However the pertinent question is whether the synchrostates consistently exist in individual subjects in the cognition related bands (β and γ), and if so, how much inter-subject variability could be expected with respect to the population average, as this is fundamental in ascertaining the possibility of classifying an individual's pathological conditions using graph-theoretic characterization of a functional brain network formed from the synchrostates. The cognitive task selected for the exploration is a set of face-perception tasks where three types of face-perception related stimuli were given to four groups of children with typical development, diagnosed with autism spectrum disorder (ASD), and with high- and low-anxiety scores.

As an illustration the topological maps for all the 3 stimulus in the β and γ bands for the low-anxiety states are shown in Figs. 12.5, 12.6, and 12.7, respectively.

The analysis described in [8] shows that there exist a small set of unique phase difference patterns at ms order time intervals amongst the EEG electrodes when

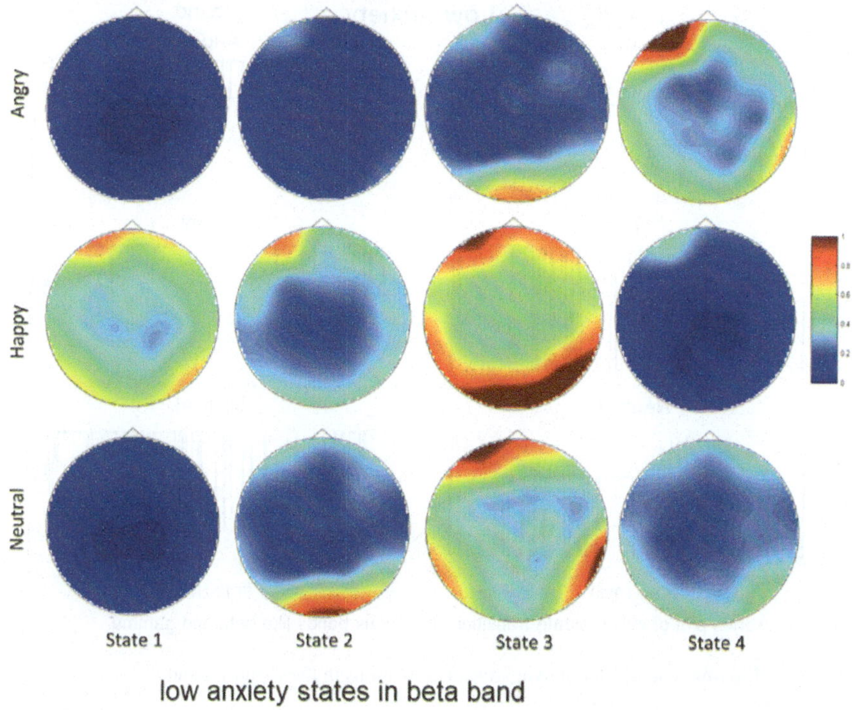

Fig. 12.5 The topographic map for all three stimuli in the β band for the low-anxiety group

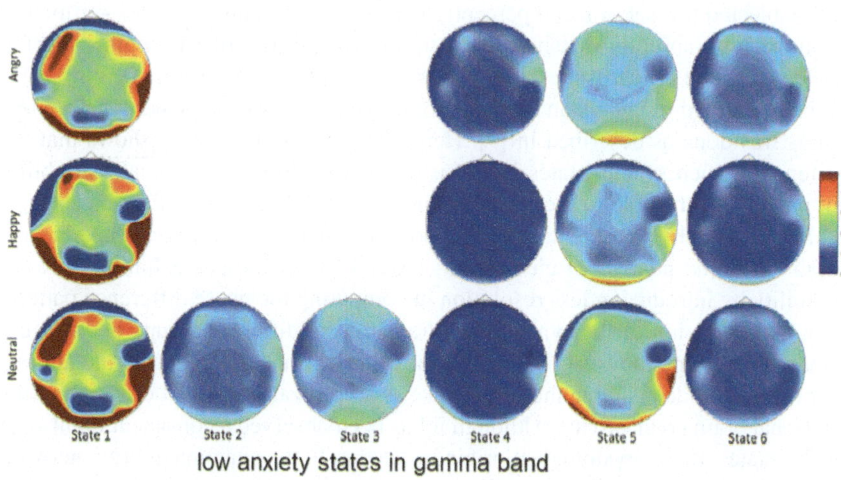

Fig. 12.6 The topographic map for all three stimuli in the γ band for the low-anxiety group

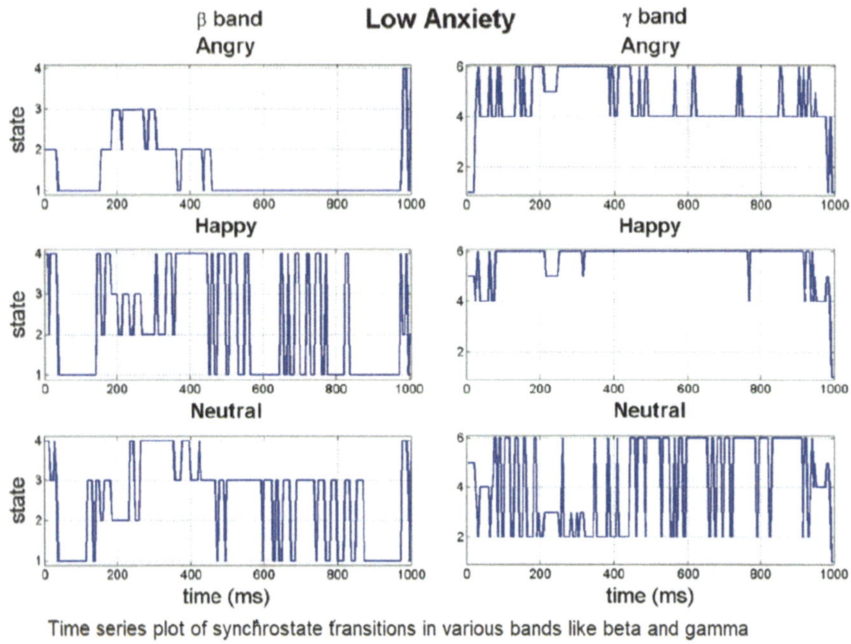

Fig. 12.7 The time-course plot of synchrostate transitions in the β and γ band

44 subjects from three different neuro-pathological groups and one healthy group were subjected to a set of facial perception tasks. These unique patterns—termed as synchrostates—abruptly switch from one to another and construct a stimulus-specific time course. The synchrostates and their transition plots can together be utilized as a generic method to understand the temporal dynamics of EEG phase synchronization as was done as described in [9]. Here, the present exploration shows that the existence of such synchrostates is consistent and exhibits only a small variability that may be attributable to the inter-person or intertrial variation often expected to be present in such experiments. Another possible factor that may contribute in such variability is the number of electrodes—fewer EEG electrodes exhibiting greater variability by introducing less resolution in computing the phase difference pattern. Also, quantification of the synchrostate transition in different groups is done in a probabilistic frame in terms of the self-transitions, which according to the authors might help in understanding the EEG phase synchronization-based derivation of the functional brain connectivity. Although it has been observed a consistent number of synchrostates, their physiological origin in relation to the anatomical brain network is yet to be established. Also it is still an open question whether the existence of synchrostates is a general phenomenon associated with active cognitive computation. However, if established as a generic phenomenon, combining the phase topographies of the synchrostates and their temporal stability from the time-course plot, one may establish a set of quantitative indices that may give a deeper understanding of the

transient phase relationship with effective connectivity in the brain, which may be useful in quantifying cognitive ability in a task-specific manner as well as classifying atypical neuropsychiatric conditions from normal brain functionality.

There is an integrated toolbox to characterize functional and effective brain connectivity, which will be discussed in the next section.

12.2.6 Integrated Toolbox to Characterize Functional and Effective Brain Connectivity

The analysis of the interdependence between time series has become an important field of research in the last years, mainly as a result of advances in the characterization of dynamical systems from the signals they produce, the introduction of concepts such as generalized and phase synchronization and the application of information theory to time series analysis. In neurophysiology, different analytical tools stemming from these concepts have added to the traditional set of linear methods, which includes the cross-correlation and the coherency function in the time and frequency domain, respectively, or more elaborated tools such as Granger Causality.

This increase in the number of approaches to tackle the existence of functional (FC) or effective connectivity (EC) between two (or among many) neural networks, along with the mathematical complexity of the corresponding time series analysis tools, makes it desirable to arrange them into a unified-easy-to-use software package. The goal is to allow neuroscientists, neurophysiologists, and researchers from related fields to easily access and make use of these analyses methods from a single integrated toolbox [12].

Here we present HERMES (http://hermes.ctb.upm.es), a toolbox for the Matlab® environment (The Mathworks, Inc) [12], which is designed to study functional and effective brain connectivity from neurophysiological data such as multivariate EEG and/or MEG records. It includes also visualization tools and statistical methods to address the problem of multiple comparisons. We believe that this toolbox will be very helpful to all the researchers working in the emerging field of brain connectivity analysis. We used this toolbox to compute the functional connectivity measures for software validation followed by hardware realization of the Graph theory measures. It's utilization will be described in detail in the next subsection.

12.2.7 EEG Processing Overview

An overview of EEG processing setup is shown in Fig. 12.1, using a wireless 8-channel EEG system for collecting data from subjects and relevant processing is undertaken to extract the related connectivity metrics. The pre-processing step involves filtering and artefact reduction, followed by feature extraction, which

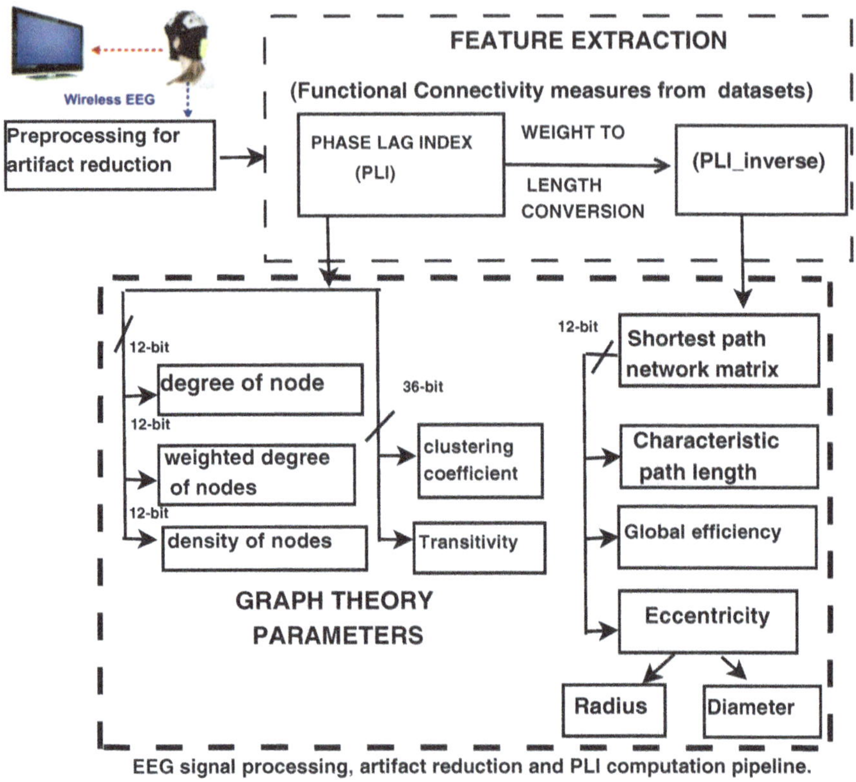

Fig. 12.8 Overview of the procedure

involves computing the FC measures using a HERMES toolbox [15], and is further quantified by a number of reduced dimensionality neuro-biological features by applying the brain connectivity toolbox (BCT) [15]. In this study, we have concentrated on the architecture and its implementation for computing one FC measure (PLI), which is quantified by 8 GT measures—(1) degree, (2) weight, (3) density, (4) clustering coefficients of the nodes, and (5) transitivity of connected network graph (all of which are computed from PLI); (6) Characteristic Path Length (CPL), (7) Global efficiency, (8) eccentricity (radius and diameter), computed from inverse of PLI (execution of 6–8 works in parallel to 1–5). The operation takes place on the phase values extracted through standard techniques (e.g., short-time Fourier transform) which is used for computing the phase lag index (PLI). The inversion (PLI INVERSE) is required to convert an input weighted connection matrix to a weighted connection-length matrix, necessary for computing the shortest path network [12] used for GT parameter extraction (Fig. 12.8).

12.2.7.1 Phase Lag Index and Inverse Computation

The sign of the phase difference for each individual element of a phase matrix, for all combination of rows are computed (viz., Fig. 12.9), followed by adding and averaging by total number of elements of the matrix. This computation corresponds to each coordinate position of the elements of a PLI matrix. Considering a matrix having M elements, θ_i and θ_j are the phase differences of the individual elements of the first two rows, the corresponding mean of the absolute value of the summation of the sign differences of the two rows is given by (1). The PLI computation has been illustrated through an example 4×4, 2D phase matrix (2), demonstrating the operation between the first two rows (3) and (4).

$$\Delta\theta_{PLI_{(i,j)}} = \frac{1}{M}\left\| \sum_{i=1}^{M} sign(\theta_i - \theta_j) \right\| \tag{12.1}$$

$$Phase = \begin{bmatrix} 1 & 2 & 5 & 6 \\ 4 & 3 & 2 & 7 \\ 8 & 9 & 13 & 5 \\ 10 & 17 & 6 & 8 \end{bmatrix} \tag{12.2}$$

$$\begin{aligned} \Delta\theta_{PLI_{(1,2)}} &= sign|[(1-4)+(2-3)+(5-2)+(6-7)]|/16 \\ &= sign|[-2]|/16 = 2/16 \end{aligned} \tag{12.3}$$

$$\Delta\theta_{PLI_{(i,j)}} = \begin{bmatrix} 1 & 2/16 & 2/16 & 4/16 \\ 2/16 & 1 & 2/16 & 4/16 \\ 2/16 & 2/16 & 1 & 2/16 \\ 4/16 & 4/16 & 2/16 & 1 \end{bmatrix} \tag{12.4}$$

$$\Delta\theta_{PLI..INVERSE_{(i,j)}} = 1 - \frac{1}{M}\left\| \sum_{i=1}^{M} sign(\theta_i - \theta_j) \right\| \tag{12.5}$$

$$= 1 - \Delta\theta_{PLI_{(i,j)}}|i \neq j \quad = \begin{bmatrix} 1 & 0.87 & 0.87 & 0.75 \\ 0.87 & 1 & 0.87 & 0.75 \\ 0.87 & 0.87 & 1 & 0.87 \\ 0.75 & 0.75 & 0.87 & 1 \end{bmatrix}$$

The PLI matrix in (12.4) represents connections among the nodes of a network where the connections (viz., edges) have weights indicating the connectivity strength whereas, its inverse represents a conversion of weight to length for the edges associated with the nodes [13]. This implies large weights correspond to strong associations and close proximity [15]. The edge length is the inverse of the connectivity, since stronger synchronization implies closer connection between two nodes, thereby

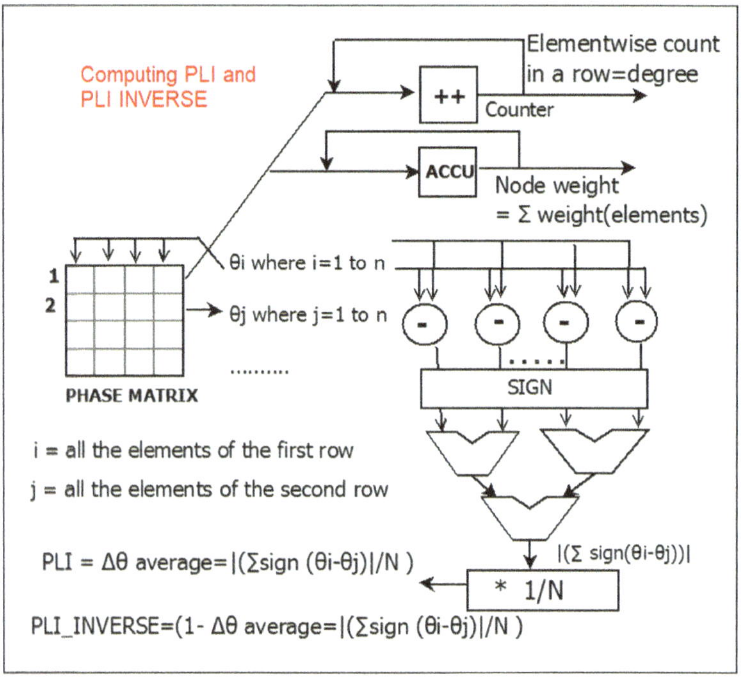

Fig. 12.9 Architecture for computing Phase Lag Index (PLI) and PLI INVERSE

requiring inverse operation. The architecture for PLI computation (cf. Fig. 12.9) illustrates the row-wise computation of the sign of the phase differences, followed by their summation and mean (division is achieved by fixed shift-add operation) for a given matrix size (8×8). The inverse computation uses the same circuit, achieved by subtracting every element of the PLI matrix from unity to produce each corresponding entry of the PLI INVERSE matrix (12.5).

12.2.7.2 Degree and Weight of Nodes—GT

A network is represented by a graph consisting of nodes and edges. The N individual components (e.g., EEG electrodes) of the network are called nodes and the k connections are called edges. The edges between nodes represent the correlation or synchronization between the activity at different neural regions, recorded by electrodes. Degree of a node (k_i) is the measure of the number of edges connected to a node/vertex given by (12.6), where $a_{i,j}$ is a particular link/edge. Similarly, the corresponding weight of the nodes is the summation of all the weights of the link connected to that node given by (12.7)

$$k_i = \sum_{j \varepsilon N} a_{i,j} \tag{12.6}$$

$$k_i^w = \sum_{j \varepsilon N} W_{i,j}. \tag{12.7}$$

Referring (4) the degree of all the nodes of the example PLI matrix $\Delta\theta_{PLI_{(i,j)}}$ is computed as [9] and weight of the nodes are computed as the summation of the individual edge weights of rows or columns connected to a particular node as $[(1 + 2/16 + 2/16 + 4/16); (2/16 + 1 + 2/16 + 4/16); (2/16 + 2/16 + 1 + 2/16); (4/16 + 4/16 + 2/16 + 1)] = [3/2, 3/2, 11/8, 13/8]$. This is achieved by a counter and accumulator which sums up the total weight of all links among nodes (similar to Fig. 12.9).

12.2.7.3 Density—GT

It denotes how dense is the physical connectivity of neuronal regions, represented by the ratio of present connections (summation of number of non-zero edges of the upper triangular PLI matrix (4)) to possible connections (number of vertices/nodes of the graph N (8)). The architecture in Fig. 12.10 computes the summation of number of non-zero elements of the upper triangular PLI matrix using a combination of a magnitude comparator (comparing with zero) and a counter module. A fully connected graph has a fixed denominator and hence division is achieved by fixed shift-add operation.

$$density_{node} = [\sum triupper(C_{i,j})|_{\neq 0}]/(N^2 - N)/2 \tag{12.8}$$

Considering the example PLI matrix $\Delta\theta_{PLI_{(i,j)}}$ in (4), density is computed as: $density_{node} = 10/[(4^2 - 4)/2]$.

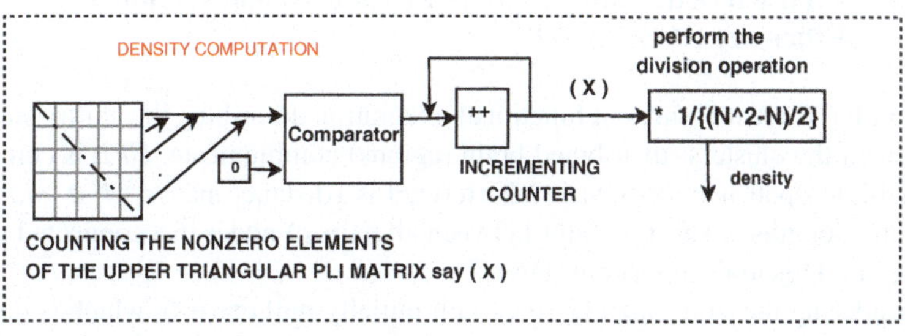

Fig. 12.10 Architecture for density estimation

12.2.7.4 Clustering Coefficient (CC) and Transitivity—GT

CC and *transitivity* are measures of functional segregation, which quantifies the division of brain network into sub-networks. They also determine the presence of clusters or modules, e.g., segregated neural groups, within the network and are based on the concept of triangles. Weighted *CC* is the average "intensity" (geometric mean) of all the triangles associated with each node. If $W_{i,j}$, $W_{j,h}$ and $W_{h,i}$ are the weights of edges of a triangle associated with a node, the *CC* is obtained as

$$t_i^w = \frac{1}{2} \sum_{j,h \varepsilon N} (W_{i,j} W_{j,h} W_{h,i})^{1/3} \tag{12.9}$$

$$CC^w = \frac{1}{n} \sum_{i \epsilon N} \frac{2t_i^w}{k_i(k_i - 1)} \tag{12.10}$$

However, it has been observed that instead of computing the cube roots, the product of three edge weight matrices results in similar morphology as the *CC* of each node. Further optimization is achieved by neglecting the denominator, which exhibits similar morphological nature for the *CC* of all nodes. Figure 12.11 illustrates the product of three PLI weighted matrices, where the diagonal elements of the resulting matrix are the *CC's* of each individual node.

Transitivity is similar to *CC* but considers a complete graph and is computed as the ratio of the summation of all values of the *CC's* of the nodes as the numerator to the summation of the total number of complete triplets attached to all the nodes in the graph as shown in (8) and the architecture in Fig. 12.11.

$$T^w = \frac{\sum_{i \epsilon N} 2t_i^w}{\sum_{i \epsilon N} k_i(k_i - 1)}. \tag{12.11}$$

12.2.7.5 Characteristic Path Length (CPL) and Weighted Global Efficiency (WGE)—GT

CPL and *WGE* are measures of functional integration describing the ability and ease with which the clusters (distributed brain regions) communicate. *CPL* is computed from a shortest path network matrix (also termed as a distance matrix-($dist\_matrix$)) containing lengths of shortest paths between all pairs of nodes in a connected graph, obtained by Dijkstra's algorithm (DA) [7].

Considering the $dist\_matrix$ row (with initially null entries), which is updated with subsequent comparisons with the PLI INVERSE matrix (the network descriptor matrix cf. Fig. 12.12). The minimum element is determined from the updated distance

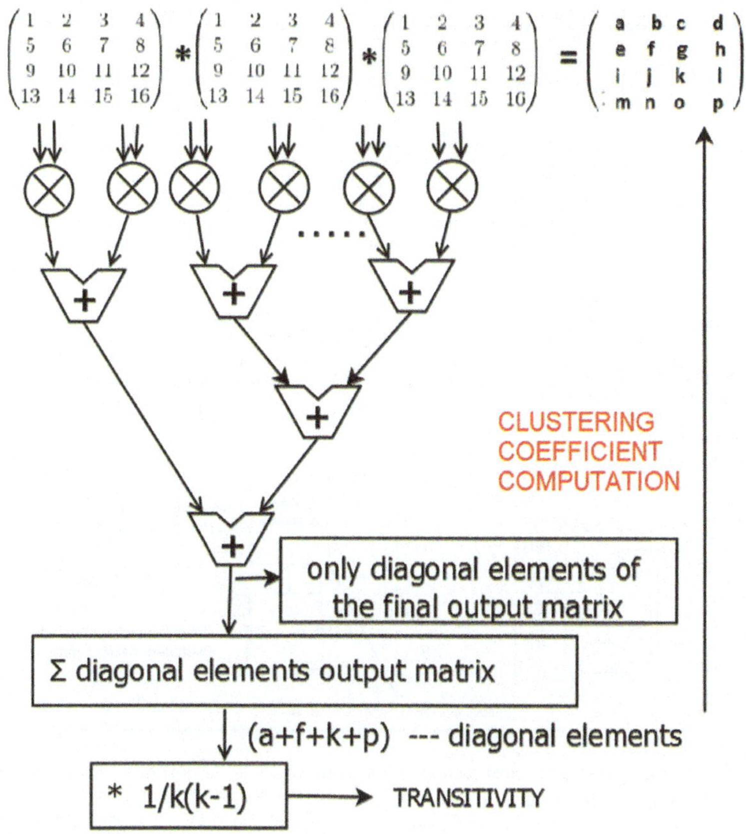

Fig. 12.11 Architecture for *Clustering Coefficient* and *Transitivity*

matrix row and the corresponding column of the minimum element is used as the row information for the next step to initiate. This minimum element is added with the corresponding coordinate of the PLI INVERSE matrix to be compared with the corresponding element of the *dist_matrix* row and the necessary computation is performed according to DA [7]. This continues until all the minimum elements of a row of the *dist_matrix* are computed (i.e., minimum distance information). This iterative update process continues for all rows of *dist_matrix* (managed by a control signal) to determine the complete shortest path network matrix (SPM). *WGE* is the mean of the inverse of each entry of the *dist_matrix*, achieved by subtracting each element of the SPM from unity. This has been illustrated with the example PLI INVERSE matrix (5), considered as input for the DA and the corresponding SPM is computed (12). The mean of SPM provides the *CPL* and accordingly, the *WGE* is computed.

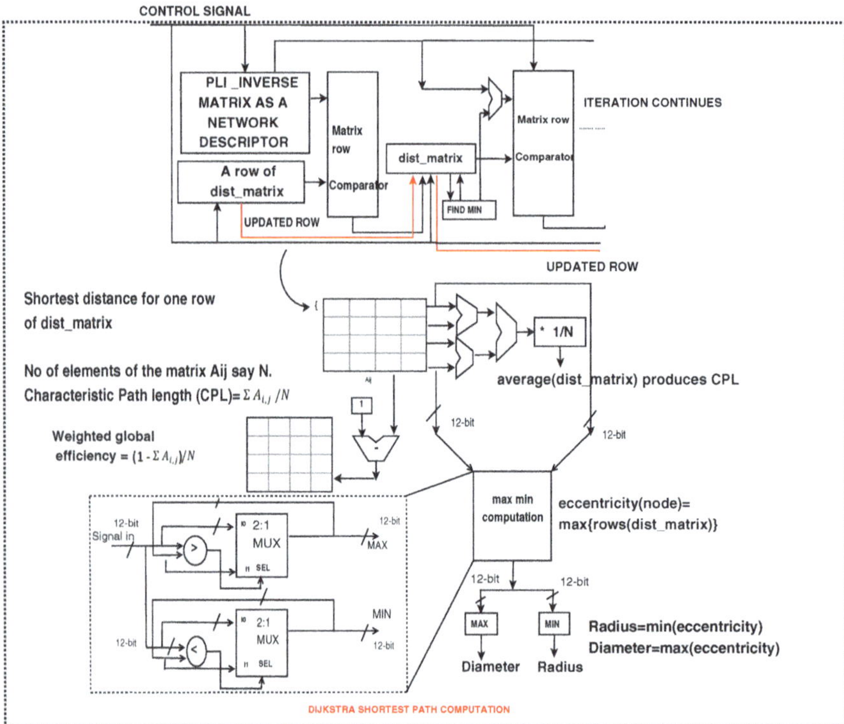

Fig. 12.12 Architecture for shortest path network using Dijkstra's algorithm for computing *CPL*, *eccentricity*, *radius* and *diameter*

$$SPM = \begin{bmatrix} 0 & 0.87 & 0.87 & 0.75 \\ 0.87 & 0 & 0.87 & 0.75 \\ 0.87 & 0.87 & 0 & 0.87 \\ 0.75 & 0.75 & 0.87 & 0 \end{bmatrix} \tag{12.12}$$

$$WGE = \frac{1}{16}\sum \begin{bmatrix} 0 & (1 - -0.87) & (1 - -0.87) & (1 - -0.75) \\ (1 - -0.87) & 0 & (1 - -0.87) & (1 - -0.75) \\ (1 - -0.87) & (1 - -0.87) & 0 & (1 - -0.87) \\ (1 - -0.75) & (1 - -0.75) & (1 - -0.87) & 0 \end{bmatrix}$$

12.2.7.6 Eccentricity, Radius and Diameter—GT

The network *Diameter* and *Radius* are the two measures to characterize the brain connectivity, which are derived from *eccentricity*. The nodal eccentricity is the maximal path length between a node and any other node in the network. Each row of the distance matrix acts as an input to a maxima-minima detection circuit (cf. Fig. 12.12) to get the maximum value for each node. This constructs the nodal eccentricity for each vertex of the network. The maximum nodal eccentricity of all the vertices forms the diameter and the minimal one forms the radius. With the given SPM (12) the maximum element of each row forms the eccentricity of each node as [.87;.87;.87;.87]. The maximum element of the eccentricity vector is the diameter and the least element forms the radius.

12.3 Results and Observation

The design entry for the architecture was coded using VHDL. In this study we have used ten 8×8 input phase matrix (8 being the number of electrodes) computed from EEG recordings of healthy adults. The RTL outputs for the FC/GT measures have been compared against MATLAB simulation and the average errors are reported in Table 12.1. A datapath of 12-bits has been used all throughout except for *CC* and *transitivity* (36 bits) due to multiplication operation. For higher resolution matrices, the error may become more significant and hence to obtain a higher accuracy, increasing the internal datapath width of the individual modules would be necessary. The design has been synthesized using STM 130-nm technology library, with a supply voltage of 1.08V and was functionally verified up to a frequency of 25 MHz. The synthesized design has an overall cell area of 2591 mm^2 (1116 K NAND2) and consumes 10 nW of dynamic power. However, for future integration onto a sensing platform the design needs to be optimized for reducing the chip area. The complexity analysis in terms of arithmetic operations and timing information obtained for each of the GT parameters is shown in Table 12.1. All the parameters are computed within 168 clock cycles (6.7 us) and works 695 times faster than software simulation (MATLAB), primarily achieved through concurrent computation of the GT parameters.

Table 12.1 Comparison of error, arithmetic complexity and timing between RTL and MATLAB simulation

Parameters	Error		Add/Sub	Mult	Clock cycles
	Average	Worst			
Phase Lag Index	0	0	1568		70
(PLI INVERSE)	0	0	64		71
Degree of node	0	0	8		88
Weight of node	0	0	64		88
Density of node	O(2^–8)	O(2^–5)	102		142
Clustering coefficient	0	0	896	1024	87
Transitivity	O(2^–10)	O(2^–9)	903	1024	88
Characteristics Path Length	O(2^–8)	O(2^–5)	21		157
Weighted Global Efficiency	O(2^–8)	O(2^–5)	21	2	158
Eccentricity	0	0	7		157
Radius / Diameter	0	0	9		168

In this study, we present an algorithm-to-architecture mapping to realize the complex network measures for brain connectivity, focusing on one FC and 8 GT parameters. The results for this *proof-of-concept* RTL implementation, in terms of error, computation time, arithmetic complexity and low power consumption could be considered favorable for future explorations-considering denser electrodes (e.g., 19/32 channel) and ASIC implementation for integration into sensor platform for automated disease prognosis and control.

In the next section we will discuss Phase Lag Index for assessment of Functional Connectivity From Multichannel EEG and MEG With Diminished Bias From Common Sources.

12.3.1 Phase Lag Index: Assessment of Functional Connectivity from Multichannel EEG and MEG With Diminished Bias from Common Sources

The objective is to address the problem of volume conduction and active reference electrodes in the assessment of functional connectivity, we propose a novel measure to quantify phase synchronization, the phase lag index (PLI), and compare its performance to the well-known phase coherence (PC), and to the imaginary component of coherency (IC). The PLI is a measure of the asymmetry of the distribution of phase differences between two signals. The performance of PLI, PC, and IC was examined in (i) a model of 64 globally coupled oscillators, (ii) an EEG with an absence seizure, (iii) an EEG data set of 15 Alzheimer patients and 13 control subjects, and (iv) two MEG data sets. Results show that PLI and PC were more sensitive than IC to increasing levels of true synchronization in the model. PC and IC were influenced stronger than PLI by spurious correlations because of common sources. All measures detected changes in synchronization during the absence seizure. In contrast to PC, PLI, and IC were barely changed by the choice of different montages. PLI and IC were superior to PC in detecting changes in beta band connectivity in AD patients. Finally, PLI and IC revealed a different spatial pattern of functional connectivity in MEG data than PC. It can be concluded that the PLI performed at least as well as the PC in detecting true changes in synchronization in model and real data but, at the same token and like-wise the IC, it was much less affected by the influence of common sources and active reference electrodes [16].

12.4 Conclusion

In this study, we present a complete survey of algorithm-to-architecture mapping to realize the complex network measures for brain connectivity, focusing on one FC and graph-theoretic parameters. The noise removal automated methodology proposed in the initial stage can remove blink and muscular artifacts real time on the hardware without the need of any extra electrode. Its reliability and robustness has also been established after exhaustive simulation study and analysis on both simulated and real data. The results for this proof-of-concept RTL implementation, in terms of error, computation time, arithmetic complexity, and low power consumption could be considered favorable for future explorations-considering denser electrodes (e.g., 32/64 channels) and ASIC implementation for integration into sensor platform for automated disease prognosis and control.

References

1. Acharyya A, Jadhav PN, Bono V, Maharatna K, Naik GR (2018) Low-complexity hardware design methodology for reliable and automated removal of ocular and muscular artifact from EEG. Comput Methods Programs Biomed 158:123–133. https://doi.org/10.1016/j.cmpb.2018.02.009
2. Aggarwal N, Agrawal R (2012) First and second order statistics features for classification of magnetic resonance brain images
3. Biswas, D, Bono, V, Scott-South, M et al. (7 more authors) (2016) Analysing wireless EEG based functional connectivity measures with respect to change in environmental factors. In: IEEE-EMBS International Conference on Biomedical and Health Informatics (BHI 2016). BHI 2016, 24-27 Feb 2016, Las Vegas, USA. IEEE , pp. 599-602. ISBN 9781509024551 https://doi.org/10.1109/BHI.2016.7455969
4. Blomstedt MHP (2015) Hardware-related complications of deep brain stimulation: a ten year experience. Acta Neurochirurgica 147:1061–1064
5. Bono V et al (2014) Artifact reduction in multichannel pervasive EEG using hybrid WPT-ICA and WPT-EMD signal decomposition techniques. In: IEEE international conference on acoustics, speech and signal processing (ICASSP), pp 5864–5868
6. Bono V, Biswas D, Das S, Maharatna K (2016) Classifying human emotional states using wireless EEG based ERP and functional connectivity measures. In: 2016 IEEE-EMBS international conference on biomedical and health informatics (BHI), Las Vegas, NV, pp 200–203. https://doi.org/10.1109/BHI.2016.7455869
7. Dijkstra E (1959) A note on two problems in connexion with graphs. Numerische Mathematik, 269–271
8. Jamal W et al (2015) Biomed Phys Eng Express 1:015002
9. Jamal W, Das S, Maharatna K, Pan I, Kuyucu D (2015) Brain connectivity analysis from EEG signals using stable phase-synchronized states during face perception tasks. Phys A Statistic Mech Appl 434:273–295, ISSN 0378-4371. https://doi.org/10.1016/j.physa.2015.03.087
10. Minati DL et al (2013) Detection of scale-freeness in brain connectivity by functional MRI: signal processing aspects and implementation of an open hardware co-processor. Med Eng Phys 35:1525–1531
11. Minati DL, Cercignani M (2013) Rapid geodesic mapping of brain functional connectivity: implementation of a dedicated co-processor in a fieldprogrammable gate array and application to resting state functional mri. Med Eng Phys 35:1532–1539
12. Niso G, Bruña R, Pereda E et al (2013) HERMES: towards an integrated toolbox to characterize functional and effective brain connectivity. Neuroinformatics 11(4):405–434. https://doi.org/10.1007/s12021-013-9186-1
13. Pal C, Biswas D, Maharatna K, Chakrabarti A (2017) Architecture for complex network measures of brain connectivity. In: IEEE international symposium on circuits and systems (ISCAS). Baltimore, MD 2017:1–4. https://doi.org/10.1109/ISCAS.2017.8050239

14. Plit AM, Barnes KA (2015) Functional connectivity classification of autism identifies highly predictive brain features but falls short of biomarker standards. NeuroImage Clinic 7:359–366
15. Rubinov OM (2010) Complex network measures of brain connectivity: uses and interpretations. NeuroImage 52:1059–1069
16. Stam CJ, Nolte G, Daffertshofer A (2007) Phase lag index: assessment of functional connectivity from multi channel EEG and MEG with diminished bias from common sources. Hum Brain Mapp 28(11):1178–1193. https://doi.org/10.1002/hbm.20346
17. VB et al (2016) Classifying human emotional states using wireless EEG based ERP and functional connectivity measures. In: International conference on biomedical and health informatics (BHI), pp 200–203
18. Xin.Di BB (2015) Dynamic brain functional connectivity modulated by resting-state networks. Brain Struct Func 220:37–46

The manufacturer's authorised representative in the EU is Springer Nature Customer Service Centre GmbH, Europaplatz 3, 69115 Heidelberg, Germany. If you have any concerns regarding our products, please contact ProductSafety@springernature.com

Printed and bound by CPI Group (UK) Ltd, Croydon, CR0 4YY
30/09/2026
02225287-0010